Mohammad Yamin

Problem Solving in Foundation Engineering using foundationPro

Springer

Mohammad Yamin
Department of Civil Engineering & Construction
Bradley University
Peoria, IL, USA

ISBN 978-3-319-17649-9 ISBN 978-3-319-17650-5 (eBook)
DOI 10.1007/978-3-319-17650-5

Library of Congress Control Number: 2015936330

Springer Cham Heidelberg New York Dordrecht London

Printed on acid-free paper

Springer International Publishing AG Switzerland is part of Springer Science+Business Media (www.springer.com)

To Allah

Preface

The main purpose of this book is to stimulate problem-solving capability and foster self-directed learning in foundation engineering subject for civil and construction engineering students and practicing professionals. It also explains the use of the foundationPro software, available at no cost, and includes a set of foundation engineering applications. Reading this or any other textbook is not enough and cannot be sufficient to perform safe and economical designs of foundations as a considerable experience and judgment are required. The overall layout of the book chapters is as follows: first, to introduce the general idea behind the title of the chapter; second, to briefly discuss the theories and methodologies and to summarize the equations and charts needed in the chapter; third, to provide a step-by-step procedure on how to deal with design problems related to the title of the chapter; fourth, to induce a number of design problems and solve these problems by hand, and then using the foundationPro software; fifth, to present a number of suggested projects to allow the reader to practice the concepts learned in the chapter; and finally, to introduce a list of references and additional useful readings about that specific chapter. In total, this book consists of four chapters. Chapter 1 deals with the design of shallow foundations resting on homogeneous soil based on bearing capacity and elastic settlement requirements. Chapter 2 presents the axial capacity of single pile foundations in homogeneous and nonhomogeneous soils based on bearing capacity and elastic settlement requirements. Chapter 3 is similar to Chap. 2 but for single drilled shaft foundations. Chapter 4 deals with the design of mechanically stabilized earth retaining walls with strip reinforcement.

Additional materials are and will be available at http://www.foundationpro.net. These materials include the following:

1. foundationPro software which includes the following applications: Shallow-1, Pile-1, Pile-2, Shaft-1, Shaft-2, and MSE Wall-1.

2. Video tutorials on how to use the various applications of foundationPro software.
3. foundationPro Forum which can be visited for general discussions about foundationPro applications. The forum can be accessed by visiting http://www.foundationpro.net/forum/.

Peoria, IL, USA Mohammad Yamin

Acknowledgements

I have taken efforts in this project. However, it would not have been possible without the help of many individuals.

Thanks are due to friends and colleagues, Drs. Enad Mahmoud and Ahmed Ibrahim.

Drs. Ehab Shatnawi, Ahmed Ashteyat, Abdullah Sharo, Diya Azzam, and Ahmed Senouci.

Thanks are also due to students, Onsel Badur, Bhagirath Reddy, Dinesh Alluri, and Robert Glod. I would like to extend my sincere thanks to all of them and to everyone else participated in bringing this project to existence.

I am grateful to Bradley University—the Office of Teaching Excellence and Faculty Development for their kind support.

I am also grateful to Lesley Poliner, Project Coordinator at Springer, and Michael Luby, Senior Publishing Editor at Springer, for their great work.

Contents

1 Shallow Foundations on Homogeneous Soil 1
1.1 Introduction 1
1.2 Theory 2
1.2.1 Bearing Capacity of Shallow Foundation 4
1.2.2 Elastic Settlement of Foundation 16
1.2.3 Foundation Loads 19
1.3 Step-by-Step Procedure 20
1.4 Design Problems 22
1.4.1 Strip/Continuous Foundation 22
1.4.2 Rectangular Foundation 30
1.4.3 Square Foundation with Horizontal Loading 37
1.4.4 Circular Foundation with One-Way Load Eccentricity . . . 41
1.4.5 Rectangular Foundation with One-Way Load Eccentricity 47
1.4.6 Square Foundation with Soil Compressibility 53
1.4.7 Rectangular Foundation with Soil Compressibility 60
1.4.8 Strip/Continuous Foundation with Horizontal Loading . . . 64
1.4.9 Square Foundation with Load Eccentricity and Horizontal Loading 70
1.4.10 Rectangular Foundation with Two Way Load Eccentricity 76
1.4.11 Rectangular Foundation with Elastic Settlement 81
1.4.12 Circular Foundation with Horizontal Loading and Elastic Settlement 89
1.4.13 Rectangular Foundation with Two-Way Eccentricity and Elastic Settlement 97
1.4.14 Circular Foundation with Load Eccentricity, Soil Compressibility, and Elastic Settlement 106
1.5 Suggested Projects 114
1.5.1 Suggested Projects: Rectangular Foundation 116

1.5.2 Suggested Projects: Circular Foundation with Load Eccentricity 117
1.5.3 Suggested Projects: Strip/Continuous Foundation with Horizontal Loading 118
1.5.4 Suggested Projects: Rectangular Foundation with Load Eccentricity 118
1.5.5 Suggested Projects: Circular Foundation with Load Eccentricity and Settlement 118
References 119

2 Axial Capacity of Single Pile Foundations in Soil 121
2.1 Introduction 121
2.2 Theory 122
2.2.1 Axial Capacity of a Single Pile (Bearing Capacity) 123
2.2.2 Axial Capacity (Elastic Settlement) 131
2.2.3 Allowable Loads 132
2.3 Step-by-Step Procedure 133
2.4 Design Problems 134
2.4.1 Circular Pile in Sandy Soil (Meyerhof's and Critical Depth Methods) 134
2.4.2 Circular Pile in Sandy Soil (Vesic and Critical Depth Methods) 144
2.4.3 Square Pile in Clay Soil (Meyerhof's and α Methods) . . . 155
2.4.4 Circular Pile in Clay Soil (Vesic's and α* Methods) 159
2.4.5 Circular Pile in Clay Soil (Meyerhof's and λ Methods) . . 164
2.4.6 Circular Pile in Sandy Soil (Three Sand Layers) 168
2.4.7 Square Pile in Sandy Soil (Two Sand Layers) 178
2.4.8 Circular Pile in Clayey Soil (Two Clay Layers) 184
2.4.9 Square Pile in Clayey Soil (Two Clay Layers) 190
2.4.10 Square Pile in Rock 196
2.4.11 Suggested Projects 198
2.4.12 Suggested Projects: Circular Pile in Sandy Soil 199
2.4.13 Suggested Projects: Circular Pile in Clayey Soil 200
2.4.14 Suggested Projects: Square Pile in Sandy Soil (Three Sand Layers) 201
References 201

3 Axial Capacity of Single Drilled Shaft Foundations in Soil 203
3.1 Introduction 203
3.2 Theory 204
3.2.1 Axial Capacity of a Single Drilled Shaft (Bearing Capacity) 206
3.2.2 Axial Capacity of a Single Drilled Shaft (Elastic Settlement) 210
3.2.3 Allowable Loads 212

3.3 Step-by-Step Procedure 212
3.4 Design Problems 213
3.4.1 Straight Shaft in Homogeneous Sandy Soil 213
3.4.2 Straight Shaft in Homogeneous Clayey Soil 218
3.4.3 Straight Shaft in Nonhomogeneous Clayey Soil 226
3.4.4 Bell Shaft in Nonhomogeneous Sandy Soil 232
3.4.5 Straight Shaft in Rock 241
3.5 Suggested Projects 243
3.5.1 Suggested Projects: Straight Shaft in Homogeneous Sandy Soil 244
3.5.2 Suggested Projects: Bell Shaft in Nonhomogeneous Sandy Soil 244
3.5.3 Suggested Projects: Straight Shaft in Rock 245
References 245

4 Design of Mechanically Stabilized Earth Retaining Walls 247
4.1 Introduction 247
4.2 Theory 248
4.2.1 Vertical and Horizontal Stresses 250
4.2.2 Internal Stability 256
4.2.3 External Stability 257
4.3 Step-by-Step Procedure 260
4.4 Design Problems 262
4.4.1 MSE Wall with Constant Length and Spacing (Horizontal and Vertical) of Strips 262
4.4.2 MSE Wall with Varying Length and Spacing (Horizontal and Vertical) of Strips 276
4.4.3 MSE Wall with Strip Surcharge Loading and Constant Length and Spacing of Strips 292
4.4.4 MSE Wall with Line Loading and Varying Strip Length 305
4.5 Suggested Projects 318
4.5.1 Suggested Projects: MSE Wall with Applied Strip Loading 318
4.5.2 Suggested Projects: MSE Wall with Embankment Loading 319
References 320

Chapter 1
Shallow Foundations on Homogeneous Soil

Abstract This chapter deals with single shallow foundations resting on homogeneous soil. Calculations of various loads (vertical, horizontal, and bending moment) a foundation can withstand are explained in details in this chapter. The calculations were performed to satisfy both bearing capacity and elastic settlement requirements. For the bearing capacity condition, the effects of many factors were considered in the analyses such as various loading conditions, foundation shapes, foundation embedment, soil compressibility, and groundwater table. Then again, the effects of several factors were considered in the elastic settlement analyses such as foundation rigidity, foundation embedment, and variation in the elastic modulus of soil with depth. Additionally, a step-by-step procedure is introduced in this chapter to develop bearing capacity and elastic settlement design charts which can be useful in the design process of shallow foundations. A number of design problems are also presented in this chapter and their solutions are explained in details. These problems were first hand-solved, and then, resolved using the Shallow-1 application of the foundationPro program. Finally, a set of design projects is suggested at the end of this chapter to allow the reader practice the concepts learned.

Keywords Shallow foundation • Bearing capacity • Elastic settlement • Shallow-1 • foundationPro

1.1 Introduction

This chapter deals with single shallow foundations on homogeneous soil. Calculations of various loads a foundation can sustain are explained in details in this chapter. Allowable and ultimate loads (vertical, horizontal, and bending moment) on a single foundation are estimated based on bearing capacity and elastic settlement requirements.

In the bearing capacity analyses, the classical bearing capacity equations for a single foundation were utilized. Various loading conditions (vertical, horizontal, and bending moments) and foundation shapes (circular, rectangular, and continuous/strip) were included in the analyses. Effect of soil compressibility on bearing capacity was also included in the analyses. Effects of the depths of foundation embedment and groundwater table were considered in the bearing capacity equation.

M. Yamin, *Problem Solving in Foundation Engineering using foundationPro*,
DOI 10.1007/978-3-319-17650-5_1

In the elastic settlement analyses, the modified settlement equation by Mayne and Poulos (1999) was utilized. This improved equation deals with circular and rectangular foundations. It considers the foundation rigidity which depends on the foundation dimensions, thickness, and elastic modulus. Also, this equation takes into account the depth of foundation embedment. Additionally, this improved elastic settlement equation considers not only a unique value for the elastic modulus of the soil underneath the foundation, but also the linearly increasing elastic modulus with depth.

A step-by-step procedure was introduced in this chapter to develop bearing capacity and elastic settlement design charts. These design charts present the relationship between various applied loads on the foundation and foundation dimensions for different shapes, depths, and allowable settlement. These charts can be useful in the foundation design process to find what will control the final design, the bearing capacity, or the elastic settlement of the foundation.

Fourteen design problems are presented in this chapter. First, these design problems were hand-solved and solution was explained in details, and then the foundationPro program was used to resolve the problems to replicate and verify the hand solution. Also, the program was used to investigate a wider and detailed solution and design alternatives for the hand-solved problems. Since the foundationPro includes a set of several applications, the Shallow-1 application of the foundationPro is the responsible application to perform bearing capacity and elastic settlement calculations for shallow foundations resting on homogeneous soil. Therefore, only Shallow-1 application will be used throughout this chapter to replicate the hand-solved problems. Five design projects are suggested at the end of this chapter to allow the reader to practice and apply the learned concepts.

1.2 Theory

This section explains how to estimate the allowable and ultimate loads that can be applied to a single shallow foundation resting on homogeneous soil. The foundation loads are estimated to meet both bearing capacity and elastic settlement requirements. All bearing capacity and elastic settlement equations are listed and all variables used in the equations are defined in the following subsections. These subsections are not meant to explain the bearing capacity and elastic settlement theories, rather to summarize the final equations from each theory which will be used in the analyses.

To determine the bearing capacity and the elastic settlement of a shallow foundation resting on homogeneous soil as shown in Fig. 1.1, soil properties (c' = cohesion, ϕ' = friction angle, μ_s = Poisson's ratio, E_s = elastic modulus of foundation soil, γ_1 and γ_2 = unit weight of the soil above and below the groundwater table, respectively) are required to perform the analysis. Depth of foundation (D_f) and depth of groundwater table (D_w) if exists are also required for the calculations. Three different foundation shapes are considered: circular (Fig. 1.2a), square/rectangular (Fig. 1.2b), and strip/continuous (Fig. 1.2c).

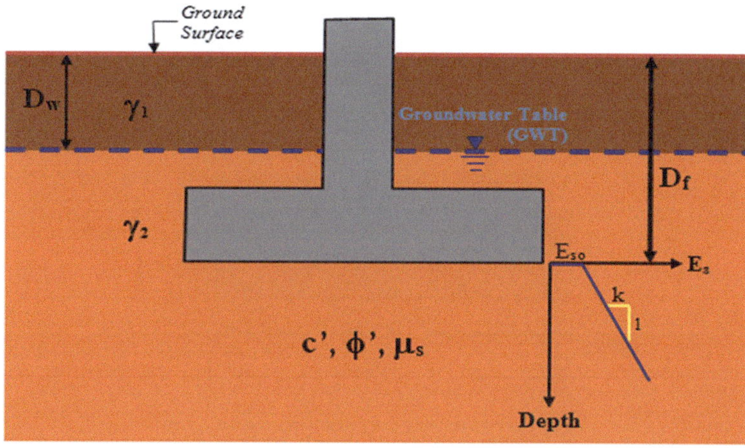

Fig. 1.1 Definition of various parameters for a shallow foundation on homogeneous soil

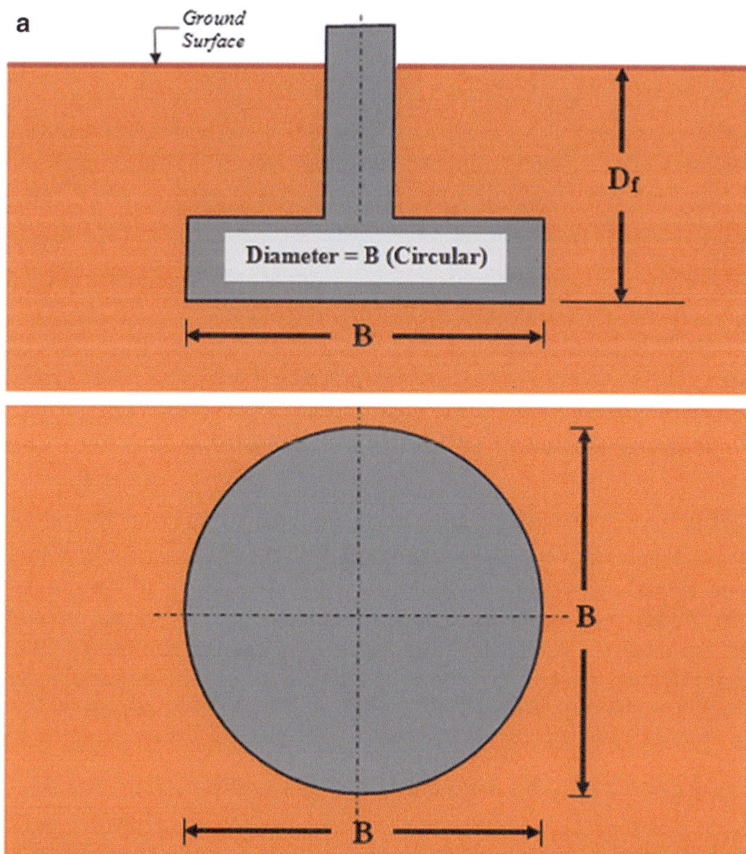

Fig. 1.2 Different foundation shapes: (**a**) Circular foundation; (**b**) square/rectangular foundation; (**c**) strip/continuous foundation

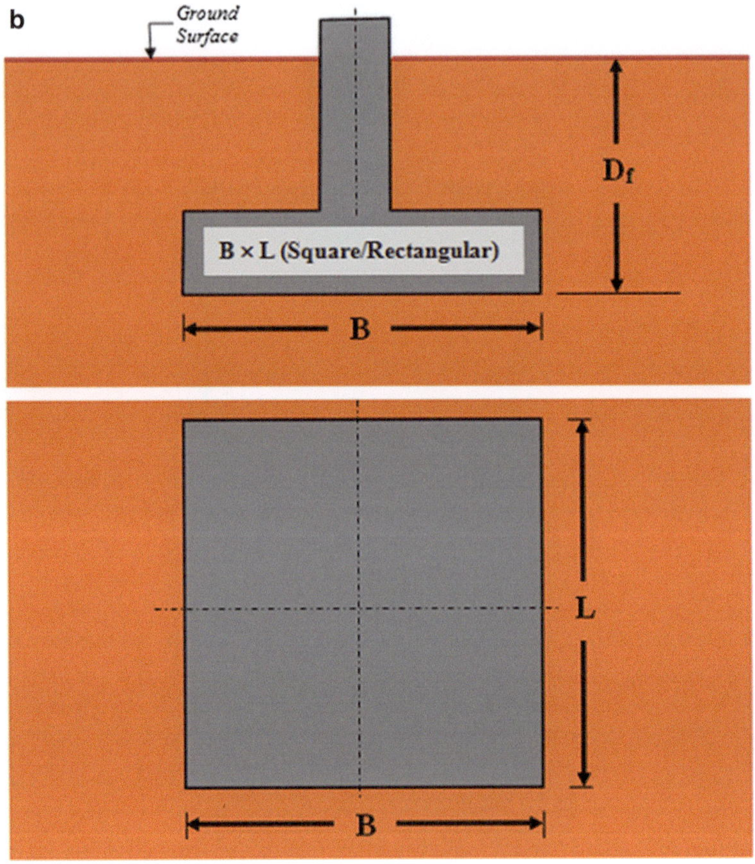

Fig. 1.2 (continued)

1.2.1 Bearing Capacity of Shallow Foundation

To estimate the allowable and ultimate loads (vertical, horizontal, and bending moment) a foundation can carry according to the bearing capacity of foundation; first, one should determine the ultimate bearing capacity of the foundation. Terzaghi (1943) was the first to present an equation for bearing capacity for different foundations with some limitations. To account for these limitations, Meyerhof (1963) suggested the following equation to estimate the ultimate bearing capacity of a single shallow foundation resting on homogeneous soil:

$$q_{\mathrm{u}} = c' N_c F_{c\mathrm{d}} F_{c\mathrm{i}} F_{c\mathrm{c}} F_{c\mathrm{s}} + q N_q F_{q\mathrm{s}} F_{q\mathrm{i}} F_{q\mathrm{c}} F_{q\mathrm{d}} + \frac{1}{2}\gamma B N_\gamma F_{\gamma\mathrm{s}} F_{\gamma\mathrm{d}} F_{\gamma\mathrm{i}} F_{\gamma\mathrm{c}} \tag{1.1}$$

In the above equation, c' represents soil cohesion, q is the effective stress at the base of the foundation, γ is the unit weight of the foundation soil, and B is the width of

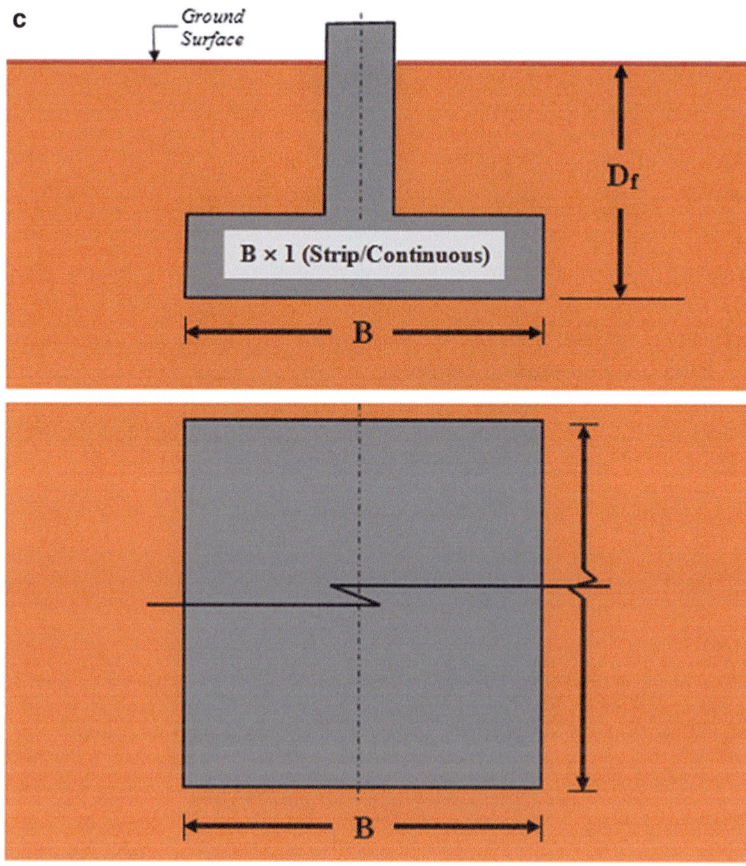

Fig. 1.2 (continued)

foundation or diameter in case of circular foundation. $F_{cs}, F_{qs}, F_{\gamma s}$ are the shape factors, $F_{cd}, F_{qd}, F_{\gamma d}$ are the depth factors, $F_{ci}, F_{qi}, F_{\gamma i}$ are the load inclination factors, N_c, N_q, N_γ are the bearing capacity factors, and F_{cc}, F_{qc}, $F_{\gamma c}$ are the soil compressibility factors.

To determine the various factors defined earlier in the bearing capacity equation, one should use the following equations which were suggested by several researchers and investigators:

1.2.1.1 Bearing Capacity Factors

Reissner (1924) derived the following equation to calculate N_q:

$$N_q = \tan^2\left(45 + \frac{\phi'}{2}\right)e^{\pi \tan(\phi')} \tag{1.2}$$

where ϕ' is the soil friction angle.

Also, Prandtl (1921) derived the following equation to determine N_c:

$$N_c = \left(N_q - 1\right) \cot\left(\phi'\right) \tag{1.3}$$

The following equation was put forth by Caquot et al. (1953) and Vesic (1973) to estimate N_γ:

$$N_\gamma = 2\left(N_q + 1\right) \tan\left(\phi'\right) \tag{1.4}$$

1.2.1.2 Effect of Foundation Depth

To account for the depth of foundation, Hansen (1970) suggested the following depth factors to be used in the ultimate bearing capacity equation:

– For a soil with $\phi' = 0$:

$$F_{cd} = 1 + 0.4\eta \tag{1.5}$$

$$F_{qd} = 1$$

$$F_{\gamma d} = 1$$

– For a soil with $\phi' > 0$, the depth factors can be calculated as follows:

$$F_{qd} = 1 + 2\tan\phi'\left(1 - \sin\left(\phi'\right)\right)^2 \eta \tag{1.6}$$

$$F_{cd} = F_{qd} - \frac{1 - F_{qd}}{N_c \tan\left(\phi'\right)} \tag{1.7}$$

$$F_{\gamma d} = 1$$

where

$$\eta = \frac{D_f}{B} \tag{1.8}$$

The ratio $\eta \leq 1$ applies for most shallow foundation cases. However, η in Eqs. (1.5) and (1.6) is replaced with η' when $\eta > 1$. η' should be in radians and can be calculated as

$$\eta' = \tan^{-1}\left(\frac{D_f}{B}\right) \tag{1.9}$$

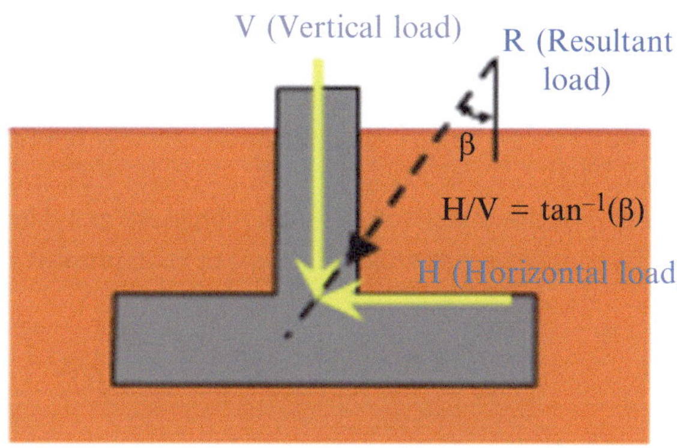

Fig. 1.3 Vertical and horizontal loads on foundation

1.2.1.3 Effect of Load Inclination

When a vertical load (V) and horizontal load (H) are applied to a foundation as shown in Fig. 1.3, the net resultant load (R) on the foundation will be inclined an angle β with the vertical. To account for this load inclination in the bearing capacity equation, Meyerhof (1963) and Hanna and Meyerhof (1981) suggested the following load inclination factors:

$$F_{ci} = F_{qi} = \left(1 - \frac{\beta^{\circ}}{90^{\circ}}\right)^2 \tag{1.10}$$

$$F_{\gamma i} = \left(1 - \frac{\beta}{\phi'}\right)^2 \tag{1.11}$$

where β is the inclination of the resultant applied load on the foundation with respect to the vertical:

$$\beta = \tan^{-1}\left(\frac{H}{V}\right) \tag{1.12}$$

1.2.1.4 Effect of Soil Compressibility

To account for the effect of the soil compressibility on the bearing capacity of shallow foundations, the derived procedure by Vesic (1973) based on the expansion of cavities in infinite soil can be followed. Hence, the soil compressibility factors F_{cc}, F_{qc}, and $F_{\gamma c}$ are estimated as follows:

$$F_{\gamma c} = F_{qc} = \exp\left\{\left(-4.4 + 0.6\frac{B}{L}\right)\tan\left(\phi'\right) + \left[\frac{\left(3.07\sin\left(\phi'\right)\right)\ \left(\log 2I_r\right)}{1 + \sin\left(\phi'\right)}\right]\right\} \tag{1.13}$$

I_r in Eq. (1.13) can be calculated using the following equation:

$$I_r = \frac{E_s}{2(1+\mu_s)\left(c' + q' \tan\left(\phi'\right)\right)} \tag{1.14}$$

where q' is the effective overburden pressure at a depth of $\left(D_f + \frac{B}{2}\right)$.

The critical rigidity index, $I_{r(cr)}$, can be expressed as

$$I_{r(cr)} = \frac{1}{2}\left\{\exp\left[\left(3.3 - 0.45\frac{B}{L}\right)\cot\left(45 - \frac{\phi'}{2}\right)\right]\right\} \tag{1.15}$$

So, if $I_r \geq I_{r(cr)}$, then the soil compressibility factors are all considered one $\left(F_{cc} = F_{qc} = F_{\gamma c} = 1\right)$

However, if $I_r < I_{r(cr)}$, then

$$F_{\gamma c} = F_{qc} = \exp\left\{\left(-4.4 + 0.6\frac{B}{L}\right)\tan\left(\phi'\right) + \left[\frac{\left(3.07\sin\left(\phi'\right)\right)\left(\log 2I_r\right)}{1+\sin\left(\phi'\right)}\right]\right\} \tag{1.16}$$

F_{cc} can be calculated depending on the soil friction angle as follows:

- For a soil with $\phi' = 0$:

$$F_{cc} = 0.32 + 0.12\frac{B}{L} + 0.60\log I_r \tag{1.17}$$

- For a soil with $\phi' > 0$:

$$F_{cc} = F_{qc} - \frac{1 - F_{qc}}{N_q \tan\phi'} \tag{1.18}$$

1.2.1.5 Effect of Groundwater Table

Sometimes, groundwater table is deep enough so that its effect on the bearing capacity calculations need not be considered. However, sometimes the groundwater table is shallow and its effect on the bearing capacity calculations is essential. For these reasons, the following cases need to be identified depending on the depth of the groundwater table to modify for its effect on the calculations.

- Case 1:
  If the depth of the groundwater table, D_w, is between zero and D_f $(0 \leq D_w \leq D_f)$, then the value of the unit weight γ and the effective stress q in the ultimate bearing capacity equation (Eq. 1.1) is taken as

$$\gamma = \gamma_2 - \gamma_w \tag{1.19}$$

$$q = \gamma_1 \times D_w + (\gamma_2 - \gamma_w) \times (D_f - D_w) \tag{1.20}$$

  where γ_1 is the unit weight of soil above the groundwater table, γ_2 is the saturated unit weight of soil below the groundwater table, and γ_w is the unit weight of water (see Fig. 1.1).
- Case 2:
  If D_w is between D_f and $D_f + B$, then Eqs. (1.21) and (1.22) can be used to calculate q and γ, respectively:

$$q = \gamma_1 D_f \tag{1.21}$$

$$\gamma = \frac{(D_w - D_f)}{B}(\gamma_1 - \gamma_2 + \gamma_w) + \gamma_2 - \gamma_w \tag{1.22}$$

- Case 3:
  If D_w is greater than $D_f + B$, then Eqs. (1.21) and (1.23) can be used to determine q and γ, respectively:

$$\gamma = \gamma_1 \tag{1.23}$$

1.2.1.6 Effect of Bending Moments on Foundation (Load Eccentricity)

To account for the effect of the applied bending moments on the bearing capacity calculations of shallow foundations, the effective area method by Meyerhof (1953) is adopted. To do so, the load eccentricities (e_B, e_L, or e_D) as a result of the applied bending moments (M_B, M_L, M_D) on the different foundation shapes (square/rectangular, circular, and strip/continuous) as shown in Fig. 1.4 are required. Load eccentricities can be calculated using the following equations:

- For square/rectangular foundations:

$$e_B = \frac{M_B}{V} \tag{1.24}$$

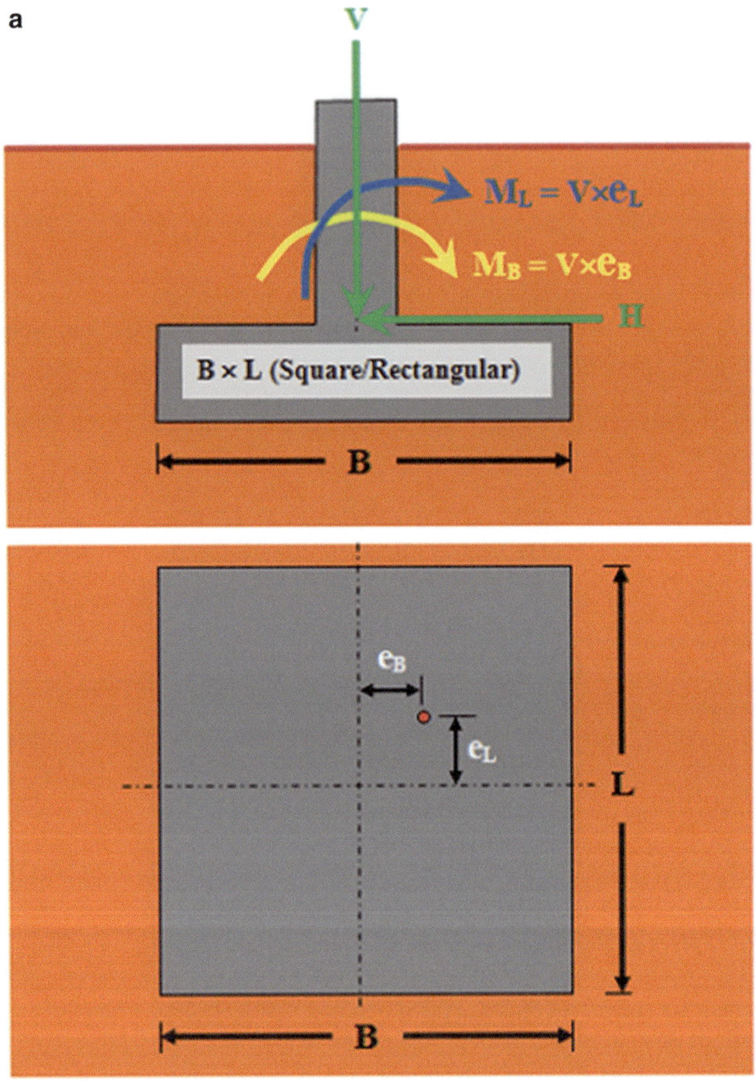

Fig. 1.4 Load eccentricity cases and application of bending moment: (**a**) Application of bending moments on square/rectangular foundation; (**b**) application of bending moments on circular foundation; (**c**) application of bending moments on strip/continuous foundation

$$e_L = \frac{M_L}{V} \tag{1.25}$$

– For circular foundations:

$$e_D = \frac{M_D}{V} \tag{1.26}$$

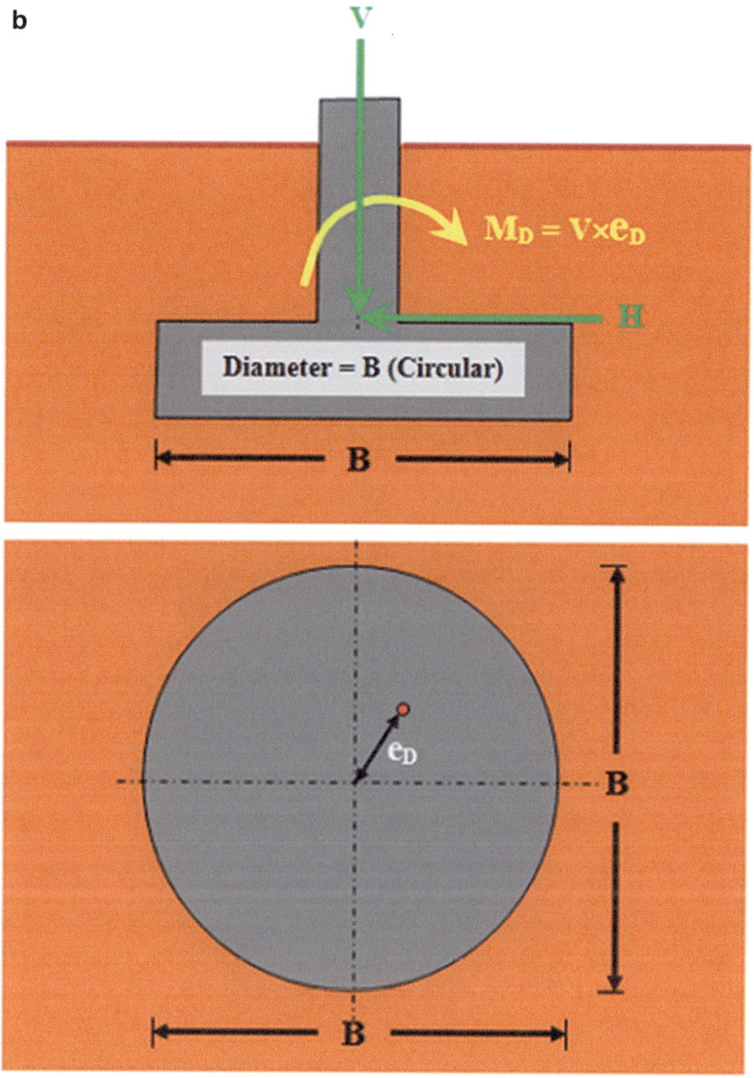

Fig. 1.4 (continued)

– For strip/continuous foundations:

Equation (1.24) can be used,
where
e_B, e_L, and e_D are the load eccentricities in the B, L, and D directions, and
M_B, M_L, and M_D are the applied bending moments on the foundation in the B, L, and D directions.

After calculating the load eccentricities, one should determine the effective dimensions of the foundation (B' and L'). The effective dimensions must be used

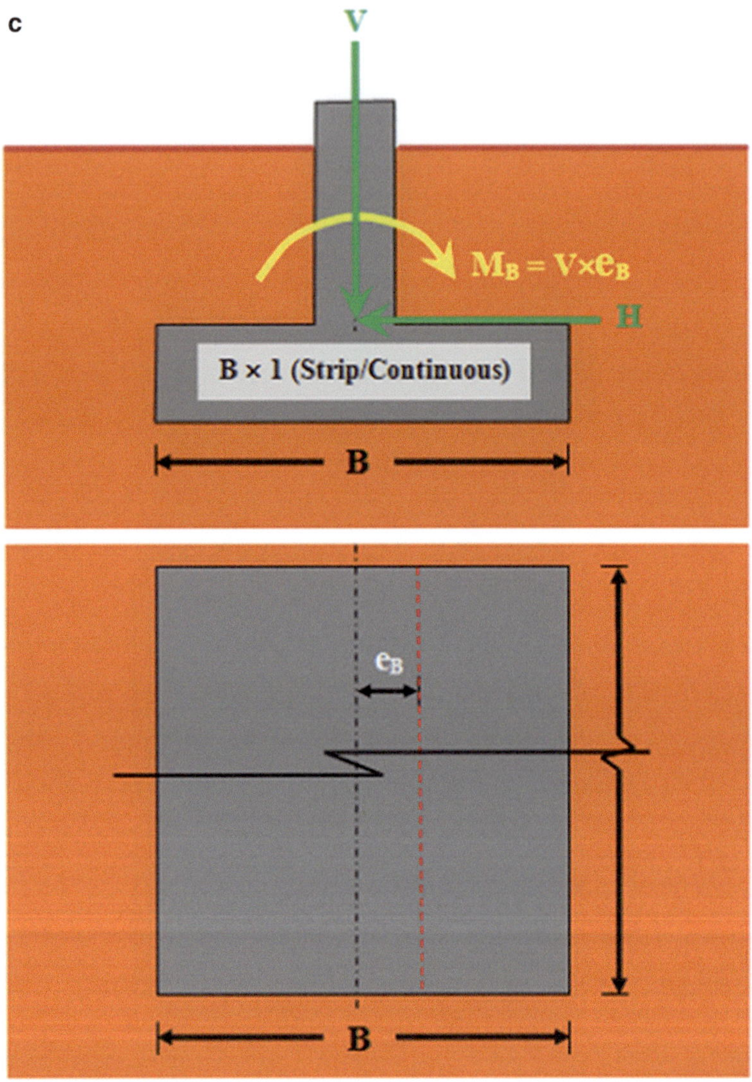

Fig. 1.4 (continued)

in the bearing capacity equation instead of the original foundation dimensions (B and L). The effective dimensions can be determined depending on the load eccentricity condition (one-way or two-way) as follows:

- One-way eccentricity
  The following equations must be used for square/rectangular foundations:
  Eccentricity in the B direction only:

$$B^{'} = B - 2e_B \tag{1.27}$$

$$L' = L \tag{1.28}$$

Eccentricity in the L direction only:

$$L' = L - 2e_L \tag{1.29}$$

$$B' = B \tag{1.30}$$

However, the following equation must be used for strip/continuous foundations:

$$B' = B - 2e_B \tag{1.31}$$

Also, the following equations must be used for circular foundations:

$$B' = f_1 D \tag{1.32}$$

$$L' = \frac{f_2 D^2}{B'} \tag{1.33}$$

where f_1 and f_2 are calculated using Eqs. (1.34) and (1.35). Equations (1.34) and (1.35) are applicable for ${}^{e_D}/_{D}$ ratio between 0.05 and 0.5:

$$f_1 = 43.473\left(\frac{e_D}{D}\right)^4 - 61.224\left(\frac{e_D}{D}\right)^3 + 32.094\left(\frac{e_D}{D}\right)^2 - 8.7505\left(\frac{e_D}{D}\right) + 1.2896 \tag{1.34}$$

$$f_2 = 1.5303\left(\frac{e_D}{D}\right)^2 - 2.438\left(\frac{e_D}{D}\right) + 0.8257 \tag{1.35}$$

- Two-way eccentricity
  Highter and Anderes (1985) suggested four two-way eccentricity cases with respect to the $\frac{e_B}{B}$ and $\frac{e_L}{L}$ ratios. To identify the appropriate case, Fig. 1.5 can be used. Then, the effective dimensions must be determined using the suggested equations for each load eccentricity case from the four cases.

Case I:

$$A' = \frac{1}{2} B_1 L_1 \tag{1.36}$$

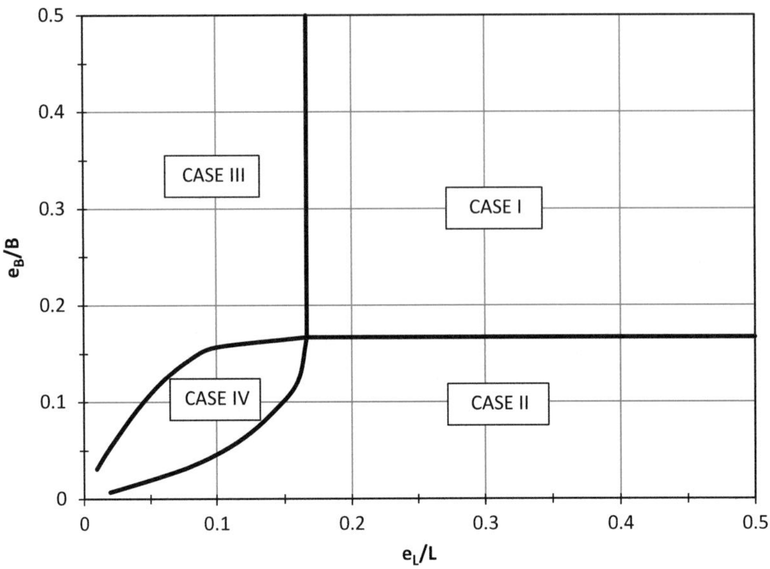

Fig. 1.5 Two-way load eccentricity cases

$$\frac{B_1}{B} = 1.5 - 3\left(\frac{e_B}{B}\right) \tag{1.37}$$

$$\frac{L_1}{L} = 1.5 - 3\left(\frac{e_L}{L}\right) \tag{1.38}$$

where A' can be calculated using Eq. (1.36) with the values of B_1 and L_1 from Eqs. (1.37) and (1.38).

Effective length can be taken as larger of L_1 or B_1, and the effective length can be expressed as L' or B', depending on the dimension that controls. Once L' or B' is found, the other unknown effective dimension can be found through the relationship in Eq. (1.36).

Case II:

$$A' = \frac{1}{2}(L_1 + L_2)B \tag{1.39}$$

$$\frac{L_1}{L} = \left(-18.8357\left(\frac{e_B}{B}\right)^2 + 6.22019\left(\frac{e_B}{B}\right) + 0.95889\right)\left(-2.0651\left(\frac{e_L}{L}\right) + 1.038\right) \tag{1.40}$$

$$\frac{L_2}{L} = \left(2.518265\left(\frac{e_B}{B}\right)^2 - 2.86483\left(\frac{e_B}{B}\right) + 0.40649\right)\ \left(-5.05047\left(\frac{e_L}{L}\right) + 2.52145\right) \tag{1.41}$$

$$L' = L_1 \text{ or } L_2 \quad \text{Whichever is larger} \tag{1.42}$$

$$B' = \frac{A'}{L'} \qquad \text{Whichever is larger} \tag{1.43}$$

where L_1 and L_2 can be used to calculate A' and L' directly from Eqs. (1.39) and (1.42). In addition, B' can be calculated from Eq. (1.43), using A' and L'.

Case III:

$$A' = \frac{1}{2}(B_1 + B_2)L \tag{1.44}$$

$$\frac{B_1}{B} = \left(-10.3897\left(\frac{e_B}{B}\right) + 5.219617\right)\ \left(0.711802\left(\frac{e_L}{L}\right) + 0.202047\right) \tag{1.45}$$

$$\frac{B_2}{B} = \left(-1.46405\left(\frac{e_B}{B}\right) + 0.731667\right)\ \left(-8.59628\left(\frac{e_L}{L}\right) + 1.37955\right) \tag{1.46}$$

$$L' = L \tag{1.47}$$

$$B' = \frac{A'}{L} \tag{1.48}$$

where B_1 and B_2 can be found through Eqs. (1.45) and (1.46), and these values can then be used to acquire A' from Eq. (1.44). In addition, A' can be used with Eq. (1.48) to yield B'.

Case IV:

$$A' = BL - \frac{1}{2}(B - B_2)(L - L_2) \tag{1.49}$$

$$\frac{B_2}{B} = \left(-6{,}470.36\left(\frac{e_B}{B}\right)^4 + 2{,}932.964\left(\frac{e_B}{B}\right)^3 - 493.417\left(\frac{e_B}{B}\right)^2 + 37.716\left(\frac{e_B}{B}\right) - 305.403\right) + \left(-461.303\left(\frac{e_L}{L}\right)^3 + 175.968\left(\frac{e_L}{L}\right)^2 - 26.045\left(\frac{e_L}{L}\right) + 305.8\right) \tag{1.50}$$

$$\frac{L_2}{L} = \left(-549.42\left(\frac{e_B}{B}\right)^3 + 202.478\left(\frac{e_B}{B}\right)^2 - 28.87\left(\frac{e_B}{B}\right)^2 + 102.43\right) + \left(355.07\left(\frac{e_L}{L}\right)^3 - 136.35\left(\frac{e_L}{L}\right)^2 + 18.05\left(\frac{e_L}{L}\right) - 101.6\right) \tag{1.51}$$

$$L' = L \tag{1.52}$$

$$B' = \frac{A'}{L} \tag{1.53}$$

where B_2 and L_2 can be determined through Eqs. (1.50) and (1.51). Then, Eq. (1.53) can be used to calculate B'.

1.2.1.7 Effect of Foundation Shape

To account for the various foundation shapes in the bearing capacity calculations, De Beer (1970) suggested the following equations to calculate the shape factors required in the ultimate bearing capacity equation (Eq. 1.1):

$$F_{cs} = 1 + \left(\frac{B'}{L'}\right)\left(\frac{N_q}{N_c}\right) \tag{1.54}$$

$$F_{qs} = 1 + \left(\frac{B'}{L'}\right)\tan\left(\phi'\right) \tag{1.55}$$

$$F_{\gamma s} = 1 - 0.4\left(\frac{B'}{L'}\right) \tag{1.56}$$

When no bending moments are applied to the foundation (no load eccentricity), the effective dimensions in Eqs. (1.54) through (1.56) are replaced with the original foundation dimensions.

1.2.2 *Elastic Settlement of Foundation*

The elastic settlement of a shallow foundation resting on a homogeneous soil with a rock layer at a depth H below the base of foundation as shown in Fig. 1.6 can be estimated using the improved equation presented by Mayne and Poulos (1999). The elastic settlement (S_e) below the center of foundation according to Mayne and Poulos can be calculated using the following equation:

$$S_e = \frac{q_{all(net)} B_e I_G I_F I_E}{E_{so}}\left(1 - \mu_s^2\right) \tag{1.57}$$

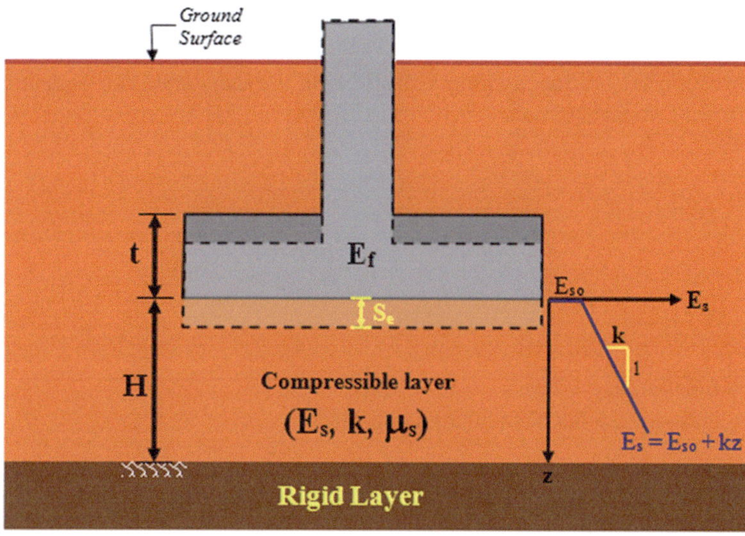

Fig. 1.6 Elastic settlement of shallow foundation

where $q_{\mathrm{all(net)}}$ is the net allowable pressure on the foundation, and I_{F} is the foundation rigidity factor, I_{E} is the foundation embedment correction factor, and μ_{s} is the Poisson's ratio.

- B_{e} of a rectangular foundation can be found using the following equation:

$$B_{\mathrm{e}} = \sqrt{\frac{4BL}{\pi}} \tag{1.58}$$

where B is the width of foundation, and L is the length of foundation

- B_{e} of a circular foundations is simply equal to B, where B is the diameter of the foundation:

$$B_{\mathrm{e}} = B \tag{1.59}$$

The modulus of elasticity of the compressible soil layer can be written as

$$E_{\mathrm{s}} = E_{\mathrm{so}} + kz \tag{1.60}$$

where k is the rate of increase in the elastic modulus of the soil with depth, E_{so} is the elastic modulus of the soil at the base of the foundation, and z is the depth.

Equation (1.57) must be used to determine the net allowable load bearing capacity, $q_{\mathrm{all(net)}}$, for an assumed maximum permissible foundation elastic settlement.

1.2.2.1 Effect of Foundation Rigidity

The correction factor to account for the rigidity of the foundation can be determined using the following equation:

$$I_F = \frac{\pi}{4} + \frac{1}{4.6 + 10\left(\frac{E_f}{E_{so} + \frac{B_e}{2}k}\right)\left(\frac{2t}{B_e}\right)^3} \tag{1.61}$$

where

t = foundation thickness
E_f = elastic modulus of foundation material

1.2.2.2 Effect of Foundation Embedment

To account for the effect of foundation embedment (D_f), the following correction factor must be calculated:

$$I_E = 1 - \left(\frac{1}{3.5\exp\left(1.22\mu_s - 0.4\left(\frac{B_e}{D_f} + 1.6\right)\right)}\right). \tag{1.62}$$

1.2.2.3 Effect of the Variation of Elastic Modulus of Soil with Depth

This correction factor, I_G, depends on the variation of elastic modulus of soil with depth as follows:

$$I_G = f(\beta_1, \beta_2) \tag{1.63}$$

where

$$\beta_1 = \frac{H}{B_e} \tag{1.64}$$

$$\beta_2 = \log\left(\frac{E_o}{kB_e}\right) \tag{1.65}$$

$$I_G = \left(-0.01189e^{-1.26658\beta_1} + 0.012608\right) \times (0.34865\beta_2^5 + 1.05867\beta_2^4 - 4.2618\beta_2^3 - 7.1333\beta_2^2 + 28.92718\beta_2 + 51.4275) \quad (1.66)$$

Equation (1.66) is only applicable for β_1 values between 0.2 and 30.

1.2.3 *Foundation Loads*

1.2.3.1 Foundation Loads Based on Bearing Capacity

Using the ultimate bearing capacity (q_u) from Eq. (1.1), one can calculate the allowable load bearing capacity q_{all} for any given factor of safety (FS) with the relationship stated in Eq. (1.67) below:

$$q_{all} = \frac{q_u}{FS} \quad (1.67)$$

Also, the vertical allowable and ultimate loads the foundation can carry are determined using the following equations:

$$V_u = q_u \cdot A' \quad (1.68)$$

$$V_{all} = \frac{V_u}{FS} \quad (1.69)$$

Similarly, the horizontal ultimate load using the following equation:

$$H_u = V_u \tan(\beta) \quad (1.70)$$

Additionally, bending moments can be determined as follows:

(a) Circular:

$$M_{u-D} = V_u \cdot e_D \quad (1.71)$$

(b) Strip/continuous:

$$M_{u-B} = V_u \cdot e_B \quad (1.72)$$

(c) Square/rectangular:

$$M_{u-L} = V_u \cdot e_L \quad (1.73)$$

1.2.3.2 Foundation Loads Based on Elastic Settlement

Using the net allowable load bearing capacity obtained from Eq. (1.57) for an assumed allowable elastic settlement value, one can calculate the allowable load bearing capacity based on elastic settlement q_{all} for any given factor of safety with the relationship stated in the equation below:

$$q_{\mathrm{all}} = q_{\mathrm{all(net)}} + \frac{q}{\mathrm{FS}} \tag{1.74}$$

Then, vertical load can be determined as follows:

– For square/rectangular foundations, the following equation must be used:

$$V_{\mathrm{all}} = \frac{q_{\mathrm{all}} \times B \times L}{1 + \frac{6e_B}{B} + \frac{6e_L}{L}} \tag{1.75}$$

– For circular foundation, the following equation must be used:

$$V_{\mathrm{all}} = \frac{q_{\mathrm{all}} \times \left(\frac{\pi}{4} D^2\right)}{1 + \frac{8e_D}{D}} \tag{1.76}$$

Also, bending moments based on allowable loads:

(a) For circular:

$$M_{\mathrm{all}-D} = V_{\mathrm{all}} e_D \tag{1.77}$$

(b) For strip/continuous:

$$M_{\mathrm{all}-B} = V_{\mathrm{all}} e_B \tag{1.78}$$

(c) For square/rectangular:

$$M_{\mathrm{all}-L} = V_{\mathrm{all}} e_L \tag{1.79}$$

1.3 Step-by-Step Procedure

To develop the bearing capacity and elastic settlement design charts, the following steps shall be followed:

Step 1:
Assume the foundation width for square/rectangular/strip foundations, or assume the foundation diameter for a circular foundation.

Step 2:
Determine the bearing capacity factors, using the known ϕ' (Eqs. 1.2–1.4).

Step 3:
Determine the depth factors using η (Eqs. 1.5–1.9).

Step 4:
Determine the load inclination factors using the assumed $\frac{H}{V}$ ratio along with ϕ' (Eqs. 1.10–1.12).

Step 5:
Determine soil compressibility factors using the foundation dimensions, c', ϕ', and μ_s (Eqs. 1.13–1.18).

Step 6:
To account for the depth of the groundwater table, determine q and γ depending on the depth of the water table and the depth of the foundation (Eqs. 1.19–1.23).

Step 7:
Determine the effective dimensions to account for the load eccentricity on the foundation. The effective dimensions can be determined using the actual dimensions of the foundation along with the moment and vertical load applied to the foundation (Eqs. 1.24–1.53 and Fig. 1.4).

Step 8:
Determine the shape factors using the effective dimensions of the foundation (Eqs. 1.54–1.56).

Step 9:
Determine the allowable and the ultimate bearing capacity of the foundation using the q_u from Eq. (1.1) along with the factor of safety for the foundation (Eq. 1.67).

Step 10:
Determine the foundation load based on bearing capacity analyses (Eqs. 1.68–1.75).

Step 11:
Determine I_G, I_E, and I_F (Eqs. 1.61–1.66).

Step 12:
Determine the net applied pressure allowable ($q_{all(net)}$) based on the assumed elastic settlement of the foundation, S_e, rigidity correction factor (I_F), embedment correction factor (I_E), influence factor (I_G), and Poisson's ratio (μ_s) (Eqs. 1.57–1.59).

Step 13:
Determine the q_{all} from Eq. (1.74).

Step 14:
Determine the foundation loads based on elastic settlement analyses (Eqs. 1.76–1.79).

Step 15:
Steps 1 through 14 can be repeated for several assumed B values to develop the required foundation capacity charts.

The above step-by-step procedure can be followed and repeated for different foundation shapes and depths. Figure 1.7 shows the flowchart that summarizes the above procedure to develop design capacity charts for shallow foundations on homogeneous soil based on bearing capacity and elastic settlement criteria.

1.4 Design Problems

In this section, a number of design problems are introduced and solved following the step-by-step procedure explained earlier in the previous section. These problems were selected to give exposure to a wide variety of challenges that can be faced while designing a shallow foundation system while helping us iron out the finer details of the theories represented above.

1.4.1 Strip/Continuous Foundation

Develop a bearing capacity design chart for a strip/continuous foundation with only vertical load (no horizontal or bending moment applied). Table 1.1 below summarizes soil properties and design parameters used in this problem. Ignore soil compressibility.

1.4.1.1 Hand Solution

To develop bearing capacity design charts, the steps below must be followed:

Step 1:
Assume foundation width, $B = 1$ m.

Step 2:
Determine bearing capacity factors.

Using Eq. (1.2):

$$N_q = \tan^2\left(45 + \frac{30}{2}\right)e^{\pi \tan(30)} = 18.414$$

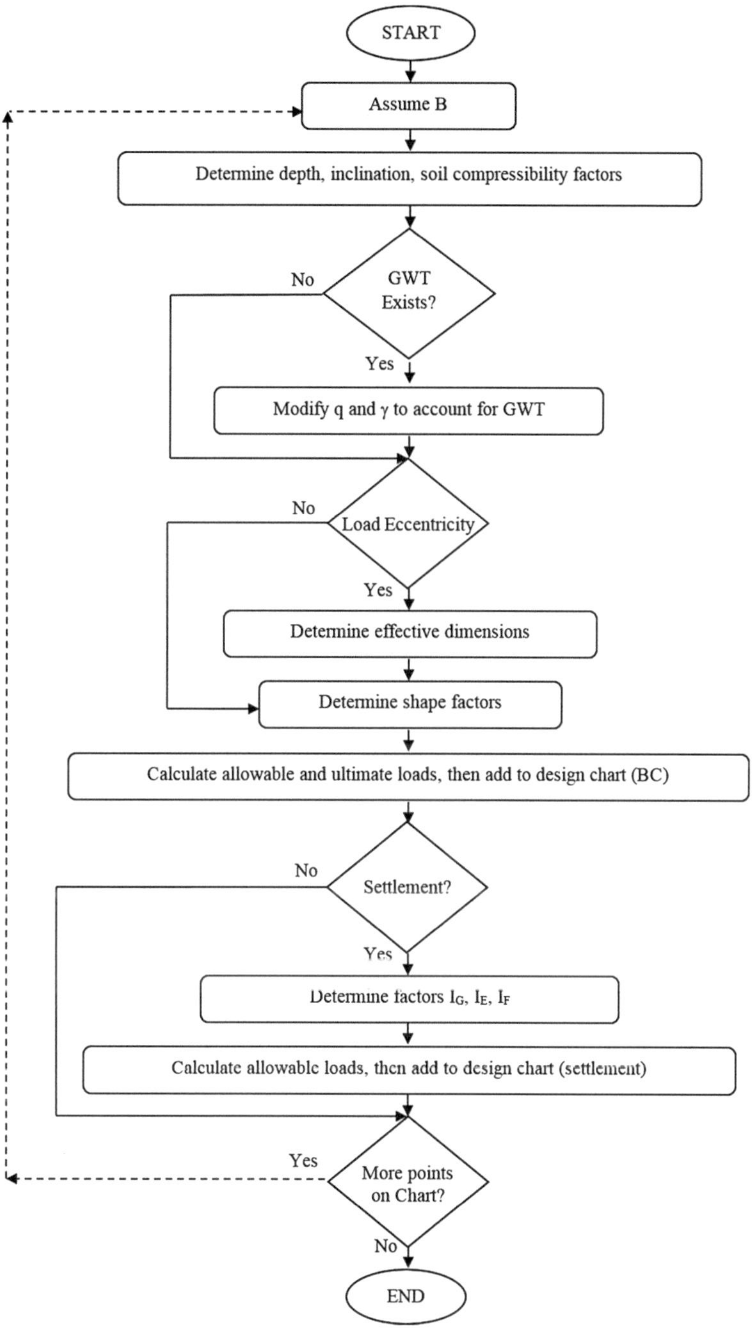

Fig. 1.7 Flowchart for development of design capacity charts

Table 1.1 Soil properties and design parameters

Property	Value	Unit
D_f	1.5	m
C	0	kN/m^2
ϕ	30	°
γ	18.85	kN/m^3
FS	3.0	–
GWT depth	Too deep	–

Using Eq. (1.3):

$$N_c = (18.414 - 1)\cot(30) = 30.135$$

Using Eq. (1.4):

$$N_\gamma = 2(18.414 + 1)\tan(30) = 22.417$$

Step 3:
Determine depth factors.

For $\phi' > 0$ and from Eq. (1.8), we find the following:

$$\eta = \frac{1.5}{1} = 1.5 > 1$$

Therefore,

Depth factors can be determined using Eqs. (1.6) and (1.7) as follows:

$$F_{qd} = 1 + 2\tan(30)(1 - \sin(30))^2 \tan^{-1}\left(\frac{1.5}{1}\right) = 1.28$$

$$F_{cd} = 1.28 - \frac{1 - 1.28}{30.135 \times \tan 30} = 1.296$$

$$F_{\gamma d} = 1$$

Step 4:
Determine load inclination factors.

Since the *H*/*V* ratio is zero (no horizontal loading is considered), the angle β can be computed using Eq. (1.12) as follows:

$$\beta = \tan^{-1}\left(\frac{0}{0}\right) = 0$$

Thus,
Using Eq. (1.10):

$$F_{ci} = F_{qi} = \left(1 - \frac{0^\circ}{90^\circ}\right)^2 = 1$$

Using Eq. (1.11):

$$F_{\gamma i} = \left(1 - \frac{0}{30}\right)^2 = 1$$

Step 5:
Determine soil compressibility factors.

Since the effect of soil compressibility on bearing capacity calculations is not required, the compressibility factors are set to 1.

Therefore,

$$F_{cc} = F_{qc} = F_{\gamma c} = 1$$

Step 6:
Determine q and γ to account for the GWT.

The groundwater table in this problem is too deep and its effect can be ignored. So, the unit weight of the foundation soil can be taken as $\gamma = 18.85\,\text{kN/m}^2$. Also, the effective stress at the foundation base can be calculated as follows:

$$q = 18.85 \times 1.5 = 28.275\,\text{kN/m}^2$$

Step 7:
Determine the effective dimensions to account for the effect of load eccentricity.

For strip/continuous foundations, Eq. (1.31) applies:

$$B' = B - 2e_B$$

where e_B can be determined from Eq. (1.24):

$$e_B = \frac{0}{V} = 0$$

Therefore:

$$B' = B = 1$$

Step 8:
Determine shape factors.

For a strip/continuous foundation, the shape factors are equal to 1. So,

$$F_{cs} = F_{qs} = F_{\gamma s} = 1$$

Step 9:
Now, one can calculate the ultimate bearing capacity of the foundation using Eq. (1.1):

$$q_u = (0)(30.135)(1.296)(1)(1)(1)(1) + (28.275)(1)(18.414)(1)(1.28)(1)(1) + \frac{1}{2}(18.85)(1)(22.417)(1)(1)(1)(1) = 877.349\,\text{kN/m}^2$$

The ultimate vertical foundation load based on bearing capacity can then be calculated from Eq. (1.68):

$$V_u = 877.349 \times 1 \times 1 = 877.349\,\text{kN/m}$$

Also, the allowable load bearing capacity is calculated using Eq. (1.67):

$$q_{all} = \frac{877.349}{3} = 292.44\,\text{kN/m}^2$$

And from Eq. (1.69):

$$V_{all} = 292.44 \times 1 \times 1 = 292.44\,\text{kN/m}$$

Step 10:
Several foundation widths (*B*) can be chosen and the above steps can be repeated to determine the foundation capacity and to be able to develop a bearing capacity design chart.

1.4.1.2 foundationPro Solution

After launching the Shallow-1 application of foundationPro program, the main sections of the applications will appear as shown in Fig. 1.8.

General Information Section

General information alongside the unit format and the factor of safety is entered as shown in Fig. 1.9.

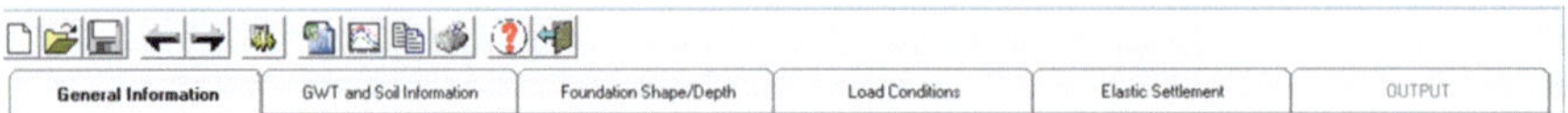

Fig. 1.8 Main sections of Shallow-1 application of foundationPro program

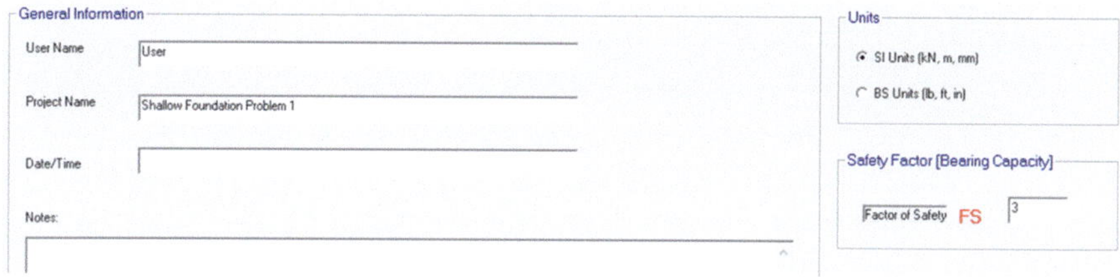

Fig. 1.9 General information

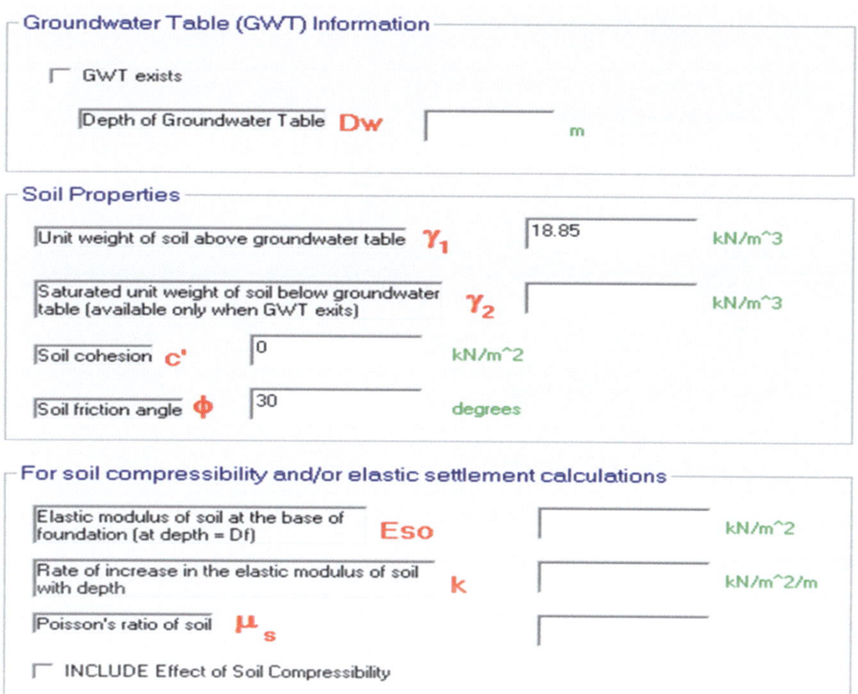

Fig. 1.10 Groundwater table and soil information section

Groundwater Table and Soil Information Section

We don't click "GWT exists" (see Fig. 1.10) since the groundwater table is very deep; we enter γ_1, c', and ϕ. We do not want to consider settlement in this problem. Therefore, no need for any soil compressibility or elastic settlement information.

Foundation Shape/Depth Section

We enter the depth of the foundation as given (see Fig. 1.11), we select the type of foundation, and we select various foundation widths to be analyzed for the given problem statement. *L*/*B* ratio is not required since this is a strip foundation.

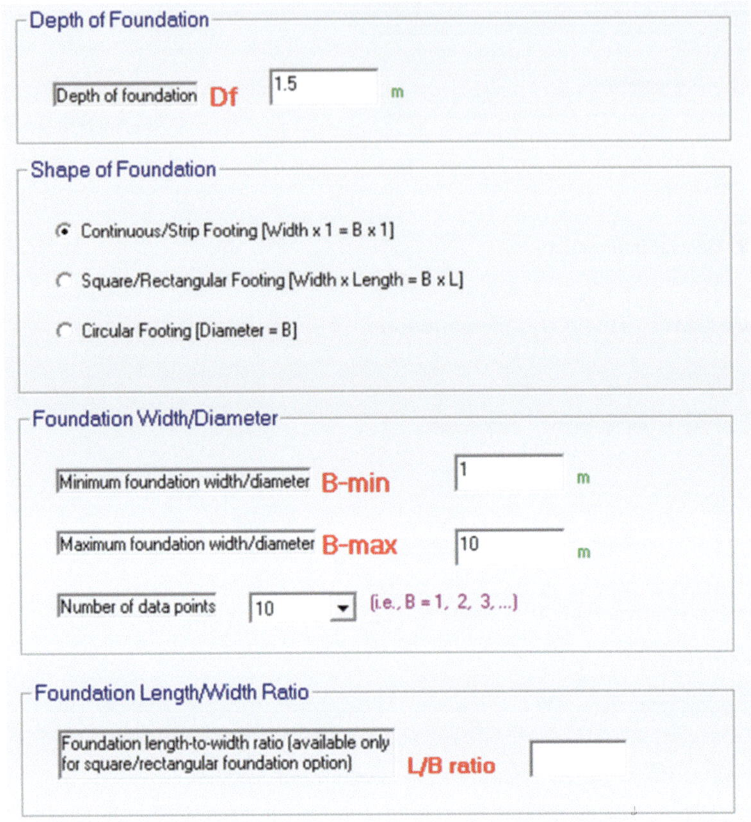

Fig. 1.11 Foundation shape/depth

Load Conditions and Elastic Settlement Sections

We can skip Load Conditions and Elastic Settlement sections in this example since they are not required in this problem. We will discuss those in future design examples. And now, we can hit the run button or the hot key F5 on the keyboard to perform the analysis! The application will check all input data before proceeding in the analyses and will also require to save the data in a file with a name of the user's choice. All saved files will be automatically given the .FPR extension to indicate that these files are used by foundationPro. Once the input check is completed and the analyses are performed, the user can view the Output section and navigate through the output results.

Output Section

One can navigate through the Output section by clicking on the tabs at the bottom of the screen as shown in Figs. 1.12 and 1.13. These tabs can give the results in terms

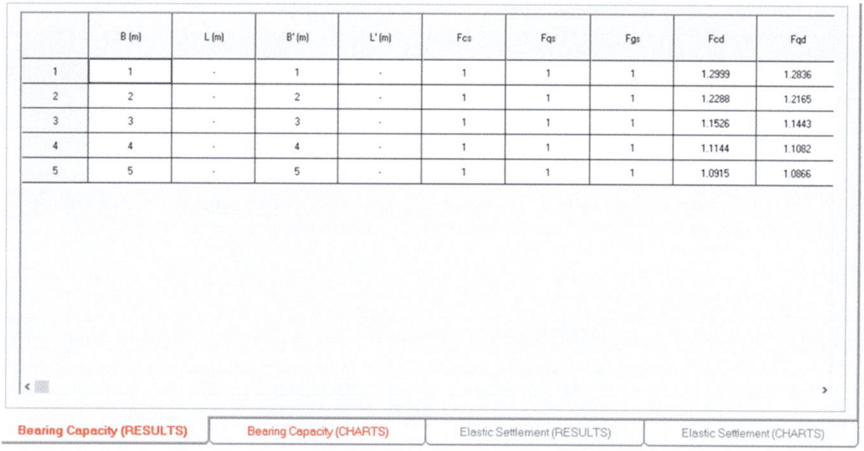

	B (m)	L (m)	B' (m)	L' (m)	Fcs	Fqs	Fgs	Fcd	Fqd
1	1	-	1	-	1	1	1	1.2999	1.2836
2	2	-	2	-	1	1	1	1.2288	1.2165
3	3	-	3	-	1	1	1	1.1526	1.1443
4	4	-	4	-	1	1	1	1.1144	1.1082
5	5	-	5	-	1	1	1	1.0915	1.0866

Fig. 1.12 Bearing capacity results (table format)

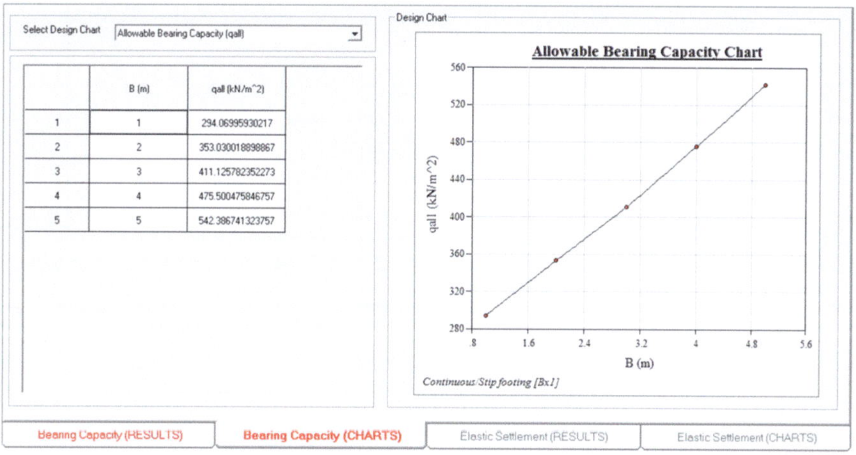

	B (m)	qall (kN/m^2)
1	1	294.06995930217
2	2	353.030018898867
3	3	411.125782352273
4	4	475.500475846757
5	5	542.386741323757

Fig. 1.13 Bearing capacity results (chart format)

of a table or a chart. One can also switch back and forth between bearing capacity results and elastic settlement results using these tabs. Excel spreadsheet button can be used to acquire the results in Excel.

One can right click on the chart and copy it directly from the program as needed. Table 1.2 summarizes some of the results for the selected foundation widths from 1 to 5 m. Allowable bearing capacity chart at various foundation widths is also shown in Fig. 1.14.

Table 1.2 Foundation loads based on bearing capacity analyses

B (m)	q (kN/m^2)	A' (m^2)	V_{all} (kN/m)	q_u (kN/m^2)	V_u (kN/m)
1	28.275	1	293.0154	879.0461	879.0461
2	28.275	2	703.4834	1,055.225	2,110.45
3	28.275	3	1,228.82	1,228.82	3,686.459
4	28.275	4	1,894.918	1,421.189	5,684.755
5	28.275	5	2,701.779	1,621.067	8,105.337

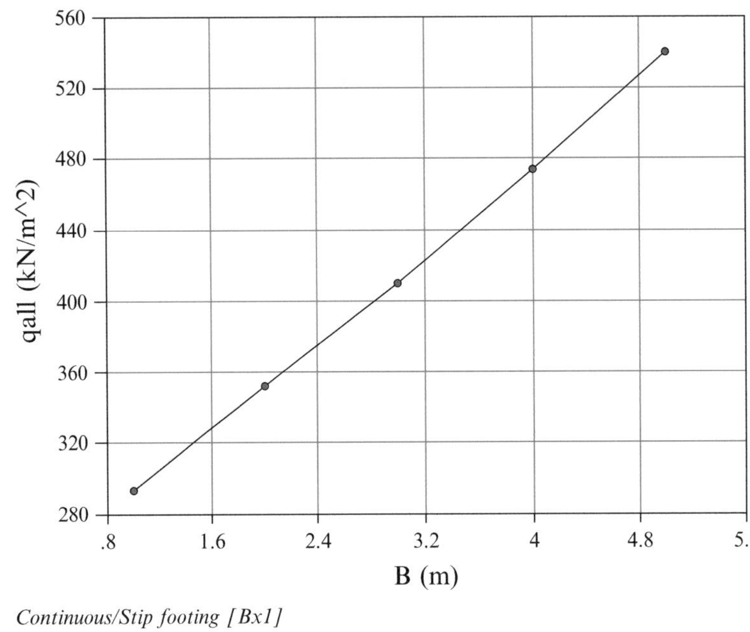

Fig. 1.14 Allowable bearing capacity chart for the continuous/strip foundation

1.4.2 Rectangular Foundation

Develop a bearing capacity design chart for a rectangular foundation with an *L*/*B* ratio of 1.5. No horizontal load and bending moments are applied to the foundation. Factor of safety is 3.0. Other information is listed in Table 1.3 below. Soil compressibility is not included.

1.4.2.1 Hand Solution

To determine the foundation load for this rectangular foundation, one must follow the steps below:

Table 1.3 Soil properties and other design parameters

Property	Value	Unit
D_f	1.3	m
c	46	kN/m^2
ϕ	4	°
γ_1	18.5	kN/m^3
γ_2	19.75	kN/m^3
GWT depth	2	m

Step 1:
Assume foundation dimensions.

So, let us assume $B = 1$ m, and then $L = 1.5$ m since $L/B = 1.5$.

Step 2:
Determine bearing capacity factors.

Bearing capacity factors can be computed from Eqs. (1.2) through (1.4) as follows:

$$N_q = \tan^2\left(45 + \frac{4}{2}\right) e^{\pi \tan(4)} = 1.43$$

$$N_c = (0.43)\cot(4) = 6.19$$

$$N_\gamma = 2(1.43 + 1)\tan(4) = 0.34$$

Step 3:
Determine depth factors.

For $\phi' > 0$ and from Eq. (1.8), η can be calculated as follows:

$$\eta = \frac{D_f}{B} = \frac{1.3}{1} = 1.3 > 1$$

Thus,

Depth factors are calculated using Eqs. (1.6) and (1.7) as below:

$$F_{qd} = 1 + 2\tan(4)(1 - \sin(4))^2 \tan^{-1}\left(\frac{1.3}{1}\right) = 1.11$$

$$F_{cd} = 1.11 - \frac{1 - 1.11}{6.19 \times \tan(4)} = 1.36$$

$$F_{\gamma d} = 1$$

Step 4:
Determine load inclination factors.

Since the H/V ratio is zero, the angle β can be determined from Eq. (1.12) as follows:

$$\beta = \tan^{-1}\left(\frac{0}{0}\right) = 0$$

Therefore,

Inclination factors are found to be 1 after using Eqs. (1.10) and (1.11):

$$F_{ci} = F_{qi} = \left(1 - \frac{0}{90°}\right)^2 = 1$$

$$F_{\gamma i} = \left(1 - \frac{0}{4}\right) = 1$$

Step 5:

Determine soil compressibility factors.

Since the effect of soil compressibility on bearing capacity calculations is not required in this problem, all compressibility factors are set to unity. Therefore,

$$F_{cc} = F_{qc} = F_{\gamma c} = 1$$

Step 6:

Determine q and γ to account for the GWT.

Given that water table is at a depth of 2 m.

Also, given that $\gamma_1 = 18.5\,\mathrm{kN/m^3}$ and $\gamma_2 = 19.75\,\mathrm{kN/m^3}$.

Since $D_f = 1.3 \le D_w = 2 \le (D_f + B) = 2.3$, Case 2 applies.

The effective stress at the foundation base can be calculated using Eq. (1.21):

$$q = 18.5 \times 1.3 = 24.05\,\mathrm{kN/m^2}$$

Also, the unit weight of the foundation soil is calculated from Eq. (1.22) as follows:

$$\gamma = \frac{(2 - 1.3)}{1}(18.5 - 19.75 + 9.81) + 19.75 - 9.81 = 15.93\,\mathrm{kN/m^3}$$

Step 7:

Determine the effective dimensions to account for the effect of load eccentricity.

No load eccentricity is considered in this example. Therefore, effective dimensions are the same as the actual assumed dimensions:

$$B' = B = 1\,\mathrm{m}$$
$$L' = L = 1.5\,\mathrm{m}$$

Step 8:

Determine shape factors.

Shape factors can be determined using Eqs. (1.54) through (1.56) as follows:

$$F_{cs} = 1 + \left(\frac{1.0}{1.5}\right)\left(\frac{1.43}{1.69}\right) = 1.5$$

$$F_{qs} = 1 + \left(\frac{1}{1.5}\right)\tan(4) = 1.04$$

$$F_{\gamma s} = 1 - 0.4\left(\frac{1}{1.5}\right) = 0.733$$

Step 9:
The ultimate bearing capacity of the foundation is now calculated using Eq. (1.1):

$$q_u = (46)(6.19)(1.36)(1)(1)(1.5) + (24.05)(1.43)(1.04)(1)(1)(1.11) + \frac{1}{2}(15.93)(1)(0.34)(0.733)(1)(1)(1) = 487.33\,\text{kN/m}^2$$

From Eq. (1.68):
The ultimate vertical load based on bearing capacity can be determined from Eq. (1.68):

$$V_u = 487.33 \times 1 \times 1.5 = 730.99\,\text{kN}$$

From Eq. (1.67):

$$q_{all} = \frac{487.33}{3} = 162.44\,\text{kN/m}^2$$

From Eq. (1.69):

$$V_{all} = 162.44 \times 1 \times 1.5 = 243.66\,\text{kN}$$

Step 10:
Several foundation dimensions can be selected and the above steps can be repeated to evaluate the impact of the foundation size on the baring capacity of the foundation.

1.4.2.2 foundationPro Solution

General Information Section

We enter the factor of safety and select the SI units as shown in Fig. 1.15.

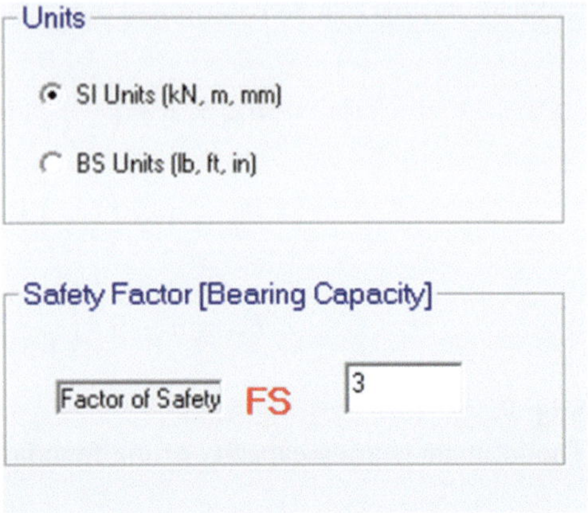

Fig. 1.15 General information section for Problem 1.2

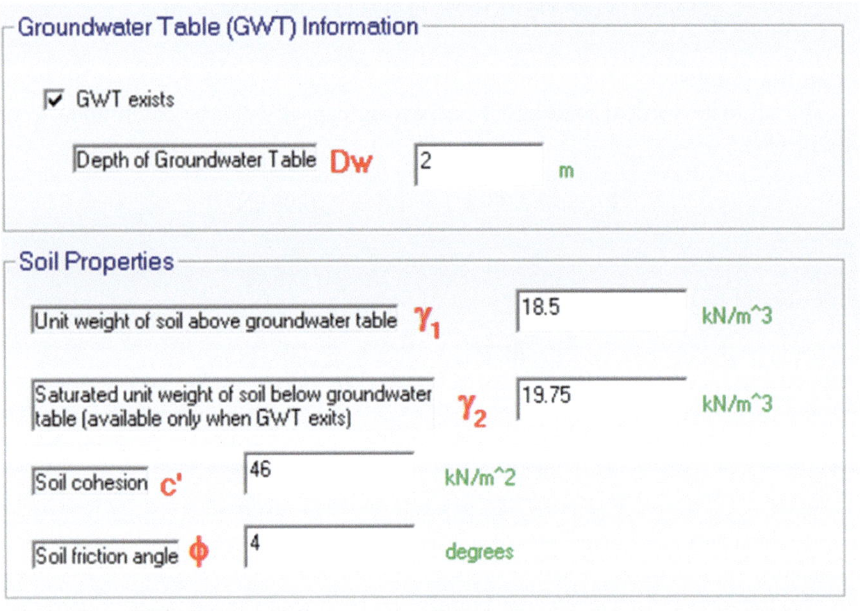

Fig. 1.16 Groundwater table

Groundwater Table and Soil Information Section

In the Groundwater Table and Soil Information section as shown in Fig. 1.16, we enter the depth of the groundwater table, saturated unit weight of the soil below the groundwater table, and unit weight of the soil above the groundwater table. We enter the soil cohesion and friction angle.

Foundation Shape/Depth Section

In this section as shown in Fig. 1.17, we enter the depth of the foundation as well as select the shape of the foundation (square/rectangular). We enter the values we would like to consider for the foundation dimensions; we also enter the *L*/*B* ratio given in the problem.

We click the run button and get the results in the Output section. Table 1.4 summarizes the foundation loads for several foundation dimensions. The allowable bearing capacity chart for this problem is shown in Fig. 1.18.

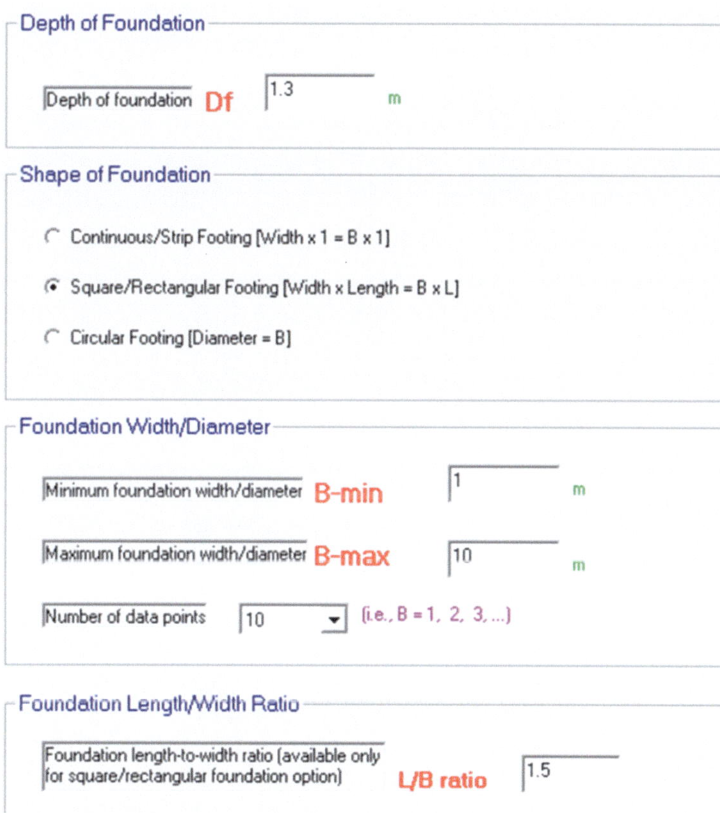

Fig. 1.17 Foundation shape section

Table 1.4 Results for the problem

B (m)	L (m)	q (kN/m^2)	A' (m^2)	q_{all} (kN/m^2)	V_{all} (kN)	q_u (kN/m^2)	V_u (kN)
1	1.5	24.05	1.5	163.6529672	245.4795	490.9589016	736.4383524
2	3	24.05	6	152.0488739	912.2932	456.1466218	2,736.879731
3	4.5	24.05	13.5	142.640744	1,925.65	427.9222321	5,776.950133
4	6	24.05	24	138.1443675	3,315.465	414.4331026	9,946.394461
5	7.5	24.05	37.5	135.6126924	5,085.476	406.8380771	15,256.42789
6	9	24.05	54	134.0633679	7,239.422	402.1901036	21,718.2656
7	10.5	24.05	73.5	133.0753866	9,781.041	399.2261599	29,343.12275
8	12	24.05	96	132.4382449	12,714.07	397.3147348	38,142.21454
9	13.5	24.05	121.5	132.0349962	16,042.25	396.1049887	48,126.75613
10	15	24.05	150	131.7954727	19,769.32	395.386418	59,307.96269

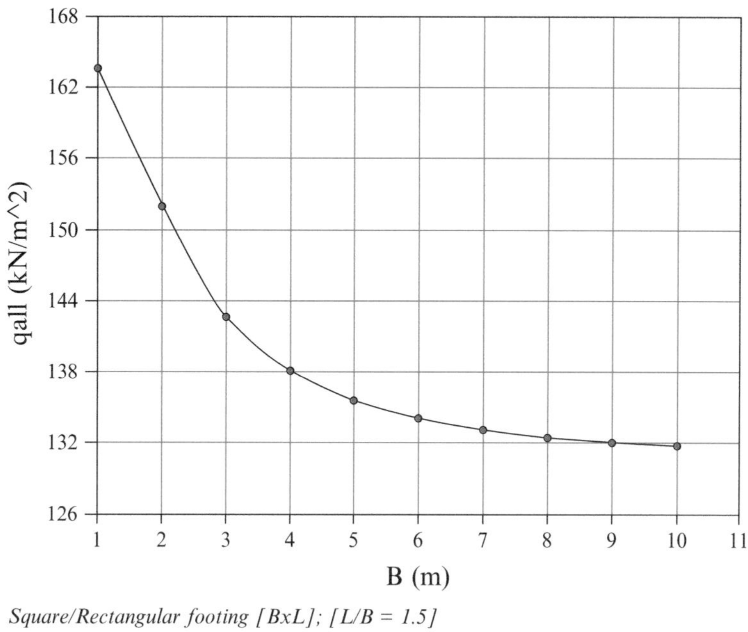

Fig. 1.18 Allowable bearing capacity chart for the problem

Table 1.5 Soil properties and design parameters

Property	Value	Unit
D_f	5	ft
c	540	lb/ft^2
ϕ	27	°
γ	104	lb/ft^3
GWT depth	Too deep	ft

1.4.3 Square Foundation with Horizontal Loading

Develop a bearing capacity design chart for a square foundation with a horizontal loading that is 25 % of the vertical load ($H/V = 0.25$). Factor of safety is 3.0. Soil compressibility is not included. Other information is listed in Table 1.5 as follows:

1.4.3.1 Hand Solution

To determine the ultimate bearing capacity q_u:

Step 1:
From the given statement $L/B = 1.0$.
So let us assume $L = 1$ ft. $B = 1$ ft.

Step 2:
Determine bearing capacity factor.
Using Eq. (1.2):

$$N_q = \tan^2\left(45 + \frac{27}{2}\right)e^{\pi \tan(27)} = 13.2$$

Using Eq. (1.3):

$$N_c = (13.2 - 1)\cot 27 = 23.94$$

Using Eq. (1.4):

$$N_\gamma = 2(13.2 + 1)\tan 27 = 14.47$$

Step 3:
Determine depth factors.
Using Eq. (1.8):

$$\eta = \frac{D_\mathrm{f}}{B} = \frac{5}{1} = 5 > 1$$

Since $\phi' > 0$ use Eq. (1.6):

$$F_{q\mathrm{d}} = 1 + 2\tan 27(1 - \sin 27)^2 \tan^{-1}\left(\frac{5}{1}\right) = 1.417$$

Use Eq. (1.7):

$$F_{\mathrm{cd}} = 1.417 - \frac{1 - 1.417}{23.94 \times \tan(27)} = 1.4511$$

$$F_{\gamma\mathrm{d}} = 1$$

Step 4:
Determine load inclination factors.
Using Eq. (1.12):

$$\beta = \tan^{-1}(0.25) = 14.19^\circ$$

Using Eq. (1.10):

$$F_{\mathrm{ci}} = F_{q\mathrm{i}} = \left(1 - \frac{14.19^\circ}{90^\circ}\right) = 0.712$$

Using Eq. (1.11):

$$F_{\gamma i} = \left(1 - \frac{14.19}{27}\right) = 0.48$$

Step 5:
Determine soil compressibility factors.
Since the soil compressibility is not included $F_{cc} = F_{qc} = F_{\gamma c} = 1$.

Step 6:
Determine q and γ to account for the GWT.
And given that the water table is very deep, Case 3 applies.
From Eq. (1.23):

$$\gamma = \gamma_1 = 104 \text{ lb/ft}^3$$

From Eq. (1.21):

$$q = 104 \times 5 = 520 \, \text{lb/ft}^2$$

Step 7:
Determine the effective dimensions (effect load eccentricity).
There is no eccentricity; therefore,
$B' = B$ and $L' = L$

Step 8:
Determine shape factors.
From Eq. (1.54):

$$F_{cs} = 1 + \left(\frac{1.0}{1.0}\right)\left(\frac{13.20}{23.94}\right) = 1.55$$

From Eq. (1.55):

$$F_{qs} = 1 + \left(\frac{1}{1}\right) \tan(27) = 1.50$$

From Eq. (1.57):

$$F_{\gamma s} = 1 - 0.4\left(\frac{1}{1}\right) = 0.6$$

Step 9:
Ultimate bearing capacity of the square foundation.
Using Eq. (1.1):

$$q_u = (540)(23.94)(1.4511)(0.712)(1)(1.55) + (520)(1.5)(13.2)(0.712)(1.417)(1) + \frac{1}{2}(104)(1)(14.47)(0.6)(1)(0.48)(1) = 31{,}305.2 \, \text{lb/ft}^2$$

From Eq. (1.68):

$$V_u = 31{,}305.2 \times 1 \times 1 = 31{,}305.2 \, \text{lb}$$

From Eq. (1.67):

$$q_{\text{all}} = \frac{31,305.2}{3} = 10,435.06\,\text{lb/ft}^2$$

From Eq. (1.69):

$$V_{\text{all}} = 14,035.06 \times 1 \times 1 = 14,035.06\,\text{lb}$$

Using the ratio given in the problem:

$$H_{\text{u}} = 31,305.2 \times 0.25 = 7,826.02\,\text{lb}$$

Step 10:
Several *B* values can be now chosen and the procedure can be repeated to evaluate the impact of the foundation size on the baring capacity of the foundation.

1.4.3.2 foundationPro Solution

General Information Section

BS units are selected for this problem as shown in Fig. 1.19.

Groundwater Table and Soil Information Sections

We don't click "GWT exists" since the groundwater table is very deep; we enter γ_1, c', and ϕ.

Foundation Shape/Depth Section

We enter the depth of the foundation as well as the foundation shape (square/rectangular). Also, we enter the values we would like to consider for the foundation dimensions. The *L*/*B* ratio of 1 for square foundation must be entered to define a square foundation as shown in Fig. 1.20.

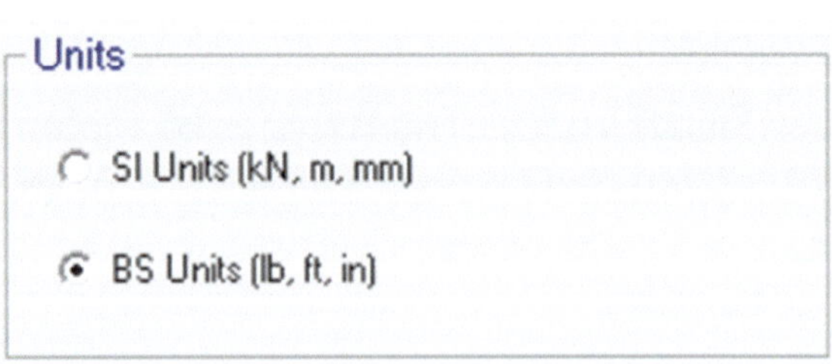

Fig. 1.19 Selected units in the General Information section

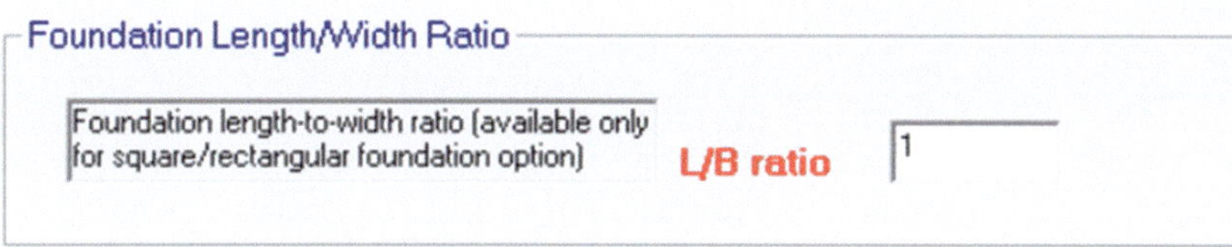

Fig. 1.20 Foundation shape and *L*/*B* ratio

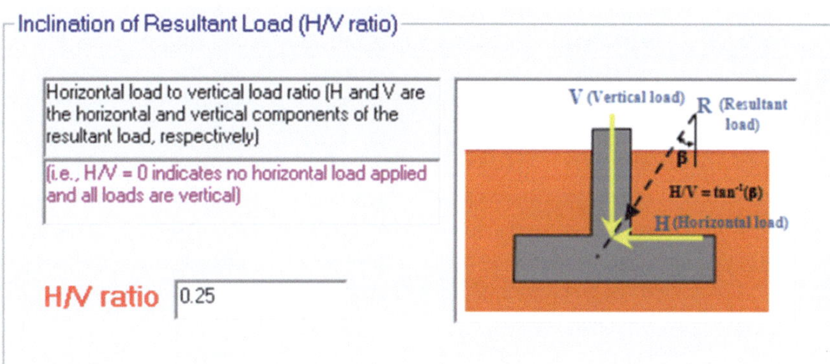

Fig. 1.21 Load conditions

Load Conditions and Elastic Settlement

In this section, we enter the *H*/*V* ratio provided in the problem as shown in Fig. 1.21.

Next, we run the program and record the results. Table 1.6 summarizes foundation loads based on bearing capacity analyses for the square foundation with dimensions ranging from 1 to 10 ft. The allowable bearing capacity chart is shown in Fig. 1.22.

1.4.4 Circular Foundation with One-Way Load Eccentricity

Develop a bearing capacity design chart for a circular footing with a vertical load and a load eccentricity ratio of 0.25. Do not consider the effect of soil compressibility in the calculations. Other information is given as Table 1.7 below:

1.4.4.1 Hand Solution

To determine the ultimate bearing capacity q_u for circular foundation:

Step 1:
Assume $D = B = 3\,\text{ft}$.

Table 1.6 Bearing capacity results

B (ft)	q (lb/ft^2)	q_{all} (lb/ft^2)	V_{all} (lb)	q_{u} (lb/ft^2)	V_{u} (lb)	H_{u} (lb)
1	520	10,472.07	10,472.07	31,416.2	31,416.2	7,854.049
1.5	520	10,290.63	23,153.91	30,871.88	69,461.73	17,365.43
2	520	10,120.78	40,483.11	30,362.34	121,449.3	30,362.34
2.5	520	9,964.597	62,278.73	29,893.79	186,836.2	46,709.05
3	520	9,823.148	88,408.33	29,469.44	265,225	66,306.25
3.5	520	9,696.654	118,784	29,089.96	356,352	89,088.01
4	520	9,584.712	153,355.4	28,754.14	460,066.2	115,016.5
4.5	520	9,486.511	192,101.9	28,459.53	576,305.6	144,076.4
5	520	9,897.392	247,434.8	29,692.18	742,304.4	185,576.1
5.5	520	9,723.244	294,128.1	29,169.73	882,384.4	220,596.1
6	520	9,584.141	345,029.1	28,752.42	1,035,087	258,771.8
6.5	520	9,471.996	400,191.8	28,415.99	1,200,576	300,143.9
7	520	9,381.033	459,670.6	28,143.1	1,379,012	344,753
7.5	520	9,307.016	523,519.6	27,921.05	1,570,559	392,639.7
8	520	9,246.766	591,793	27,740.3	1,775,379	443,844.8
8.5	520	9,197.855	664,545	27,593.57	1,993,635	498,408.8
9	520	9,158.393	741,829.8	27,475.18	2,225,489	556,372.4
9.5	520	9,126.887	823,701.6	27,380.66	2,471,105	617,776.2
10	520	9,102.145	910,214.5	27,306.43	2,730,643	682,660.8

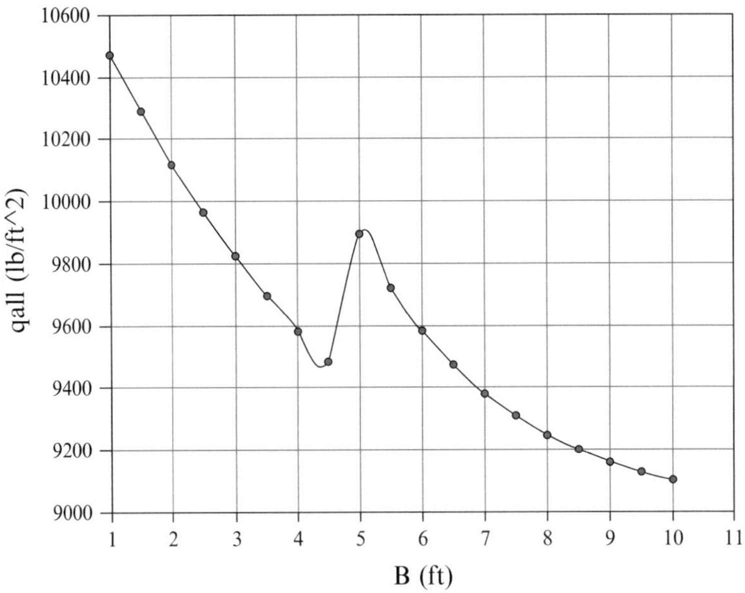

Square/Rectangular footing [BxL]; [L/B = 1]

Fig. 1.22 Allowable bearing capacity chart

Table 1.7 Soil properties and other design parameters

Property	Value	Unit
D_f	6	ft
c	320	lb/ft^2
ϕ	28	$^\circ$
γ	108.5	lb/ft^3
$\frac{e_D}{D}$ ratio	0.25	–
FS	3.5	–
GWT depth	Too deep	ft

Step 2:
Determine bearing capacity factor.
Using Eq. (1.2):

$$N_q = \tan^2\left(45 + \frac{28}{2}\right)e^{\pi \tan(28)} = 14.72$$

Using Eq. (1.3):

$$N_c = (14.72 - 1)\cot(28) = 25.80$$

Using Eq. (1.4):

$$N_\gamma = 2(14.72 + 1)\tan(28) = 16.72$$

Step 3:
Determine depth factors.
Using Eq. (1.8):

$$\eta = \frac{6}{3} = 2 > 1$$

Since $\phi' > 0$ use Eq. (1.6):

$$F_{qd} = 1 + 2\tan 28(1 - \sin 28)^2 \tan^{-1}\left(\frac{6}{3}\right) = 1.33$$

Use Eq. (1.7)

$$F_{cd} = 1.33 - \frac{1 - 1.33}{25.80 \times \tan 28} = 1.355$$

$$F_{\gamma d} = 1$$

Step 4:
Determine load inclination factors.
Using Eq. (1.12):

$$\beta = \tan^{-1}\left(\frac{0}{0}\right) = 0$$

Using Eq. (1.10):

$$F_{ci} = F_{qi} = \left(1 - \frac{0°}{90°}\right)^2 = 1$$

Using Eq. (1.11):

$$F_{\gamma i} = \left(1 - \frac{0}{30}\right) = 1$$

Step 5:
Determine soil compressibility factors.
Since the soil compressibility is not included $F_{cc} = F_{qc} = F_{\gamma c} = 1$.

Step 6:
Determine q and γ to account for the GWT.
Given that the water table is very deep, Case 3 applies.
From Eq. (1.23):

$$\gamma = \gamma_1 = 108.5\,\text{lb/ft}^3$$

From Eq. (1.21):

$$q = 108.5 \times 6 = 651\,\text{lb/ft}^2$$

Step 7:
Determine the effective dimensions.
Given that, the load eccentricity for this circular foundation is $\frac{e_D}{D} = 0.25$.
Using Eqs. (1.34) and (1.35), f_1 and f_2 can be determined as follows:

$$f_1 = 43.473(0.25)^4 - 61.224(0.25)^3 + 32.094(0.25)^2 - 8.7505(0.25) + 1.2896 = 0.321$$
$$f_2 = 1.5303(0.25)^2 - 2.438(0.25) + 0.8257 = 0.3125$$

The effective dimensions to account for the effect of load eccentricity can now be calculated using Eq. (1.32):

$$B' = 0.321 \times 3 = 0.963$$

Using Eq. (1.33):

$$L^{'} = \frac{0.3125 \times 3^2}{0.963} = 2.919$$

Step 8:
Determine shape factors.
From Eq. (1.54):

$$F_{cs} = 1 + \left(\frac{0.963}{2.919}\right)\left(\frac{14.72}{25.80}\right) = 1.188$$

From Eq. (1.55):

$$F_{qs} = 1 + \left(\frac{0.963}{2.919}\right)\tan 28 = 1.175$$

From Eq. (1.56):

$$F_{\gamma s} = 1 - 0.4\left(\frac{0.963}{2.919}\right) = 0.86$$

Step 9:
The ultimate bearing capacity of the circular footing from Eq. (1.1):

$$\begin{aligned} q_{\text{u}} = {} & 320 \times 25.80 \times 1.355 \times 1 \times 1 \times 1.188 + 651 \times 14.72 \times 1.175 \times 1 \times 1 \times 1.33 \\ & + 0.5 \times 108.5 \times 0.963 \times 16.72 \times 0.86 \times 1 \times 1 \times 1 = 29{,}016.62\,\text{lb/ft}^2 \end{aligned}$$

From Eq. (1.68):

$$V_u = 29{,}151.09 \times 0.963 \times 2.919 = 81{,}943.32\,\text{lb}$$

From Eq. (1.67):

$$q_{\text{all}} = \frac{29{,}151.09}{3.5} = 8{,}328.88\,\text{lb/ft}^2$$

From Eq. (1.69):

$$V_{\text{all}} = 8{,}328.88 \times 0.963 \times 2.919 = 23{,}412.46\,\text{lb}$$

From the ratio given in the problem statement:

$$M_{\text{u}_D} = 81{,}943.32 \times 0.25 = 20{,}485.83\,\text{ft lb}$$

Step 10: Several foundation diameters can be chosen and the above steps can be followed again to evaluate the impact of the foundation diameter on the baring capacity of the foundation.

1.4.4.2 foundationPro Solution

General Information Section

We enter the factor of safety and select the unit type we wish to use (Fig. 1.23).

Groundwater Table Section

We don't click "GWT exists" since the groundwater table is very deep; we enter γ_1, c', and ϕ.

Foundation Shape/Depth Section

We enter the depth of the foundation as well as foundation shape (circular). We enter the values we would like to consider for the foundation dimensions; we also leave the *L*/*B* ratio empty due to the fact that this is a circular foundation (Fig. 1.24).

Load Conditions Section

In this section, we enter the load eccentricity ratio given in the problem after checking the appropriate box for circular foundation as shown in Fig. 1.25.

Fig. 1.23 Factor of safety

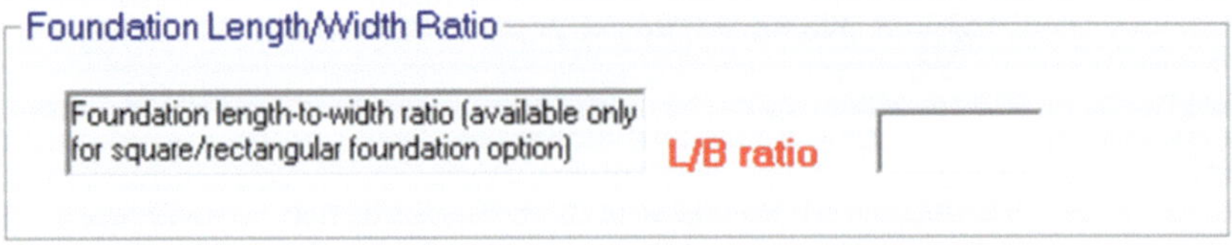

Fig. 1.24 Foundation shape

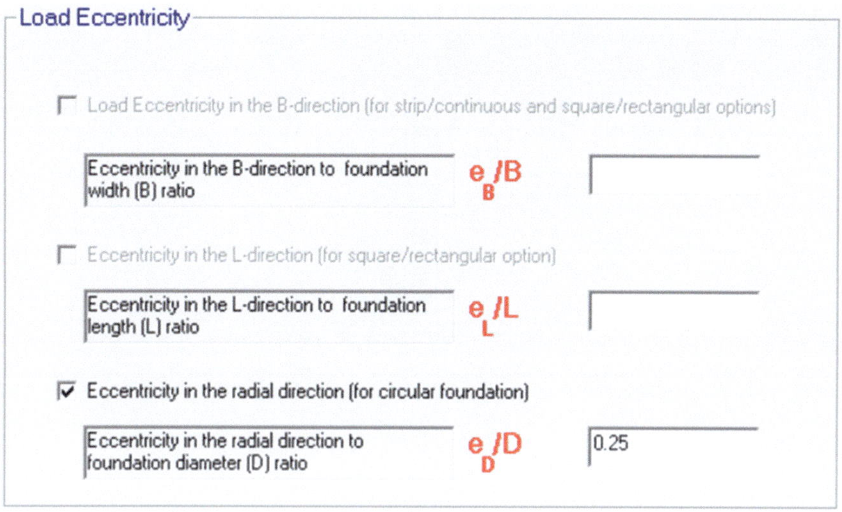

Fig. 1.25 Load conditions

After running the program, the output will appear in the Output section of the program. Table 1.8 shows the bearing capacity results for this circular foundation with load eccentricity. The allowable bearing capacity chart is also shown in Fig. 1.26.

1.4.5 Rectangular Foundation with One-Way Load Eccentricity

Develop a bearing capacity design chart for a rectangular foundation with L/B ratio of 1.2, and eccentricity ratio of 0.25 in the B direction. Factor of safety is 4.0. Soil compressibility is not included. Other required parameters are listed in Table 1.9.

1.4.5.1 Hand Solution

To determine the ultimate bearing capacity q_u:

Step 1:
From the given statement $L/B = 1.2$.

So, let us assume $L = 1.2$ m. Therefore, $B = 1$ m.

Step 2:
Determine bearing capacity factors.

Table 1.8 Foundation loads at various diameters

B (ft)	B' (ft)	q (lb/ft^2)	q_{all} (lb/ft^2)	V_{all} (lb)	q_{u} (lb/ft^2)	V_{u} (lb)	$M_{\text{u-}D}$ (lb ft)
3	0.963124	651	8,303.214073	23,303.75	29,061.24926	81,563.12	20,390.78013
4.5	1.444686	651	8,076.263118	51,000.28	28,266.92091	178,501	44,625.242
6	1.926248	651	8,320.060807	93,404.12	29,120.21283	326,914.4	81,728.60732

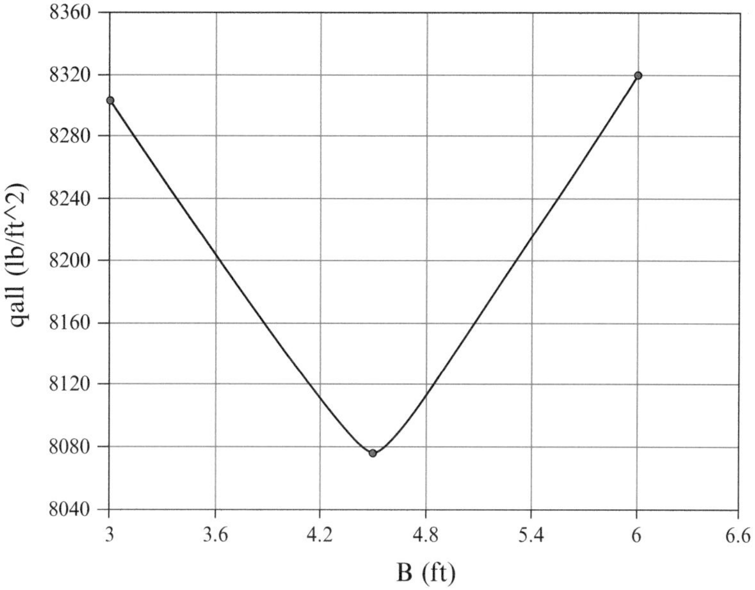

Circular footing [B]

Fig. 1.26 Allowable bearing capacity chart

Table 1.9 Soil properties and other design parameters

Property	Value	Unit
D_f	1.3	m
c	0	kN/m^2
ϕ	32	°
γ_1	18.1	kN/m^3
γ_2	19.25	kN/m^3
e_B/B	0.25	–
GWT depth	0.5	m

Using Eq. (1.2):

$$N_q = \tan^2\left(45 + \frac{32}{2}\right)e^{\pi \tan(32)} = 23.18$$

Using Eq. (1.3):

$$N_c = (23.18 - 1)\cot(32) = 35.49$$

Using Eq. (1.4):

$$N_\gamma = 2(23.18 + 1)\tan(32) = 30.22$$

Step 3:
Determine depth factors.

Given $\phi' > 0$ and using Eq. (1.8):

$$\eta = \frac{D_f}{B} = \frac{1.3}{1} = 1.3 > 1$$

Using Eq. (1.6):

$$F_{qd} = 1 + 2\tan(32)\,(1 - \sin(32))^2 \tan^{-1}\left(\frac{1.3}{1}\right) = 1.252$$

Using Eq. (1.7):

$$F_{cd} = 1.252 - \frac{1 - 1.252}{35.49 \times \tan(32)} = 1.264$$

$$F_{\gamma d} = 1$$

Step 4:
Determine load inclination factors.

No horizontal load is applied to the foundation. Therefore, the *H*/*V* ratio is zero.
Using Eq. (1.12):

$$\beta = \tan^{-1}\left(\frac{0}{0}\right) = 0$$

Using Eq. (1.10)

$$F_{ci} = F_{qi} = \left(1 - \frac{0}{90^\circ}\right)^2 = 1$$

Using Eq. (1.11):

$$F_{\gamma i} = \left(1 - \frac{0}{32}\right) = 1$$

Step 5:
Determine soil compressibility factors.

Since the soil compressibility is not included in the given problem

$$F_{cc} = F_{qc} = F_{\gamma c} = 1$$

Step 6:
Determine q and γ to account for the GWT.

Given water table is at depth 0.5 m.

Given that, $\gamma_1 = 18.1$ kN/m^3 and $\gamma_2 = 19.25$ kN/m^3.
Here $0 \leq D_w \leq D_f$, then Case 1 applies.
Using Eq. (1.19):

$$\gamma = 19.25 - 9.81 = 9.44\,\text{kN/m}$$

Using Eq. (1.20):

$$q = 18.1(0.5) + (19.25 - 9.81) \times (1.3 - 0.5) = 16.602\,\text{kN/m}^2$$

Step 7:
Determine the effective dimensions (effect of load eccentricity).
Given $e_B/B = 0.25$ (eccentricity in the B)
For rectangular footings:
Use Eq. (1.27):

$$B' = B - 2e_B = 1 - (2 \times 0.25) = 0.5$$

From Eq. (1.25):

$$e_L = \frac{0}{V} = 0$$

From Eq. (1.29):

$$L' = L - 2e_L = L = 1.2\,\text{m}$$

Step 8:
Determine shape factors.
Using Eq. (1.54):

$$F_{cs} = 1 + \left(\frac{0.5}{1.2}\right)\left(\frac{23.18}{35.49}\right) = 1.27$$

Using Eq. (1.55):

$$F_{qs} = 1 + \left(\frac{0.5}{1.2}\right)\tan(32) = 1.26$$

Using Eq. (1.56):

$$F_{\gamma s} = 1 - 0.4\left(\frac{0.5}{1.2}\right) = 0.8$$

Step 9:
The ultimate bearing capacity of the rectangular foundation from Eq. (1.1):

$$q_u = (0)\,(35.49)\,(1.25)\,(1)\,(1)\,(1.277) + (16.802)\,(23.18)\,(1.26)\,(1)\,(1)\,(1.24) + \frac{1}{2}(9.44)\,(0.5)\,(30.22)\,(0.8)\,(1)\,(1)\,(1) = 666.94\,\mathrm{kN/m^2}$$

From Eq. (1.68):

$$V_u = 666.94 \times 1.2 \times 0.5 = 400.17\,\mathrm{kN}$$

From Eq. (1.67):

$$q_{all} = \frac{666.94}{4} = 166.7\,\mathrm{kN/m^2}$$

From Eq. (1.69):

$$V_{all} = 166.7 \times 1.2 \times 0.5 = 100\,\mathrm{kN}$$

From the ratio given in the problem statement:

$$M_{u_B} = 400.17 \times 0.25 = 100.04\,\mathrm{kN\,m}$$

Step 10:
Different values of B can be chosen and the procedure can be repeated to evaluate the impact of the size of the foundation on the baring capacity of the foundation.

1.4.5.2 foundationPro Solution

General Information Section

We enter the factor of safety and select the unit type we wish to use.

Groundwater Table and Soil Information Section

We enter the depth of the groundwater table, unit weight of the soil under the groundwater table, and unit weight of the soil above the groundwater table. We enter the soil cohesion and friction angle.

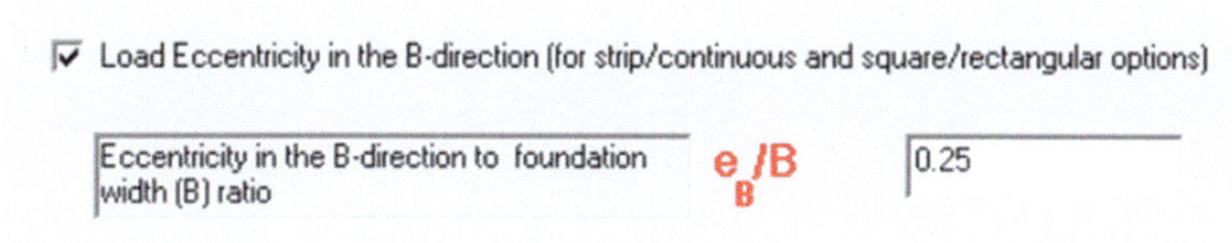

Fig. 1.27 Load eccentricity ratio in the *B* direction

Foundation Shape/Depth Section

We enter the depth of the foundation, and then select the foundation shape as square/rectangular.

We enter the values we would like to consider for the foundation dimensions; we also enter the *L*/*B* ratio given in the problem.

Load Conditions Section

In this section, we check the appropriate load eccentricity box to enable the textbox and then we enter the load eccentricity ratio given in the problem as shown in Fig. 1.27.

Elastic Settlement Section

We do not consider elastic settlement for this problem

Now, we can click the run button and then view the output results. Bearing capacity results from foundationPro are shown in Table 1.10 for the selected foundation dimensions with eccentricity in one direction. The allowable bearing capacity versus foundation widths are shown in Fig. 1.28.

1.4.6 Square Foundation with Soil Compressibility

Develop a bearing capacity design chart for a square foundation under vertical loading. Use a factor of safety of 3.0. Include the effect of soil compressibility on bearing capacity calculations. Soil properties and other design parameters are listed in Table 1.11.

1.4.6.1 Hand Solution

To determine the ultimate bearing capacity q_u:

Step 1:
From the given statement $L/B = 1.0$.
So, let us assume $L = 1$ m and $B = 1$ m.

Table 1.10 Bearing capacity results

B (m)	L (m)	B' (m)	L' (m)	q_{all} (kN/m^2)	V_{all} (kN)	q_u (kN/m^2)	V_u (kN)	$M_{u\text{-}B}$ (kN m)
1	1.2	0.5	1.2	166.7359	100.04152	666.9434346	400.1660607	100.0415152
2	2.4	1	2.4	172.7153	414.51663	690.8610512	1,658.066523	414.5166307
3	3.6	1.5	3.6	180.3163	973.70824	721.2653644	3,894.832968	973.708242
4	4.8	2	4.8	191.5446	1,838.8286	766.1785965	7,355.314526	1,838.828632
5	6	2.5	6	204.2238	3,063.3577	816.8953961	12,253.43094	3,063.357735

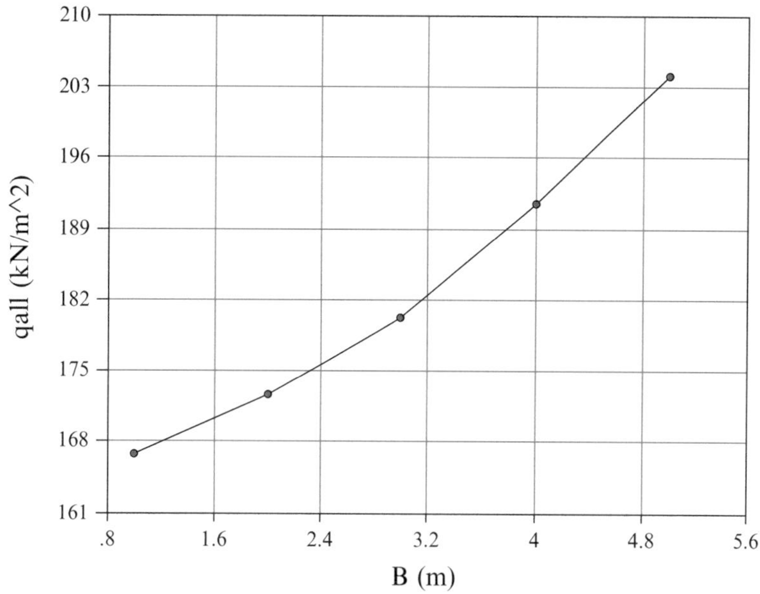

Fig. 1.28 Allowable bearing capacity chart

Table 1.11 Soil properties and design parameters

Property	Value	Unit
D_f	1.3	m
c	0	kN/m^2
ϕ	30	°
γ_1	18.1	kN/m^3
γ_2	19.25	kN/m^3
GWT Depth	0.5	m
E_s	12,400	kN/m^2
μ_s	0.35	–

Step 2:
Determine bearing capacity factors.
Using Eq. (1.2):

$$N_q = \tan^2\left(45 + \frac{30}{2}\right)e^{\pi \tan(30)} = 18.40$$

Using Eq. (1.3):

$$N_c = (18.40 - 1)\cot(30) = 30.14$$

Using Eq. (1.4):

$$N_\gamma = 2(18.40 + 1)\tan(30) = 22.40$$

Step 3:
Determine depth factors.
Given $\phi' > 0$ and using Eq. (1.8):

$$\eta = \frac{D_\mathrm{f}}{B} = \frac{1.3}{1} = 1.3 > 1$$

Using Eq. (1.6):

$$F_{q\mathrm{d}} = 1 + 2\tan(30)\,(1 - \sin(30))^2 \tan^{-1}\left(\frac{1.3}{1}\right) = 1.26$$

Using Eq. (1.7):

$$F_{\mathrm{cd}} = 1.26 - \frac{1 - 1.26}{30.14 \times \tan(30)} = 1.274$$

$$F_{\gamma\mathrm{d}} = 1$$

Step 4:
Determine load inclination factors.
Since H/V ratio is zero and using Eq. (1.12):

$$\beta = \tan^{-1}\left(\frac{0}{0}\right) = 0$$

Using Eq. (1.10):

$$F_{\mathrm{ci}} = F_{q\mathrm{i}} = \left(1 - \frac{0}{90^\circ}\right)^2 = 1$$

Using Eq. (1.11):

$$F_{\gamma\mathrm{i}} = \left(1 - \frac{0}{30}\right) = 1$$

Step 5:
Determine soil compressibility factors.
q' at a depth of $D_\mathrm{f} + B/2$ can be calculated as

$$q' = 18.1 \times 0.5 + 1.3(19.75 - 9.81) = 21.972\,\mathrm{kN/m^2}$$

Using Eq. (1.14):

$$I_{\mathrm{r}} = \frac{12,400}{2 \times (1 + 0.35)\ (0 + 21.972 \times \tan(30))} = 21.972$$

Using Eq. (1.15):

$$I_{\mathrm{r(cr)}} = \frac{1}{2}\left\{\exp\left[\left(3.3 - 0.45\frac{1}{1}\right)\cot\left(45 - \frac{30}{2}\right)\right]\right\} = 69$$

$I_{\mathrm{r}} \geq I_{\mathrm{r(cr)}}$, then

$$F_{cc} = F_{qc} = F_{\gamma c} = 1$$

Step 6:
Determine q and γ to account for the GWT.
Given groundwater table is at a depth of 0.5 m.
Given that, $\gamma_1 = 18.1$ kN/m^3 and $\gamma_2 = 19.25$ kN/m^3.
$0 \leq D_{\mathrm{w}} \leq D_{\mathrm{f}}$, then Case 1 applies.
Using Eq. (1.19):

$$\gamma = 19.25 - 9.81 = 9.44\,\mathrm{kN/m^3}$$

Using Eq. (1.20):

$$q = 18.1(0.5) + (19.25 - 9.81) \times (1.3 - 0.5) = 16.602\,\mathrm{kN/m^2}$$

Step 7:
Determine the effective dimensions (effect of load eccentricity).
There is no eccentricity; therefore, effective dimensions are the same as the actual dimensions:
$B' = B$ and $L' = L$

Step 8:
Determine shape factors.
Using Eq. (1.54):

$$F_{cs} = 1 + \left(\frac{1.0}{1.0}\right)\left(\frac{18.4}{30.14}\right) = 1.6$$

Using Eq. (1.55):

$$F_{qs} = 1 + \left(\frac{1.0}{1}\right)\tan(30) = 1.577$$

Using Eq. (1.56):

$$F_{\gamma s} = 1 - 0.4\left(\frac{1.0}{1.0}\right) = 0.6$$

Step 9:
The ultimate bearing capacity of the square footing from Eq. (1.1):

$$q_u = (0)(30.14)(1.274)(1)(1)(1.6) + (16.602)(18.4)(1.577)(1)(1)(1.26) + \frac{1}{2}(9.44)(1.0)(22.4)(0.6)(1)(1)(1) = 670.42\,\text{kN/m}^2$$

From Eq. (1.68):

$$V_u = 670.42 \times 1 \times 1 = 670.42\,\text{kN}$$

From Eq. (1.67):

$$q_{\text{all}} = \frac{670.42}{3} = 223.47\,\text{kN/m}^2$$

From Eq. (1.69):

$$V_{\text{all}} = 223.47 \times 1 \times 1 = 223.47\,\text{kN}$$

Step 10:
Different values of *B* can be chosen and the procedure can be repeated to evaluate the impact of the size of the foundation on the baring capacity of the foundation.

1.4.6.2 foundationPro Solution

General Information Section

We enter the factor of safety and select the appropriate units.

Groundwater Table and Soil Information Section

We enter the depth of the groundwater table, unit weight of the soil below the groundwater table, and unit weight of the soil above the groundwater table. We enter the soil cohesion and friction angle as well in this section. We enter the elastic modulus of soil at the base of foundation, rate of increase in the elastic modulus of soil, and Poisson's ratio of soil to account for the effect of soil compressibility.

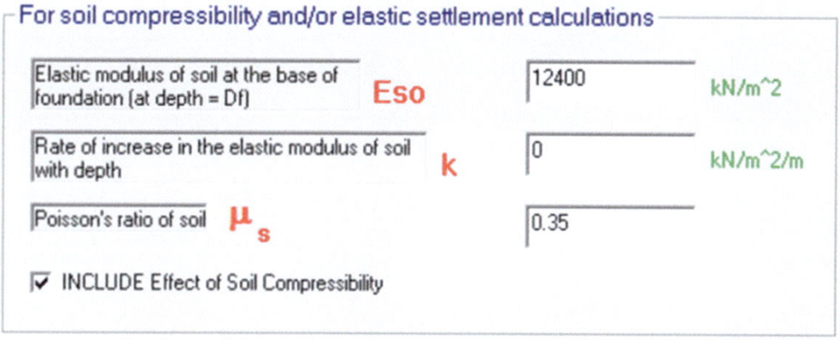

Fig. 1.29 Soil compressibility information

Table 1.12 Foundation loads based on bearing capacity results

B (m)	q_{all} (kN/m^2)	V_{all} (kN)	q_u (kN/m^2)	V_u (kN)
1	224.2040263	224.2040263	672.6120788	672.6121
2	233.0597098	932.2388393	699.1791294	2,796.717
3	244.1611962	2,197.450766	732.4835886	6,592.352
4	260.2859129	4,164.574607	780.8577388	12,493.72
5	278.4199218	6,960.498045	835.2597654	20,881.49

Make sure to check the box at the bottom as shown in Fig. 1.29 to include the effect of soil compressibility in the bearing capacity analyses.

Foundation Shape/Depth Section

We enter the depth of the foundation, and then we select the foundation shape as square. We enter the values we would like to consider for the foundation dimensions; we also enter the L/B ratio of 1 to specify square foundation.

Load Conditions and Elastic Settlement Sections

We can skip both sections since no load eccentricity and settlement analyses are required in this problem. And now, we can hit the run button! Table 1.12 summarizes the foundation load results for foundation widths ranging from 1 to 5 m. Also, the allowable bearing capacity chart is shown in Fig. 1.30.

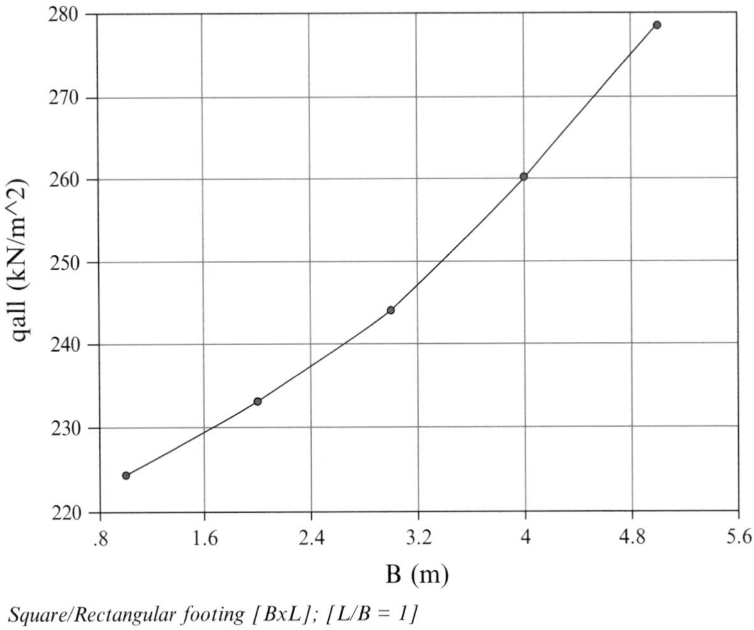

Fig. 1.30 Allowable bearing capacity chart

Table 1.13 Soil properties and other design parameters

Property	Value	Unit
D_f	0.6	m
c	38	kN/m^2
ϕ	12	°
γ	18.1	kN/m^3
FS	3.0	
GWT depth	Too deep	–
E_s	1,600	kN/m^2
μ_s	0.3	–

1.4.7 Rectangular Foundation with Soil Compressibility

Develop a bearing capacity design chart for a rectangular foundation with *L*/*B* ratio of 2; no bending moments or horizontal load is applied. Groundwater table is too deep and can be ignored. Include the effect of soil compressibility on bearing capacity calculations and use the soil parameters as provided in Table 1.13.

1.4.7.1 Hand Solution

To determine the ultimate bearing capacity q_u:

Step 1:
From $L/B = 2.0$.
So, let us assume $L = 1$ m and then $B = 0.5$ m.

Step 2: Determine bearing capacity factors.
Using Eq. (1.2):

$$N_q = \tan^2\left(45 + \frac{12}{2}\right) e^{\pi \tan(12)} = 2.97$$

Using Eq. (1.3):

$$N_c = (2.97 - 1)\cot(12) = 9.28$$

Using Eq. (1.4):

$$N_\gamma = 2(2.97 + 1)\tan(12) = 1.69$$

Step 3:
Determine depth factors.
Given $\phi' > 0$ and using Eq. (1.8):

$$\mu = \frac{D_f}{B} = \frac{0.6}{0.5} = 1.2 > 1$$

Using Eq. (1.6):

$$F_{qd} = 1 + 2\tan(12)(1 - \sin(12))^2 \tan^{-1}(1.2) = 1.26$$

Using Eq. (1.7):

$$F_{cd} = 1.26 - \frac{1 - 1.26}{9.28 \times \tan(12)} = 1.39$$

$$F_{\gamma d} = 1$$

Step 4:
Determine load inclination factors.
Using Eq. (1.12), since H/V ratio is zero:

$$\beta = \tan^{-1}\left(\frac{0}{0}\right) = 0$$

Using Eq. (1.10):

$$F_{ci} = F_{qi} = \left(1 - \frac{0}{90^\circ}\right)^2 = 1$$

Using Eq. (1.11):

$$F_{\gamma i} = \left(1 - \frac{0}{12}\right) = 1$$

Step 5:
Determine soil compressibility factors.
q' at a depth of $D_f + B/2$ can be calculated as

$$q' = 18.1\left(0.6 + \left(\frac{0.5}{2}\right)\right) = 15.385\,\text{kN/m}^2$$

Using Eq. (1.14):

$$I_r = \frac{1,600}{2 \times (1 + 0.33)\,(38 + 15.385 \times \tan(12))} = 14.57$$

Using Eq. (1.15):

$$I_{r(cr)} = \frac{1}{2}\left\{\exp\left[\left(3.3 - 0.45\left(\frac{0.5}{1}\right)\right)\cot\left(45 - \frac{12}{2}\right)\right]\right\} = 22.29$$

$I_r < I_{r(cr)}$, then
Using Eq. (1.16):

$$F_{\gamma c} = F_{qc} = \exp\left\{\left(-4.4 + 0.6\left(\frac{0.5}{1}\right)\right)\tan(12) + \left[\frac{(3.07\sin(12))\,(\log(2(14.57)))}{1 + \sin(12)}\right]\right\}$$
$$= 0.907$$

Using Eq. (1.18):

$$F_{cc} = 0.907 - \frac{1 - 0.907}{2.97 \times \tan(12)} = 0.759$$

Step 6:
Determine q and γ to account for the GWT.
Given groundwater table is deep and can be ignored and using Eq. (1.23):

$$\gamma_1 = 18.1\ \text{kN/m}^3 = \gamma.$$

Also, using Eq. (1.21):

$$q = 18.1 \times 0.6 = 10.86\,\text{kN/m}^2$$

Step 7:
Determine the effective dimensions (effect of load eccentricity).
Since there is no load eccentricity; $B^{'} = B$ and $L^{'} = L$.

Step 8:
Determine shape factors.
Using Eq. (1.54):

$$F_{cs} = 1 + \left(\frac{0.5}{1.0}\right)\left(\frac{2.97}{9.28}\right) = 1.16$$

Using Eq. (1.55):

$$F_{qs} = 1 + \left(\frac{0.5}{1}\right)\tan(12) = 1.106$$

Using Eq. (1.56):

$$F_{\gamma s} = 1 - 0.4\left(\frac{0.5}{1.0}\right) = 0.8$$

Step 9:
The ultimate bearing capacity of the square foundation can be then calculated from Eq. (1.1):

$$q_u = (38)(9.28)(1.39)(1)(0.759)(1.16) + (10.86)(2.97)(1.106)(1)(0.907)(1.26) + \frac{1}{2}(18.1)(0.5)(1.69)(0.8)(1)(1)(0.907) = 477.83\,\text{kN/m}^2$$

From Eq. (1.68):

$$V_u = 477.83 \times 1 \times 0.5 = 238.915\,\text{kN}$$

From Eq. (1.67):

$$q_{all} = \frac{477.83}{3} = 159.27\,\text{kN/m}^2$$

From Eq. (1.69):

$$V_{\text{all}} = 223.47 \times 1 \times 0.5 = 79.6\,\text{kN}$$

Step 10:
Several foundation widths can be chosen and the procedure can be repeated to evaluate the effect of foundation dimensions on the baring capacity of the foundation.

1.4.7.2 foundationPro Solution

General Information Section

We enter the factor of safety and the units to be considered throughout the analyses.

Groundwater Table and Soil Information Section

We don't click "GWT exists" since the groundwater table is very deep; we enter γ_1, c', and ϕ. Enter the elastic modulus of soil at the base of foundation, rate of increase in the elastic modulus of soil with depth (can be considered zero since a unique value is provided in the problem), and the Poisson's ratio of soil.

Foundation Shape/Depth Section

We enter the depth of the foundation, and then select the foundation shape as rectangular.

We enter the values we would like to consider for the foundation dimensions; we also enter the *L*/*B* ratio of 2 given in the problem.

Load Conditions and Elastic Settlement Sections

We can skip both sections since these are not required for this problem. And now, we can hit the run button and view the Output section. Foundation loads this rectangular foundation can sustain including the effect of soil compressibility are listed in Table 1.14 for many sets of foundation dimensions. The allowable bearing capacity chart is also shown in Fig. 1.31.

1.4.8 Strip/Continuous Foundation with Horizontal Loading

Develop a bearing capacity design chart for a strip/continuous foundation with a horizontal loading that is 30 % of the vertical loading (i.e., *H*/*V* ratio = 0.3).

Table 1.14 Bearing capacity results

B (m)	L (m)	q_{all} (kN/m^2)	V_{all} (kN)	q_u (kN/m^2)	V_u (kN)
0.5	1	157.7269573	78.86347864	473.1808718	236.5904
1	2	144.9075039	289.8150078	434.7225117	869.445
1.5	3	135.8447918	611.3015632	407.5343755	1,833.905
2	4	131.4055593	1,051.244474	394.2166778	3,153.733
2.5	5	128.8307925	1,610.384906	386.4923775	4,831.155
3	6	127.1999418	2,289.598953	381.5998255	6,868.797
3.5	7	126.1177981	3,089.886054	378.3533944	9,269.658
4	8	125.3862059	4,012.358588	376.1586176	12,037.08
4.5	9	124.8946297	5,058.232504	374.6838892	15,174.7
5	10	124.5763773	6,228.818864	373.7291319	18,686.46

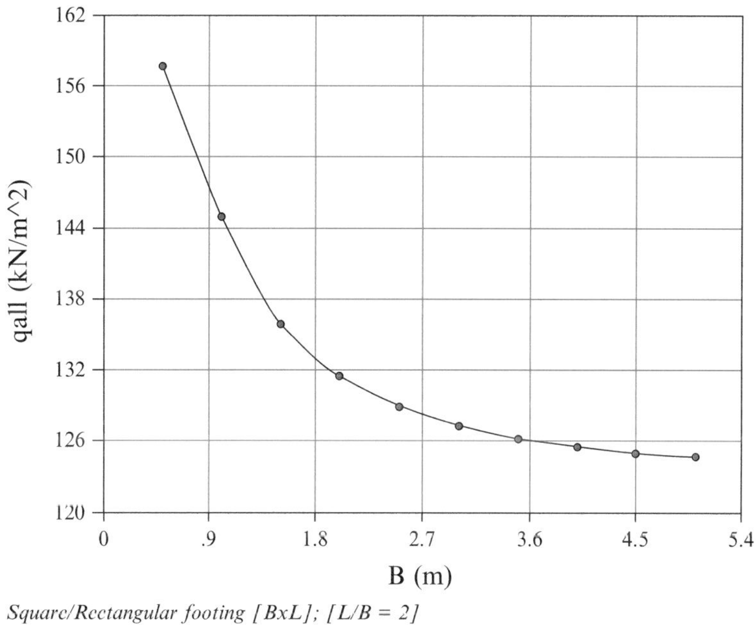

Fig. 1.31 Allowable bearing capacity chart

Groundwater table is at the ground surface. Ignore the effect of soil compressibility in the analysis and use a factor of safety of 3.0. Other given information is summarized in Table 1.15 below.

Table 1.15 Soil properties and other design parameters

Property	Value	Unit
D_f	3.5	ft
c	150	lb/ft^2
ϕ	28	°
γ_1	124	lb/ft^3
γ_2	124	lb/ft^3
GWT depth	Ground surface	ft

1.4.8.1 Hand Solution

To develop a design chart, one must follow the steps below:

Step 1:
Assume $B = 1$ ft.

Step 2:
Determine bearing capacity factors.
Using Eq. (1.2):

$$N_q = \tan^2\left(45 + \frac{\phi'}{2}\right) e^{\pi \tan(\phi)'} = 14.72$$

Using Eq. (1.3):

$$N_c = (N_q - 1)\cot\phi' = 25.80$$

Using Eq. (1.4):

$$N_\gamma = 2(N_q + 1)\tan\phi' = 16.72$$

Step 3:
Determine depth factors.
Using Eq. (1.8):

$$\mu = \frac{D_f}{B} = \frac{3.5}{1} = 3.5 > 1$$

Since $\phi' > 0$ and from Eq. (1.6):

$$F_{qd} = 1 + 2\tan(28)(1 - \sin(28))^2 \tan^{-1}\left(\frac{3.5}{1}\right) = 1.386$$

Use Eq. (1.7)

$$F_{cd} = 1.386 - \frac{1 - 1.386}{25.80 \times \tan(28)} = 1.41$$

$$F_{\gamma d} = 1$$

Step 4:
Determine load inclination factors.
For the given *H*/*V* ratio and from Eq. (1.12):

$$\beta = \tan^{-1}(0.3) = 16.7$$

Using Eq. (1.10):

$$F_{ci} = F_{qi} = \left(1 - \frac{16.7^\circ}{90^\circ}\right)^2 = 0.66$$

Using Eq. (1.10):

$$F_{\gamma i} = \left(1 - \frac{16.7}{28}\right) = 0.403$$

Step 5:
Determine soil compressibility factors.
Since the effect of soil compressibility is not required $F_{cc} = F_{qc} = F_{\gamma c} = 1$.

Step 6:
Determine q and γ to account for the GWT.
Since the groundwater table is at the ground surface ($D_W = 0$ ft), Case I applies.
Use Eq. (1.19):

$$\gamma = 124 - 62.4 = 61.6 \text{ lb/ft}^3.$$

Use Eq. (1.20):

$$q = 0 + (61.6)(3.5) = 215.6 \text{lb/ft}^2$$

Step 7:
Determine the effective dimensions (effect of load eccentricity).
There is no load eccentricity. Therefore,

$$B' = B = 1 \text{ ft}.$$

Step 8: Determine shape factors.
For a strip footing the shape factors are equal to 1. So,

$$F_{cs} = F_{qs} = F_{\gamma s} = 1$$

Step 9:
The ultimate bearing capacity of the strip footing can be now calculated from Eq. (1.1) as

$$q_{\mathrm{u}}(150)(25.8)(1.41)(0.66)(1)(1)+(215.6)(14.72)(1)(0.66)(1)(1.38)$$
$$+\frac{1}{2}(61.6)(1)(16.72)(1)(1)(0.403)(1)=6,699.50\,\mathrm{lb/ft^2}$$

From Eq. (1.68):

$$V_{\mathrm{u}}=6,660.49\times 1\times 1=6,660.49\,\mathrm{lb/ft}$$

From Eq. (1.67):

$$q_{\mathrm{all}}=\frac{6,660.49}{3}=2,220.16\,\mathrm{lb/ft^2}$$

From Eq. (1.69):

$$V_{\mathrm{all}}=2,220.16\times 1\times 1=2,220.16\,\mathrm{lb/ft}$$

Using the ratio given in the problem:

$$H_{\mathrm{u}}=6,660.49\times 0.3=1,998.417\,\mathrm{lb/ft}$$

Step 10:
Several B values can be selected and the procedure can be repeated to evaluate the effect of the foundation dimension on the bearing capacity of the strip/continuous foundation.

1.4.8.2 foundationPro Solution

General Information Section

We enter the given factor of safety and select the BS units in this section to use throughout the analyses.

Groundwater Table and Soil Information Section

We click “GWT exists” and since the groundwater table is at surface we enter “0” for the depth of the groundwater table; we enter γ_1, c', and ϕ. We do not enter any values for the elastic modulus of soil since it is not required in the problem.

Foundation Shape/Depth Section

We enter the depth of the foundation as well as select the type of the foundation (continuous).

We enter the values we would like to consider for the foundation dimensions; we leave the *L*/*B* ratio section empty due to the fact that this is a continuous footing.

Load Conditions and Elastic Settlement Sections

We enter the *H*/*V* ratio provided in the problem. We do not consider load eccentricity or elastic settlement. And now, we can hit the run button! Ultimate horizontal and vertical loads the strip/continuous foundation can sustain are summarized in Table 1.16. The ultimate horizontal load versus foundation width is shown in Fig. 1.32.

Table 1.16 Allowable and ultimate foundation load results

B (ft)	q_{all} (lb/ft^2)	V_{all} (lb/ft)	q_u (lb/ft^2)	V_u (lb/ft)	H_u (lb/ft)
1	2,253.472481	2,253.472481	6,760.417444	6,760.417444	2,028.125233
2	2,205.964799	4,411.929598	6,617.894397	13,235.78879	3,970.736638
3	2,183.362334	6,550.087003	6,550.087003	19,650.26101	5,895.078303
4	2,258.850982	9,035.403929	6,776.552947	27,106.21179	8,131.863536
5	2,243.269298	11,216.34649	6,729.807894	33,649.03947	10,094.71184

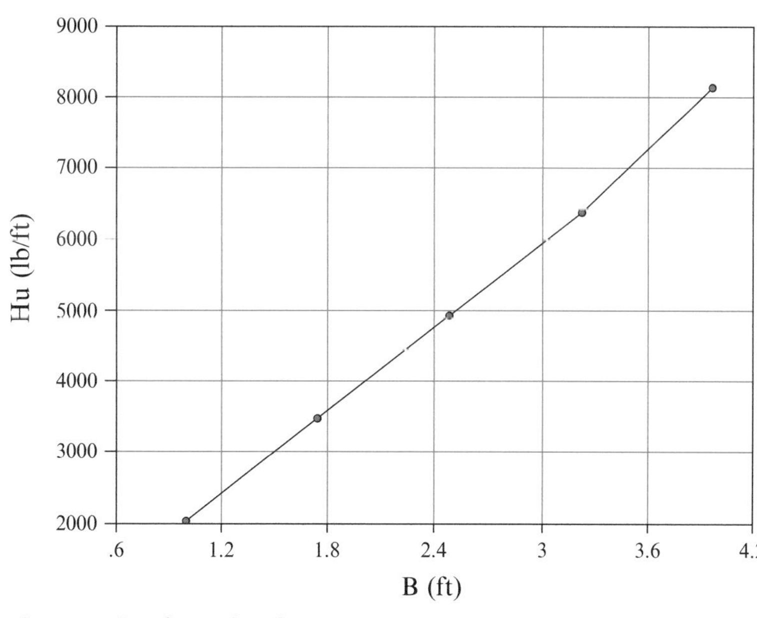

Fig. 1.32 Ultimate horizontal load chart

Table 1.17 Soil properties and other design parameters

Property	Value	Unit
D_f	4	ft
c	459	lb/ft^2
ϕ	32	°
γ_1	115	lb/ft^3
γ_2	122	lb/ft^3
FS	4.0	–
GWT depth	1.5	ft

1.4.9 *Square Foundation with Load Eccentricity and Horizontal Loading*

Develop a bearing capacity design chart for a square foundation with load eccentricity ratios in both directions as follows: $e_B/B = 0.2$ and $e_L/L = 0.2$, and a horizontal loading of 15 % of the vertical loading (i.e., H/V ratio $= 0.15$). Ignore the effect of soil compressibility. Other required parameters are listed in Table 1.17.

1.4.9.1 Hand Solution

To determine the ultimate bearing capacity (q_u), one must follow the steps below:

Step 1:
Assume foundation dimensions.

So, let us assume $L = 1$ ft, and then $B =$ ft ($L/B = 1$).

Step 2:
Determine bearing capacity factors.
Using Eq. (1.2):

$$N_q = \tan^2\left(45 + \frac{32}{2}\right) e^{\pi \tan (32)'} = 23.176$$

Using Eq. (1.3)

$$N_c = (23.176 - 1) \cot (32) = 35.490$$

Using Eq. (1.4):

$$N_\gamma = 2(23.176 + 1) \tan (32) = 30.22$$

Step 3:
Determine depth factors.
Given $\phi' > 0$ and

Using Eq. (1.8):

$$\eta = \frac{D_{\mathrm{f}}}{B} = \frac{4}{1} = 4.0 > 1$$

Use Eq. (1.6):

$$F_{q\mathrm{d}} = 1 + 2\tan(32)\,(1 - \sin(32))^2 \tan^{-1}\left(\frac{4}{1}\right) = 1.3661$$

Use Eq. (1.7):

$$F_{\mathrm{cd}} = 2.517 - \frac{1 - 2.517}{35.49 \times \tan(32)} = 1.38$$

$$F_{\gamma\mathrm{d}} = 1$$

Step 4:
Determine load inclination factors.
Using Eq. (1.12) and with $H/V = 0.15$:

$$\beta = \tan^{-1}(0.15) = 8.53$$

Using Eq. (1.10):

$$F_{\mathrm{ci}} = F_{q\mathrm{i}} = \left(1 - \frac{8.53^\circ}{90^\circ}\right)^2 = 0.819$$

Using Eq. (1.11):

$$F_{\gamma\mathrm{i}} - \left(1 - \frac{8.53}{32}\right) - 0.733$$

Step 5:
Determine soil compressibility factors.

Since the effect of soil compressibility is not required in this problem soil compressibility factors are taken as $F_{\mathrm{cc}} = F_{q\mathrm{c}} = F_{\gamma\mathrm{c}} = 1$.

Step 6:
Determine q and γ to account for the GWT which is at a depth of 1.5 ft.

Given that, $\gamma_1 = 115$ lb/ft^3; $\gamma_2 = 122$ lb/ft^3.
Since $D_{\mathrm{w}} = 1.5$ ft is less than $D_{\mathrm{f}} = 4$ ft,
γ can be calculated from Eq. (1.19):

$$\gamma = (\gamma_2 - \gamma_{\mathrm{w}}) = 59.6\,\mathrm{lb/ft^3}$$

And q from Eq. (1.20):

$$q = 115(1.5) + (122 - 62.4) \times (4 - 1.5) = 321.5\,\text{lb/ft}^2$$

Step 7:
Determine the effective dimensions (effect of load eccentricity).

Since load eccentricity is in two ways, Fig. 1.5 must be used to determine the applicable two-way eccentricity case. With the given load eccentricity ratios, one can find that Case I applies.

Therefore, Eq. (1.37) must be used to determine B_1:

$$B_1 = 1.5 - 3(0.2) = 0.9\,\text{ft}$$

Using Eq. (1.38):

$$L_1 = 1.5 - 3(0.2) = 0.9\,\text{ft}$$

Using Eq. (1.36):

$$A' = \frac{1}{2}B_1L_1 = 0.405\ \text{ft}^2$$

Using Eq. (1.48):

$$L' = 0.9\,\text{ft}$$

$$B' = \frac{0.405}{0.9} = 0.45\,\text{ft}$$

Step 8:
Determine shape factors.

Using Eq. (1.54):

$$F_{cs} = 1 + \left(\frac{0.45}{0.9}\right)\left(\frac{23.18}{35.49}\right) = 1.32$$

Using Eq. (1.55):

$$F_{qs} = 1 + \left(\frac{0.45}{0.9}\right)\tan(32) = 1.31$$

Using Eq. (1.56):

$$F_{\gamma s} = 1 - 0.4\left(\frac{0.45}{0.9}\right) = 0.8$$

Step 9:
The ultimate bearing capacity of the square foundation can be now calculated from Eq. (1.1) as follows:

$$q_u = (459)(35.49)(1.38)(0.819)(1)(1.32) + (321.5)(23.18)(1.31)(0.819)(1)(1.366) + \frac{1}{2}(59.6)(0.9)(30.22)(0.8)(1)(0.733)(1) = 35,752.79\,\mathrm{lb/ft^2}$$

From Eq. (1.68):

$$V_u = 35,752.79 \times 0.45 \times 0.9 = 14,479.88\,\mathrm{lb}$$

From Eq. (1.67):

$$q_{all} = \frac{35,752.79}{4} = 8,938.19\,\mathrm{lb/ft^2}$$

From Eq. (1.69):

$$V_{all} = 8,938.19 \times 0.45 \times 0.9 = 3,619.96\,\mathrm{lb}$$

From the *H*/*V*, $e_B/_B$, and $e_L/_L$ ratios given in the problem statement, one can find the ultimate loads the 1 ft × 1 ft square foundation can sustain as follows:

$$H_u = 14,479.88 \times 0.15 = 2,171.982\,\mathrm{lb}$$
$$M_{u_B} = 14,479.88 \times 0.2 \times 1 = 2,895.976\,\mathrm{ft\,lb}$$
$$M_{u_L} = 14,479.88 \times 0.2 \times 1 = 2,895.976\,\mathrm{ft\,lb}$$

Step 10:
Different values of *B* can be chosen and the steps can be repeated to determine the bearing capacity of the square foundation in terms of the foundation dimensions.

1.4.9.2 foundationPro Solution

General Information Section

We enter the factor of safety and select the BS units to be used throughout the analysis.

Groundwater Table and Soil Information Section

We click "GWT exists" and enter the depth of the groundwater table as 1.5 ft; we enter γ_1, c', and ϕ. We do not enter any values for the elastic modulus of soil since settlement analysis is not required in the problem.

Foundation Shape/Depth Section

We enter the depth of the foundation, and then select the foundation shape as square/rectangular. We enter the values we would like to consider for the foundation dimensions; we enter the L/B ratio of 1 to indicate square foundation.

Load Conditions Section

We enter the H/V ratio provided in the problem. We also enter the load eccentricity values stated in the problem statement. See Fig. 1.33 for details. We can now skip the elastic settlement section since settlement analysis is not required and hit the run button. Foundation loads for this square foundation under horizontal loading and two-way load eccentricities are summarized in Table 1.18 (Fig. 1.34).

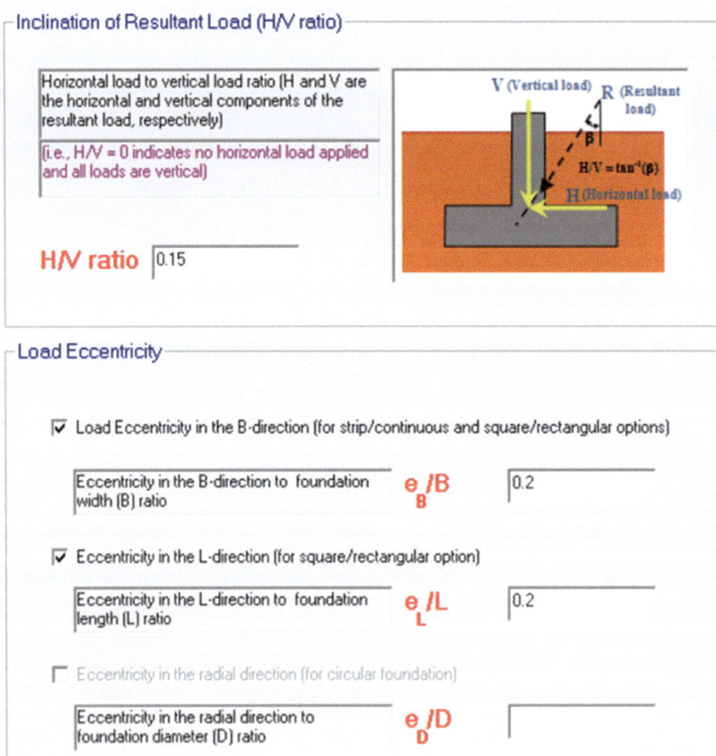

Fig. 1.33 Load conditions data

Table 1.18 Foundation loads at various foundation dimensions

B (ft)	B' (ft)	q_{all} (lb/ft^2)	V_{all} (lb)	q_u (lb/ft^2)	V_u (lb)	H_u (lb)	$M_{u\text{-}B}$ (lb ft)	$M_{u\text{-}L}$ (lb ft)
1	0.45	8,916.836133	3,611.318634	35,667.34453	14,445.27454	2,166.79118	2,889.054907	2,889.054907
2	0.9	8,575.91792	13,892.98703	34,303.67168	55,571.94812	8,335.792218	11,114.38962	11,114.38962
3	1.35	8,306.065118	30,275.60735	33,224.26047	121,102.4294	18,165.36441	24,220.48588	24,220.48588
4	1.8	8,498.609634	55,070.99043	33,994.43854	22,0283.9617	33,042.59426	44,056.79234	44,056.79234
5	2.25	8,191.871534	82,942.69928	32,767.48614	331,770.7971	49,765.61957	66,354.15942	66,354.15942

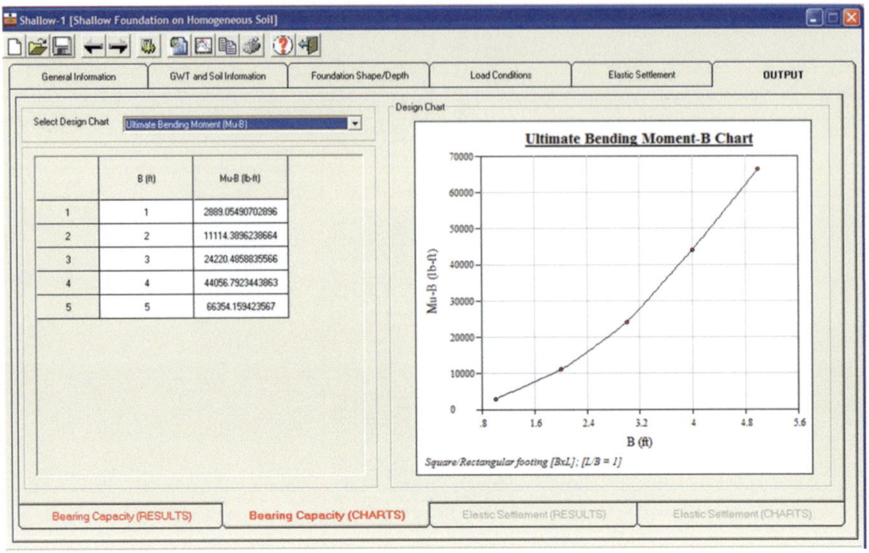

Fig. 1.34 Ultimate bending moment in the *B*-direction

Table 1.19 Soil properties and other parameters

Property	Value	Unit
D_f	1.25	m
c	55	kN/m^2
ϕ	20	°
γ_1	17.9	kN/m^3
γ_2	18.95	kN/m^3
FS	3.0	–
GWT depth	2	m
E_s	1,850	kN/m^2
μ_s	0.33	–

1.4.10 Rectangular Foundation with Two-Way Load Eccentricity

Develop a bearing capacity design chart for a rectangular foundation with L/B ratio of 1.4, and load eccentricity ratios of $e_B/B = 0.1$ and $e_L/L = 0.25$. No horizontal load is applied to the foundation. Include the effect of soil compressibility on bearing capacity. Groundwater table is at a depth of 2 m below the ground surface. Other required parameters are summarized in Table 1.19.

1.4.10.1 Hand Solution

To determine the foundation loads, one must follow the steps below:

Step 1:
Assume foundation dimensions.
So, let us assume $L = 1.4$ m. Therefore, $B = 1$ m since $L/B = 1.4$.

Step 2:
Determine bearing capacity factors.
Using Eq. (1.2):

$$N_q = \tan^2\left(45 + \frac{20}{2}\right)e^{\pi \tan(20)} = 6.40$$

Using Eq. (1.3)

$$N_c = (6.40 - 1)\cot(20) = 14.48$$

Using Eq. (1.4):

$$N_\gamma = 2(6.40 + 1)\tan(20) = 5.39$$

Step 3:
Determine depth factors.
For $\phi' > 0$ and from Eq. (1.8):

$$\eta = \frac{D_\mathrm{f}}{B} = \frac{1.25}{1} = 1.25 > 1$$

Use Eq. (1.6):

$$F_{q\mathrm{d}} = 1 + 2\tan(20)(1 - \sin(20))^2 \tan^{-1}\left(\frac{1.25}{1}\right) = 1.28$$

Use Eq. (1.7):

$$F_{c\mathrm{d}} = 1.28 - \frac{1 - 1.28}{14.48 \times \tan(20)} = 1.33$$

$$F_{\gamma\mathrm{d}} = 1$$

Step 4:
Determine load inclination factors.
Since H/V ratio is 0, all load inclination factors are set to 1:

$$F_{c\mathrm{i}} = F_{q\mathrm{i}} = F_{\gamma\mathrm{i}} = 1$$

Step 5:
Determine soil compressibility factors.
q' at a depth of $(D_f + B/2)$ can be calculated as

$$q' = 17.9 \times (1.25 + 0.5) = 31.25\,\text{kN/m}^2$$

Using Eq. (1.14):

$$I_r = \frac{1{,}850}{2 \times (1 + 0.33)\,(55 + 31.25 \times \tan(20))} = 10.57$$

Using Eq. (1.15):

$$I_{r(cr)} = \frac{1}{2}\left\{\exp\left[\left(3.3 - 0.45\left(\frac{1}{1.4}\right)\right)\cot\left(45 - \frac{20}{2}\right)\right]\right\} = 35.25$$

$I_r < I_{r(cr)}$, then
Using Eq. (1.16):

$$F_{\gamma c} = F_{qc} = \exp\left\{\left(-4.4 + 0.6\left(\frac{1}{1.4}\right)\right)\tan(20) + \left[\frac{(3.07 \sin(20))\log(2(10.57)))}{1 + \sin(20)}\right]\right\}$$
$$= 0.664$$

And using Eq. (1.18):

$$F_{cc} = 0.664 - \frac{1 - 0.664}{6.40 \times \tan(20)} = 0.519$$

Step 6:
Determine q and γ to account for the effect of the groundwater table.

Given that groundwater table is at depth of 2 m, and $\gamma_1 = 17.9$ kN/m^3 and $\gamma_2 = 18.95$ kN/m^3.

Also, we know that $D_f = 1.25\,\text{m} \leq D_w = 2\,\text{m} \leq D_f + B = 2.25\,\text{m}$. Hence,

$$q = 17.9 \times 1.25 = 22.37\,\text{kN/m}^2$$

$$\gamma = \frac{(2 - 1.25)}{1}(17.9 - 18.95 + 9.81) + 18.95 - 9.81 = 15.702\,\text{kN/m}^3$$

Step 7:
Determine the effective dimensions (effect of load eccentricity).

Given $e_B/B = 0.1$ (eccentricity in the B) and $e_L/L = 0.25$ (eccentricity in the L).

Therefore, one can find from Fig. 1.5 that Case II applies to this two-way eccentricity problem.

The effective area can be calculated using Eq. (1.39):

$$A' = \frac{1}{2}(1.017 + 0.255)1 = 0.6364\,\text{m}^2$$

The dimensions L_1 can be computed from Eq. (1.40):

$$L_1 = \left(-18.8357(0.1)^2 + 6.22019(0.1) + 0.95889\right)\ (-2.0651(0.25) + 1.038) \times 1.4 = 1.017\,\text{m}$$

Similarly, L_2 from Eq. (1.41):

$$L_2 = \left(2.518265(0.1)^2 - 2.86483(0.1) + 0.40649\right)\ (-5.05047(0.25) + 2.52145) \times 1.4 = 0.255\,\text{m}$$

The effective width is then calculated using Eq. (1.43):

$$B' = \frac{0.6364}{1.017} = 0.625\,\text{m}$$

Also, the effective length is then determined from Eq. (1.42):

$$L' = 1.017\,\text{m}$$

Step 8:
Determine shape factors.
Use Eq. (1.54):

$$F_{cs} = 1 + \left(\frac{0.625}{1.017}\right)\left(\frac{6.40}{14.48}\right) = 1.271$$

Use Eq. (1.55):

$$F_{qs} = 1 + \left(\frac{0.625}{1.017}\right)\tan(20) = 1.223$$

Use Eq. (1.56):

$$F_{\gamma s} = 1 - 0.4\left(\frac{0.625}{1.017}\right) = 0.754$$

Step 9:
The ultimate bearing capacity of the rectangular footing can be now calculated using Eq. (1.1) as follows:

$$q_u = (55)(14.48)(1.33)(1)(0.519)(1.271) + (22.37)(6.4)(1.22)(1)(0.664)(1.28) + \frac{1}{2}(15.702)(0.625)(5.39)(0.754)(1)(1)(0.664) = 860.40\,\text{kN/m}^2$$

The ultimate vertical load is calculated from Eq. (1.68) as follows:

$$V_u = 860.40 \times 1.017 \times 0.625 = 546.89\,\text{kN}$$

From Eq. (1.67):

$$q_{all} = \frac{860.40}{3} = 286.94\,\text{kN/m}^2$$

From Eq. (1.69):

$$V_{all} = 286.94 \times 1.017 \times 0.625 = 182.29\,\text{kN}$$

From the ratios given in the problem statement:

$$M_{u_B} = 546.89 \times 0.1 \times 1 = 54.6891\,\text{m kN}$$
$$M_{u_L} = 546.89 \times 0.25 \times 1 = 136.77\,\text{m kN}$$

Step 10:
Now, several other foundation dimensions can be selected to develop design charts following the steps above.

1.4.10.2 foundationPro Solution

General Information Section

We enter the factor of safety and select the SI units to be used throughout the analysis.

Groundwater Table and Soil Information Section

We click "GWT exists" and enter the depth of the groundwater table in the provided textbox; we enter γ_1, γ_2, c', and ϕ. Then, we enter the elastic modulus of soil at the base of foundation, rate of increase in the elastic modulus of soil, and the Poisson's ratio of soil.

Foundation Shape/Depth Section

We enter the depth of the foundation, and then select the foundation shape as required by the problem statement. Then, we enter the values we would like to consider for the foundation dimensions; we enter the *L*/*B* ratio of 1.4 given in the problem.

Load Conditions Section

We do not consider the horizontal loading for this problem; therefore we enter "0" for *H*/*V* ratio. We enter the load eccentricity values as stated in the problem statement.

Elastic Settlement Section

We can skip this section since settlement analysis is not required in this problem. So, now we can hit the run button and view the Output section. Allowable and ultimate foundation loads at various dimensions under two-way eccentricity condition are summarized in Table 1.20. Also, the allowable bearing capacity chart is shown in Fig. 1.35.

1.4.11 Rectangular Foundation with Elastic Settlement

Develop bearing capacity and elastic settlement design charts for a rectangular foundation with *L*/*B* ratio of 1.5. The maximum elastic foundation settlement should not exceed 25 mm. The elastic modulus of the soil below the foundation increases with depth at a rate of 160 kN/m^2/m. Top of rock is at 5.5 m below the base of the foundation. Ignore the effect of soil compressibility on the bearing capacity calculations. Use a safety factor of 3.0. Soil properties and other parameters are provided in Table 1.21.

1.4.11.1 Hand Solution

To develop design capacity charts based on bearing capacity and elastic settlement conditions, one must follow the steps below:

Step 1:
Assume $B = 1$ m; therefore, $L = 1.5$ m since $L/B = 1.5$.

Step 2:
Determine bearing capacity factors.

Table 1.20 Foundation loads at various foundation dimensions

B (m)	L (m)	B' (m)	L' (m)	q_{all} (kN/m^2)	V_{all} (kN)	q_u (kN/m^2)	V_u (kN)	$M_{u\text{-}B}$ (kN m)	$M_{u\text{-}L}$ (kN m)
1	1.4	0.625783	1.017141	291.7515	185.7024	875.2544	557.1073	55.71073	139.2768
2	2.8	1.251565	2.034282	267.3844	680.7703	802.1531	2,042.311	204.2311	510.5777
3	4.2	1.877348	3.051423	250.4298	1,434.608	751.2895	4,303.823	430.3823	1,075.956
4	5.6	2.50313	4.068563	241.7897	2,462.421	725.369	7,387.263	738.7263	1,846.816
5	7	3.128913	5.085704	236.5033	3,763.412	709.5098	11,290.23	1,129.023	2,822.559
6	8.4	3.754696	6.102845	232.9153	5,337.097	698.7458	16,011.29	1,601.129	4,002.822
7	9.8	4.380478	7.119986	230.3149	7,183.279	690.9447	21,549.84	2,154.984	5,387.459
8	11.2	5.006261	8.137127	228.3456	9,302.02	685.0369	27,906.06	2,790.606	6,976.515
9	12.6	5.632043	9.154268	226.8085	11,693.62	680.4254	35,080.85	3,508.085	8,770.213
10	14	6.257826	10.17141	225.5834	14,358.58	676.7501	43,075.75	4,307.575	10,768.94

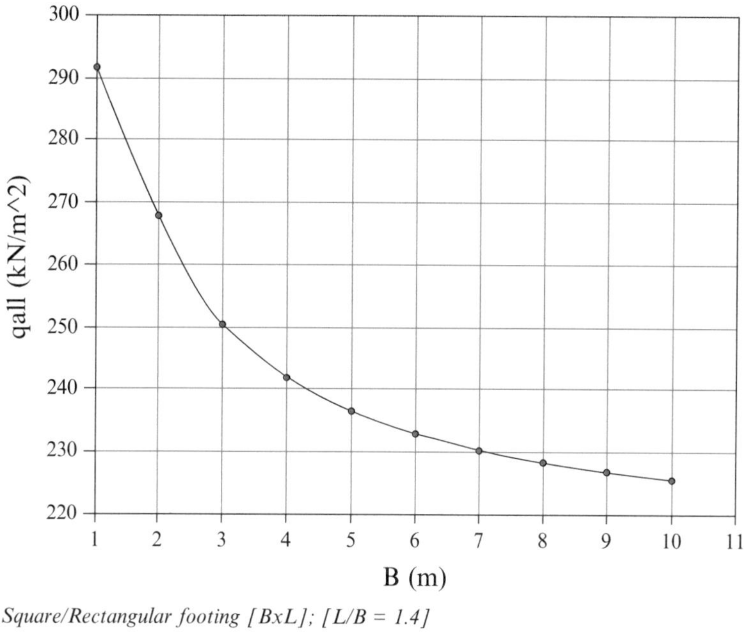

Fig. 1.35 Allowable bearing capacity chart

Table 1.21 Soil properties and other parameters

Property	Value	Unit
D_f	1.3	m
c	16	kN/m^2
ϕ	30	°
γ	18.6	kN/m^3
E_f	25,000,000	kN/m^2
t	400	mm
E_s	11,000	kN/m^2
μ_s	0.35	–
k	160	kN/m^2/m
S_e	25	mm
H	5.5	m
GWT depth	Too deep	–

Using Eq. (1.2):

$$N_q = \tan^2\left(45 + \frac{30}{2}\right)e^{\pi \tan(30)} = 18.14$$

Using Eq. (1.3):

$$N_c = (18.14 - 1)\cot(30) = 30.14$$

Using Eq. (1.4):

$$N_\gamma = N_c = \left(N_q - 1\right)\cot(\phi) = 22.40$$

Step 3:
Determine depth factors.
For $\phi^{'} > 0$ and from Eq. (1.8):

$$\eta = \frac{D_\text{f}}{B} = \frac{1.3}{1} = 1.5 > 1$$

Use Eq. (1.6):

$$F_{q\text{d}} = 1 + 2\tan(30)(1 - \sin(30))^2 \tan^{-1}\left(\frac{1.3}{1}\right) = 1.2642$$

Use Eq. (1.7):

$$F_{c\text{d}} = 1.28 - \frac{1 - 1.28}{30.14 \times \tan 30} = 1.296$$

$$F_{\gamma\text{d}} = 1$$

Step 4:
Determine load inclination factors.
Since no horizontal load is considered (H/V ratio = 0), all load inclination factors are set to 1 throughout this problem.

From Eq. (1.12):

$$\beta = \tan^{-1}\left(\frac{0}{0}\right) = 0$$

From Eq. (1.10):

$$F_{c\text{i}} = F_{q\text{i}} = \left(1 - \frac{0°}{90°}\right)^2 = 1$$

From Eq. (1.11):

$$F_{\gamma\text{i}} = \left(1 - \frac{0}{30}\right) = 1$$

Step 5:
Determine soil compressibility factors.

Since the effect of soil compressibility is not required in the given problem, all compressibility factors are set to 1.

Therefore $F_{cc} = F_{qc} = F_{\gamma c} = 1$

Step 6:
Determine q and γ to account for the GWT.
Given that groundwater table is very deep:
Unit weight of the foundation soil is determined using Eq. (1.23):

$$\gamma = \gamma_1 = 18.6\,\mathrm{kN/m^3}$$

Also, the effective stress is calculated using Eq. (1.21) as

$$q = 18.6 \times 1.3 = 24.18\,\mathrm{kN/m^2}$$

Step 7:
Determine the effective dimensions to account for the effect of load eccentricity.
Since there is no eccentricity,

$$B^{'} = B = 1\,\mathrm{m}$$

$$L_L^{'} = L = 1.5\,\mathrm{m}$$

Step 8:
Determine shape factors.
Use Eq. (1.54):

$$F_{cs} = 1 + \left(\frac{1}{1.5}\right)\left(\frac{18.14}{30.14}\right) = 1.401$$

Use Eq. (1.55):

$$F_{qs} = 1 + \left(\frac{1}{1.5}\right)\tan(30) = 1.38$$

Use Eq. (1.56):

$$F_{\gamma s} = 1 - 0.4\left(\frac{1}{1.5}\right) = 0.7333$$

Step 9:
The ultimate bearing capacity of the rectangular foundation is now computed from Eq. (1.1) as follows:

$$q_{\mathrm{u}} = (16)\,(30.14)\,(1.296)\,(1)\,(1)\,(1)\,(1.401) + (24.18)\,(1.38)\,(18.14)\,(1)\,(1.26)\,(1) + \frac{1}{2}(18.6)\,(1)\,(22.4)\,(0.733)\,(1)\,(1)\,(1) = 1,790.98\,\mathrm{kN/m^2}$$

The ultimate vertical load is found from Eq. (1.68) as follows:

$$V_{\mathrm{u}} = 1,790.98 \times 1 \times 1.5 = 2,686.47\,\mathrm{kN}$$

From Eq. (1.67), the allowable load bearing capacity is calculated as

$$q_{\mathrm{all}} = \frac{1,790.98}{3} = 596.99\,\mathrm{kN/m^2}$$

Also, the allowable vertical load is computed from Eq. (1.69):

$$V_{\mathrm{all}} = 596.99 \times 1 \times 1.5 = 895.49\,\mathrm{kN}$$

Step 10:
Different values of B can be chosen and the procedure can be repeated to evaluate the impact of the foundation dimensions on the bearing capacity of the foundation.

Step 11:
To perform elastic settlement analysis:

The effective dimension can be found from Eq. (1.58) as follows:

$$B_{\mathrm{e}} = \sqrt{\frac{4 \times 1 \times 1.5}{\pi}} = 1.38$$

Given $E_{\mathrm{s}} = 11{,}000$ kN/m^2.

Since I_{G} is influence factor for the variation of E_{s} with depth $= f(\beta_1, \beta_2)$, one must find β_1 and β_2 as follows:

From Eq. (1.64):

$$\beta_1 = \frac{5.5}{1.38} = 3.978$$

From Eq. (1.65):

$$\beta_2 = \log\left(\frac{11,000}{160 \times 1.38}\right) = 1.69$$

I_G is then computed from Eq. (1.66):

$$I_G = \left(-0.01189e^{-1.26658\beta_1} + 0.012608\right) \times \left(0.34865\beta_2^5 + 1.05867\beta_2^4 - 4.2618\beta_2^3 - 7.1333\beta_2^2 + 28.92718\beta_2 + 51.4275\right) = 0.912$$

The foundation rigidity correction factor can be computed from Eq. (1.61) as follows:

$$I_F = \frac{\pi}{4} + \frac{1}{4.6 + 10\left(\frac{25,000,000}{11,000 + \frac{1.38}{2} \times 160}\right)\left(\frac{2 \times 0.4}{1.38}\right)^3} = 0.785$$

Also, foundation embedment correction factor is computed using Eq. (1.62):

$$I_E = 1 - \left(\frac{1}{3.5\exp(1.22 \times 0.35 - 0.4)\left(\frac{1.38}{1.3} + 1.6\right)}\right) = 0.895$$

Given $S_e = 0.025$ m and from Eq. (1.57), one can calculate the net allowable bearing capacity $q_{All(net)}$ as follows:

$$0.025 = \frac{q_{All(net)} \times 1.38 \times 0.91 \times 0.785 \times 0.895}{11,000}\left(1 - 0.3^2\right)$$

Therefore,

$$q_{All(net)} = 342.51\,\text{kN/m}^2$$

From Eq. (1.67):

$$q_{all} = 342.51 + \frac{24.18}{3} = 350.57\,\text{kN/m}^2$$

Hence,

The allowable vertical load can be solved for using Eq. (1.75) as follows:

$$350.57 = V_{all}\left(\left(\frac{1}{1 \times 1.5}\right) + \frac{0}{1.5 \times 1^2} + \frac{0}{1 \times 1.5^2}\right)$$

Thus,

$$V_{all} = 525.855\,\text{kN}$$

1.4.11.2 foundationPro Solution

General Information Section

In this section, we enter the factor of safety and select the SI units to be used throughout the analysis.

Groundwater Table and Soil Information Section

We do not click "GWT exists" since the groundwater table is too deep to consider. We enter γ_1, c', and ϕ. Then, we enter the elastic modulus of soil at the base of foundation, rate of increase in the elastic modulus of soil, and the Poisson's ratio of soil.

Foundation Shape/Depth Section

In this section, we enter the depth of the foundation and select the foundation shape as required. We enter the values we would like to consider for the foundation dimensions. Also, the *L*/*B* ratio of 1.5 is used as required in the problem.

Load Conditions Section

This section can be skipped and no information is required since horizontal load and bending moments are not applied to the foundation.

Elastic Settlement Section

We enter the required parameters to be used for the elastic settlement analysis in this problem as shown in Fig. 1.36.

And now, we can hit the run button and view the Output section. Foundation loads based on elastic settlement analyses are shown in Figs. 1.37 and 1.38. Also, bearing capacity results at various foundation dimensions are summarized in Table 1.22. The bearing capacity results are also shown in Fig. 1.39. Foundation capacity results (allowable load bearing capacity and vertical load) are summarized in Table 1.23. The allowable vertical foundation loads based on settlement analyses at various foundation dimensions are also shown in Fig. 1.40.

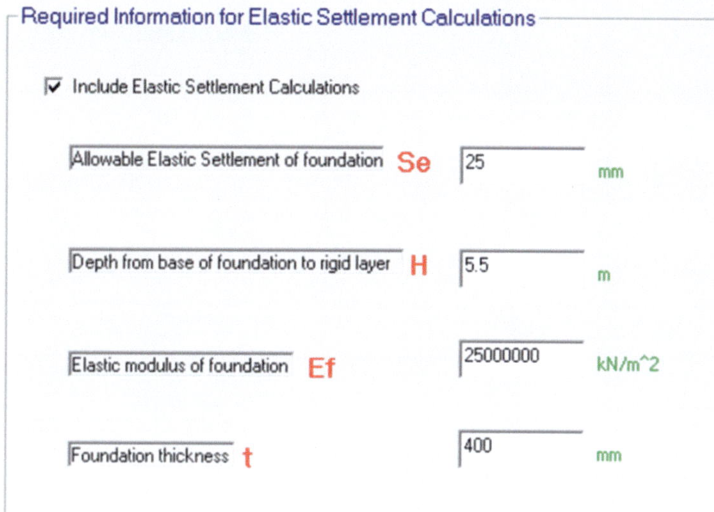

Fig. 1.36 Required elastic settlement information

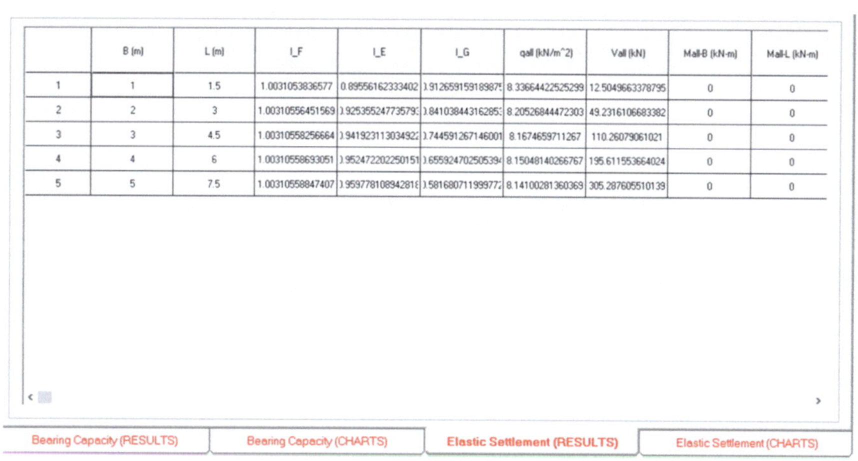

	B (m)	L (m)	I_F	I_E	I_G	qall (kN/m^2)	Vall (kN)	Mall-B (kN-m)	Mall-L (kN-m)
1	1	1.5	1.0031053836577	0.89556162333402	).912659159189875	8.33664422525299	12.5049663378795	0	0
2	2	3	1.00310556451569	).92535524773579	).841038443162853	8.20526844472303	49.2316106683382	0	0
3	3	4.5	1.00310558256664	).941923113034922	).744591267146001	8.1674659711267	110.26079061021	0	0
4	4	6	1.00310558693051	).952472202250151	).655924702505394	8.15048140266767	195.611553664024	0	0
5	5	7.5	1.00310558847407	).959778108942818	).581680711999772	8.14100281360369	305.287605510139	0	0

Fig. 1.37 Allowable foundation load results based on elastic settlement analyses

1.4.12 *Circular Foundation with Horizontal Loading and Elastic Settlement*

Develop a bearing capacity and elastic settlement design charts for circular foundation with a horizontal loading of 10 % of the vertical loading (i.e., $H/V = 0.1$). Use a safety factor of 3.2. Ignore the effect of soil compressibility on bearing capacity calculations. Soil properties and other design parameters are provided in Table 1.24.

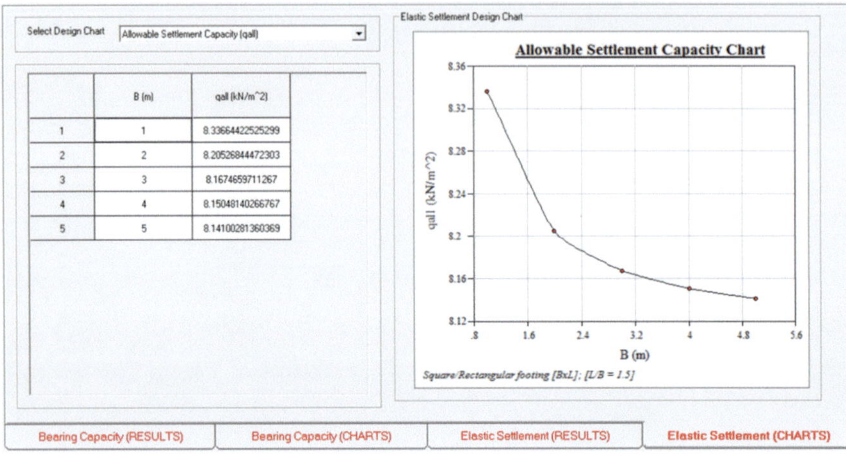

Fig. 1.38 Allowable settlement capacity results

Table 1.22 Bearing capacity results at various foundation dimensions

B (m)	L (m)	q_{all} (kN/m^2)	V_{all} (kN)	q_{u} (kN/m^2)	V_{u} (kN)
1	1.5	599.9378643	899.9068	1,799.813593	2,699.72
2	3	616.8443573	3,701.066	1,850.533072	11,103.2
3	4.5	639.9666553	8,639.55	1,919.899966	25,918.65
4	6	645.8035776	15,499.29	1,937.410733	46,497.86
5	7.5	658.0136522	24,675.51	1,974.040956	74,026.54
6	9	673.3832546	36,362.7	2,020.149764	109,088.1
7	10.5	690.4496025	50,748.05	2,071.348807	152,244.1
8	12	708.4839547	68,014.46	2,125.451864	204,043.4
9	13.5	727.0856764	88,340.91	2,181.257029	265,022.7
10	15	746.0188364	111,902.8	2,238.056509	335,708.5

1.4.12.1 Hand Solution

To develop design capacity charts (bearing capacity and elastic settlement), one must follow the steps as below:

Step 1:
Assume foundation diameter, $B = D = 2$ ft.

Step 2:
Determine bearing capacity factors.
Using Eq. (1.2):

$$N_q = \tan^2\left(45 + \frac{8}{2}\right)e^{\pi\tan(8)} = 2.06$$

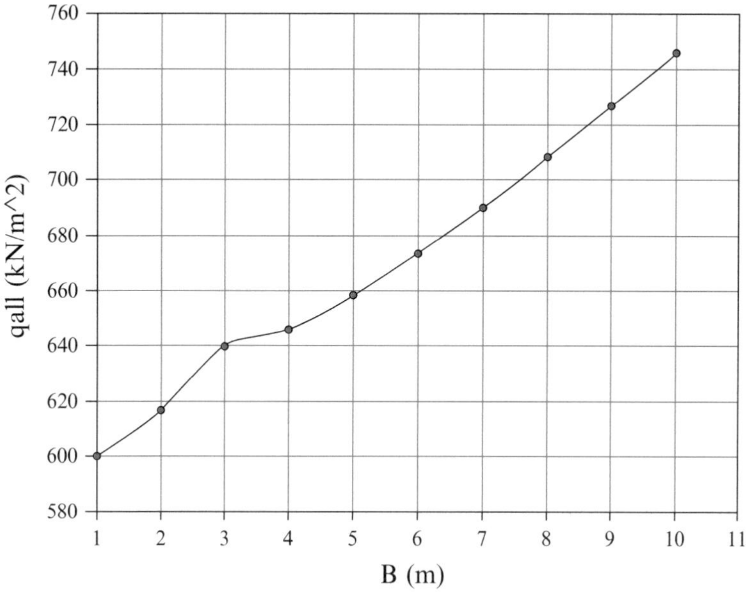

Square/Rectangular footing [BxL]; [L/B = 1.5]

Fig. 1.39 Allowable bearing capacity chart

Table 1.23 Foundation capacity results based on elastic settlement analyses

B (m)	L (m)	q_{all} (kN/m^2)	V_{all} (kN)
1	1.5	361.2143795	541.8215693
2	3	193.1326161	1,158.795697
3	4.5	144.2336947	1,947.154878
4	6	121.5691412	2,917.659388
5	7.5	108.1685785	4,056.321694
6	9	98.98223983	5,345.040951
7	10.5	92.10770596	6,769.916388
8	12	86.7195085	8,325.072816
9	13.5	82.40459696	10,012.15853
10	15	78.91533192	11,837.29979

Using Eq. (1.3):

$$N_c = (2.06 - 1)\cot(8) = 7.53$$

Using Eq. (1.4):

$$N_\gamma = N_c = (2.06 - 1)\cot(8) = 0.86$$

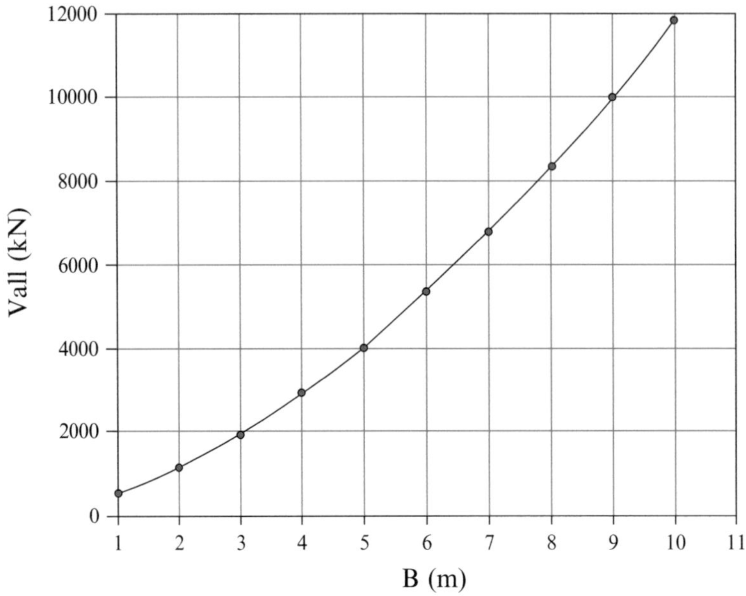

Square/Rectangular footing [BxL]; [L/B = 1.5]

Fig. 1.40 Allowable vertical load based on elastic settlement analyses

Table 1.24 Soil properties and other design parameters

Property	Value	Unit
D_f	4	ft
c	1,400	lb/ft^2
ϕ	8	°
γ_1	119	lb/ft^3
γ_2	135	lb/ft^3
E_f	522,000,000	lb/ft^2
t	16.8	in.
E_s	250,000	lb/ft^2
μ_s	0.3	–
k	0	lb/ft^2/ft
S_e	1	in.
H	10	ft
GWT depth	Too deep	–

Step 3: Determine depth factors.

For $\phi' > 0$ and from Eq. (1.8):

$$\eta = \frac{D_f}{B} = \frac{4}{2} = 2 > 1$$

Use Eq. (1.6):

$$F_{qd} = 1 + 2\tan(8)\,(1 - \sin(8))^2 \tan^{-1}\left(\frac{4}{2}\right) = 1.23$$

Use Eq. (1.7):

$$F_{cd} = 1.23 - \frac{1 - 1.23}{7.53 \times \tan(8)} = 1.447$$

$$F_{\gamma d} = 1$$

Step 4:
Determine load inclination factors.

For *H*/*V* ratio of 0.1 and from Eq. (1.12):

$$\beta = \tan^{-1}(0.1) = 5.71$$

From Eq. (1.10):

$$F_{ci} = F_{qi} = \left(1 - \frac{5.71^\circ}{90^\circ}\right)^2 = 0.87$$

From Eq. (1.11):

$$F_{\gamma i} = \left(1 - \frac{5.71}{8}\right) = 0.286$$

Step 5:
Determine soil compressibility factors.

Since the effect of soil compressibility is not required in this problem, all factors are set to 1.

Thus,

$$F_{cc} = F_{qc} = F_{\gamma c} = 1$$

Step 6:
Determine q and γ to account for the GWT.

Since groundwater table is very deep, its effect can be ignored. Therefore, the unit weight of the foundation soil can be found using Eq. (1.23) as follows:

$$\gamma = \gamma_1 = 119\,\text{lb/ft}^3$$

Also, the effective stress at the base of the foundation can be computed from Eq. (1.21):

$$q = 119 \times 4 = 476\,\text{lb/ft}^2$$

Step 7:
Determine the effective dimensions to account for the effect of load eccentricity. Since there is no load eccentricity,

$$B^{'} = B = 2\,\text{ft.}$$

Step 8:
Determine shape factors.
Use Eq. (1.54):

$$F_{cs} = 1 + (1)\left(\frac{2.06}{7.53}\right) = 1.273$$

Use Eq. (1.55):

$$F_{qs} = 1 + (1)\tan(8) = 1.14$$

Use Eq. (1.56):

$$F_{\gamma s} = 1 - 0.4(1) = 0.6$$

Step 9:
The ultimate bearing capacity of the circular foundation can be now computed from Eq. (1.1) as follows:

$$q_u = (1,400)(7.53)(1.447)(0.87)(1)(1.273) + (476)(1.14)(2.06)(0.93)(1)(1.23) \\ + \frac{1}{2}(119)(2.57)(0.86)(1)(0.6)(0.286)(1) = 18,545.78\,\text{lb/ft}^2$$

The ultimate vertical load based on bearing capacity can then be calculated from Eq. (1.68):

$$V_u = 18,545.78 \times 3.14 = 38,233.7\,\text{kN}$$

Also, from Eq. (1.67):

$$q_{all} = \frac{18,545.78}{3.2} = 5,795.55\,\text{kN/m}^2$$

The allowable vertical load is then determined using Eq. (1.69):

$$V_{\text{all}} = 5,795.55 \times 3.14 = 18,198.0\,\text{kN}$$

Since the *H*/*V* ratio is 0.1, the ultimate horizontal load is computed as follows:

$$H_{\text{u}} = 538,233.7 \times 0.1 = 53,823.37\,\text{kN}$$

Step 10:
Several foundation diameters can be selected and the above nine steps can be repeated to evaluate the bearing capacity of the foundation at various diameters.

To determine foundation loads based on elastic settlement analyses, one must follow the steps below:

Step 11:
For the effective dimension for circular foundations, one can use Eq. (1.67):

$$B_{\text{e}} = B = 2\,\text{ft}$$

From Eq. (1.64):

$$\beta_1 = \frac{10}{2} = 5$$

Given that $k = 0$ lb/ft^2/ft.

From Eq. (1.65) and since k is 0, β_2 can be assumed 2 (i.e., log (100)):

Now, one can determine I_{G} from Eq. (1.66) as follows:

$$I_{\text{G}} = \left(-0.01189\text{e}^{-1.26658\beta_1} + 0.012608\right) \times (0.34865\beta_2^5 + 1.05867\beta_2^4 - 4.2618\beta_2^3 - 7.1333\beta_2^2 + 28.92718\beta_2 + 51.4275) = 0.91$$

The foundation rigidity correction factor can also be determined using Eq. (1.61):

$$I_{\text{F}} = \frac{\pi}{4} + \frac{1}{4.6 + 10\left(\frac{522 \times 10^6}{250,000 + \frac{2}{2} \times 0}\right)\left(\frac{2 \times 1.4}{2}\right)^3} = 0.785$$

Also, foundation embedment correction factor is determined from Eq. (1.62) as follows:

$$I_{\text{E}} = 1 - \left(\frac{1}{3.5\exp(1.22 \times 0.3 - 0.4)\left(\frac{2}{4} + 1.6\right)}\right) = 0.859$$

Given $S_e = 1$ in. (which is the allowable elastic settlement the foundation can experience) and using Eq. (1.57):

$$0.0833 = \frac{q_{\mathrm{All(net)}} \times 2 \times 0.91 \times 0.785 \times 0.859}{250,000}\left(1 - 0.3^2\right)$$

One can solve for the net allowable load bearing capacity, $q_{\mathrm{All(net)}}$.

Thus,

$$q_{\mathrm{All(net)}} = 18,647.01\,\mathrm{lb/ft^2}$$

Then, the allowable load bearing capacity is calculated from Eq. (1.67):

$$q_{\mathrm{all}} = 18,647.01 + \frac{476}{3.2} = 18,795.76\,\mathrm{lb/ft^2}$$

From Eq. (1.75):

$$22,476.1 = V_{\mathrm{all}}\left(\frac{4}{\pi \times 2^2} + \frac{0}{\pi \times 2^3}\right)$$

One can solve for the vertical allowable load based on settlement:

$$V_{\mathrm{all}} = 59,018.71\,\mathrm{lb}$$

1.4.12.2 foundationPro Solution

General Information Section

We enter the factor of safety and select the BS units to be used throughout the analysis.

Groundwater Table and Soil Information Section

In this section, we do not click “GWT exists” since the groundwater table is too deep to consider. We enter γ_1, c', and ϕ. We enter the elastic modulus of soil at the base of foundation, rate of increase in the elastic modulus of soil, and the Poisson’s ratio of soil.

Foundation Shape/Depth Section

In this section, we enter the depth of the foundation and then set the foundation shape as circular. Then, we enter the values we would like to consider for the foundation dimensions; we do not enter anything for the *L*/*B* ratio since this is a circular foundation.

Load Conditions Section

We consider the horizontal loading for this problem by entering an *H*/*V* ratio of 0.1 given in the problem statement. We do not need to consider the load eccentricity since this does not apply in this problem.

Elastic Settlement Section

We enter the "S_e," "*H*," "E_f," and "*t*" for the elastic settlement analyses in this problem.

And now, we can hit the run button and view the Output section. Bearing capacity results (allowable and ultimate) at various foundation diameters are summarized in Table 1.25. Also, allowable bearing capacity design chart is shown in Fig. 1.41. Foundation load based on elastic settlement analyses are summarized in Table 1.26 and also shown in Fig. 1.42.

1.4.13 Rectangular Foundation with Two-Way Eccentricity and Elastic Settlement

Develop bearing capacity and elastic settlement design charts for rectangular foundation with *L*/*B* ratio of 1.5, and load eccentricity of $e_B/B = 0.1$ and $e_L/L = 0.1$. The maximum elastic settlement of foundation must not exceed 25 mm and top of rock is at 8 m below ground surface. Table 1.27 lists the soil properties and other required design parameters.

1.4.13.1 Hand Solution

The following steps must be followed to develop capacity design charts (elastic settlement and bearing capacity):

Step 1:
Assume $B = 1$ m; then $L = 1.5$ m since $L/B = 1.5$.

Step 2:
Determine bearing capacity factors.
Using Eq. (1.2):

$$N_q = \tan^2\left(45 + \frac{14}{2}\right) e^{\pi \tan(14)} = 3.59$$

Table 1.25 Bearing capacity results at various foundation diameters

B (ft)	q_{all} (lb/ft^2)	V_{all} (lb)	q_u (lb/ft^2)	V_u (lb)	H_u (lb)
1	6,047.732779	4,749.878218	19,352.74489	15,199.6103	1,519.96103
2	5,710.637252	17,940.49604	18,274.03921	57,409.58733	5,740.958733
3	5,433.865922	38,409.73484	17,388.37095	122,911.1515	12,291.11515
4	5,549.602652	69,738.36369	17,758.72849	223,162.7638	22,316.27638
5	5,241.521068	102,917.0255	16,772.86742	329,334.4817	32,933.44817
6	5,037.048061	142,419.1787	16,118.5538	455,741.3718	45,574.13718
7	4,891.779955	188,257.7547	15,653.69586	602,424.8149	60,242.48149
8	4,783.514912	240,445.6849	15,307.24772	769,426.1918	76,942.61918
9	4,699.918577	298,995.901	15,039.73945	956,786.8831	95,678.68831
10	4,633.590339	363,921.3343	14,827.48908	1,164,548.27	116,454.827
11	4,579.820716	435,234.9163	14,655.42629	1,392,751.732	139,275.1732
12	4,535.470054	512,949.5786	14,513.50417	1,641,438.651	164,143.8651
13	4,498.364747	597,078.2525	14,394.76719	1,910,650.408	191,065.0408
14	4,466.95222	687,633.8697	14,294.2471	2,200,428.383	220,042.8383
15	4,440.093915	784,629.3615	14,208.30053	2,510,813.957	251,081.3957

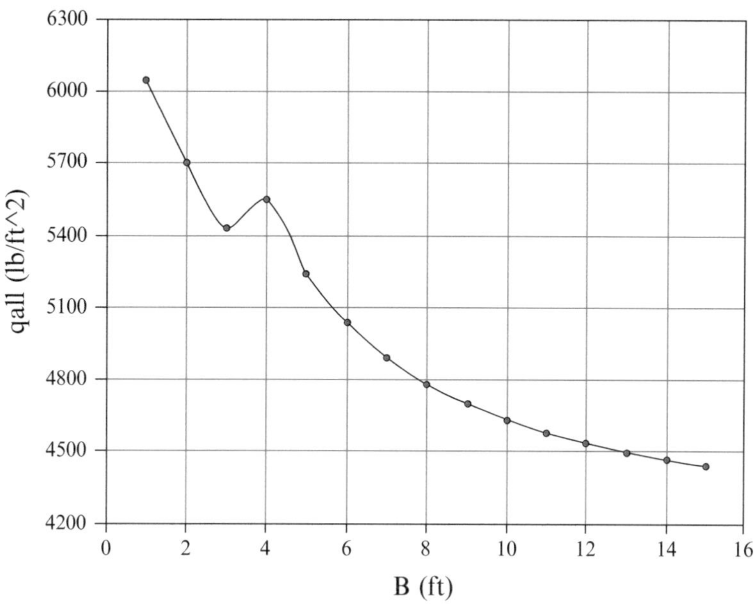

Circular footing [B]

Fig. 1.41 Allowable bearing capacity design chart

Table 1.26 Allowable loads based on elastic settlement analyses at various foundation diameters

B (ft)	q_{all} (lb/ft^2)	V_{all} (lb)
1	36,960.75429	29,028.90854
2	18,177.10465	57,105.05843
3	12,106.86226	85,578.36649
4	9,232.880254	116,023.7951
5	7,606.970614	149,362.5187
6	6,578.441141	186,001.0413
7	5,875.008218	226,096.8126
8	5,365.3734	269,693.0826
9	4,979.430272	316,777.6667
10	4,676.721378	367,308.8382
11	4,432.423864	421,227.3251
12	4,230.534969	478,462.2329
13	4,060.309722	538,934.206
14	3,914.284624	602,557.3032
15	3,787.125278	669,240.2785

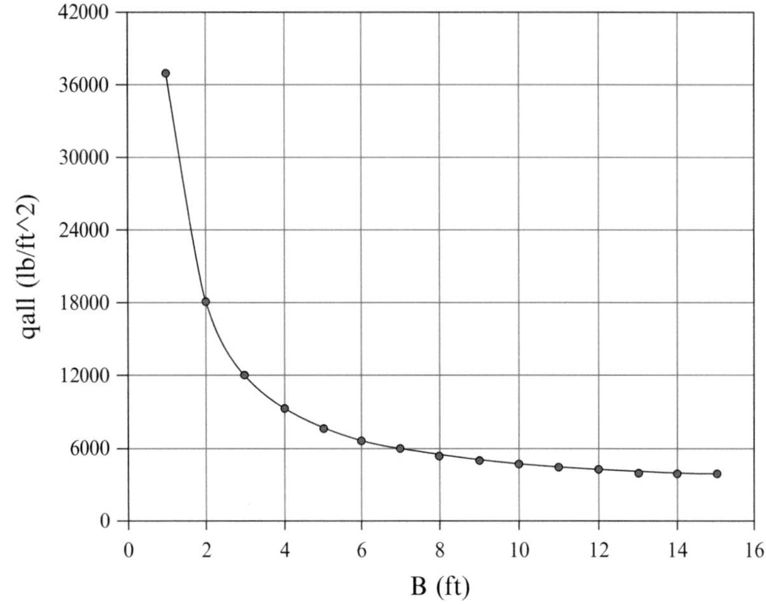

Circular footing [B]

Fig. 1.42 Allowable settlement design capacity chart

Table 1.27 Soil properties and other design parameters

Property	Value	Unit
D_f	1.5	m
c	55	kN/m^2
ϕ	14	°
γ	19.1	kN/m^3
FS	3.0	–
E_f	25,000,000	kN/m^2
t	400	mm
E_s	20,000	kN/m^2
μ_s	0.3	–
k	50	kN/m^2/m
S_e	25	mm
H	6.5	m
GWT depth	Too deep	–

Using Eq. (1.3):

$$N_c = (3.59 - 1)\cot(14) = 10.37$$

Using Eq. (1.4):

$$N_\gamma = N_c = (3.59 - 1)\cot(14) = 2.29$$

Step 3:
Determine depth factors.
For $\phi' > 0$ and from Eq. (1.8), one can find η as follows:

$$\eta = \frac{D_\mathrm{f}}{B} = \frac{1.5}{1} = 1.5 > 1$$

Use Eq. (1.6):

$$F_{q\mathrm{d}} = 1 + 2\tan(14)(1 - \sin(14))^2 \tan^{-1}\left(\frac{1.5}{1}\right) = 1.281$$

Use Eq. (1.7):

$$F_{c\mathrm{d}} = 1.281 - \frac{1 - 1.281}{10.37 \times \tan(14)} = 1.39$$

$$F_{\gamma\mathrm{d}} = 1$$

Step 4:
Determine load inclination factors.
All load inclination factors are set to 1 since the H/V ratio is zero:

$$F_{c\mathrm{i}} = F_{q\mathrm{i}} = F_{\gamma\mathrm{i}} = 1$$

Step 5:
Determine soil compressibility factors.
The effect of soil compressibility on bearing capacity calculations is not required; therefore,

$$F_{c\mathrm{c}} = F_{q\mathrm{c}} = F_{\gamma\mathrm{c}} = 1$$

Step 6:
Determine q and γ to account for the GWT.
In this problem, the groundwater table is too deep and its effect can be ignored. Therefore,

The unit weight of the foundation soil can be determined using Eq. (1.23):

$$\gamma = \gamma_1 = 19.1\,\mathrm{kN/m^2}$$

Also, the effective stress at the base of the foundation can be computed from Eq. (1.21) as follows:

$$q = 19.1 \times 1.5 = 28.65\,\mathrm{kN/m^2}$$

Step 7:
Determine the effective dimensions to account for the effect of load eccentricity.

Given,

Figure 1.5 can be used to determine the two-way load eccentricity case that applies. Given $e_B/B = 0.1$ (eccentricity in the B) and $e_L/L = 0.1$ (eccentricity in the L), one can find that Case IV controls.

So, B_2 dimension can be computed from Eq. (1.49):

$$\frac{B_2}{B} = \left(-6,470.36(0.1)^4 + 2,932.964(0.1)^3 - 493.417(0.1)^2 + 37.716(0.1) - 305.403\right) + \left(-461.303(0.1)^3 + 175.968(0.1)^2 - 26.045(0.1) + 305.8\right) = 0.214$$

Similarly, L_2 dimension can be computed from Eq. (1.51):

$$\frac{L_2}{L} = \left(-549.42(0.1)^3 + 202.478(0.1)^2 - 28.87(0.1) + 102.43\right) + \left(355.07(0.1)^3 - 136.35(0.1)^2 + 18.05(0.1) - 101.6\right) = 0.203$$

Therefore,

The effective area is determined using Eq. (1.49):

$$A' = 1 \times 1.5 - \frac{1}{2}(1 - 0.214)\,(1.5 - 0.304) = 1.029$$

Using Eq. (1.52):

$$B' = \frac{1.029}{1.5} = 0.686\,\mathrm{m}$$

Using Eq. (1.52):

$$L' = 1.5\,\mathrm{m}$$

Step 8:
Determine shape factors.
Use Eq. (1.54):

$$F_{cs} = 1 + \left(\frac{0.686}{1.5}\right)\left(\frac{2.59}{10.37}\right) = 1.15$$

Use Eq. (1.55):

$$F_{qs} = 1 + \left(\frac{0.686}{1.5}\right)\tan(14) = 1.115$$

Use Eq. (1.56):

$$F_{\gamma s} = 1 - 0.4\left(\frac{0.686}{1.5}\right) = 0.817$$

Step 9:
The ultimate bearing capacity of the rectangular foundation can be calculated from Eq. (1.1):

$$q_u = (55)(10.37)(1.39)(1)(1)(1.15) + (28.65)(2.59)(1.115)(1)(1)(1.28) + \frac{1}{2}(19.1)(0.686)(2.29)(0.817)(1)(1)(1) = 1,121.5\,\text{kN/m}^2$$

The ultimate vertical load based on bearing capacity can be calculated from Eq. (1.68):

$$V_u = 1,121.5 \times 1 \times 0.686 = 769.349\,\text{kN}$$

Then, the allowable load bearing capacity can be calculated from Eq. (1.67):

$$q_{all} = \frac{1,121.5}{3} = 373.833\,\text{kN/m}^2$$

From Eq. (1.69):

$$V_{all} = 373.833 \times 1 \times 0.686 = 256.44\,\text{kN}$$

Thus, the ultimate bending moments in both directions can be calculated as follows:

$$M_{u_B} = 769.349 \times 0.1 \times 1 = 76.949\,\text{m kN}$$
$$M_{u_L} = 769.349 \times 0.1 \times 1 = 76.949\,\text{m kN}$$

Step 10:
Several foundation dimensions can be chosen and the procedure can be repeated to study the effect of foundation dimensions on the bearing capacity of the foundation.

Step 11:
To find V_{all} for given allowable settlement:
From Eq. (1.58)

$$B_{\text{e}} = \sqrt{\frac{4 \times 1 \times 1.5}{\pi}} = 1.38\,\text{m}$$

From Eq. (1.64):

$$\beta_1 = \frac{6.5}{1.38} = 4.71$$

From Eq. (1.65):

$$\beta_2 = \log\left(\frac{20,000}{50 \times 1.38}\right) = 2.46 > 2$$

So, I_{G} can be computed from Eq. (1.66) as follows:

$$\begin{aligned} I_{\text{G}} =& \left(-0.01189\text{e}^{-1.26658\beta_1} + 0.012608\right) \times \left(0.34865\beta_2^5 + 1.05867\beta_2^4 - 4.2618\beta_2^3\right. \\ & \left. - 7.1333\beta_2^2 + 28.92718\beta_2 + 51.4275\right) = 0.94 \end{aligned}$$

The foundation rigidity correction factor can also be determined from Eq. (1.61):

$$I_{\text{F}} = \frac{\pi}{4} + \frac{1}{4.6 + 10\left(\dfrac{25,000,000}{20,000 + \dfrac{1.38}{2} \times 55}\right)\left(\dfrac{2 \times 0.4}{1.38}\right)^3} = 0.789$$

And the embedment correction factor is determined using Eq. (1.62):

$$I_{\text{E}} = 1 - \left(\frac{1}{3.5\exp(1.22 \times 0.3 - 0.4)\left(\dfrac{1.38}{1.5} + 1.6\right)}\right) = 0.88$$

Given $S_{\text{e}} = 0.025$ m and from Eq. (1.57), one can solve for the net allowable bearing capacity:

$$0.025 = \frac{q_{\text{All(net)}} \times 1.38 \times 0.94 \times 0.789 \times 0.88}{20,000}\left(1 - 0.3^2\right)$$

So,

$q_{\text{All(net)}} = 610.04\ \text{kN/m}^2$.

The allowable bearing capacity can then be calculated from Eq. (1.67):

$$q_{\text{all}} = 610.04 + \frac{28.65}{3} = 619.59\,\text{kN/m}^2$$

Therefore,

Equation (1.75) can be used to solve for the allowable vertical load based on elastic settlement as follows:

$$619.59 = V_{\text{all}}\left(\left(\frac{1}{1 \times 1.5}\right) + \frac{0.6}{1.5 \times 1^2} + \frac{0.9}{1 \times 1.5^2}\right)$$

Thus,

$$V_{\text{all}} = 422.447\,\text{kN}$$

1.4.13.2 foundationPro Solution

General Information Section

In this section, we enter the factor of safety and select the SI units to be used throughout the analysis.

Groundwater Table and Soil Information Section

Groundwater table and soil information are entered in this section. We do not click "GWT exists" since the groundwater table is too deep to consider. We enter γ_1, c', and ϕ. We enter the elastic modulus of soil at the base of foundation, rate of increase in the elastic modulus of soil, and the Poisson's ratio of soil.

Foundation Shape/Depth Section

We enter the depth of the foundation and then select the foundation shape. We enter the values we would like to consider for the foundation dimensions; we also enter the *L*/*B* ratio given in the problem statement.

Load Conditions Section

In this section, we define the load conditions as we do not consider any horizontal loading for this problem. However, we enter the e_B/B and e_L/L values given in the problem to account for the load eccentricity on the foundation.

Elastic Settlement Section

We enter the "S_e," "H," "E_f," and "t" to allow for elastic settlement analysis in this problem.

Now, we can hit the run button and view the Output section. Various foundation loads based on bearing capacity and elastic settlement analyses are summarized in Tables 1.28 and 1.29. Additionally, design capacity charts are shown in Figs. 1.43 and 1.44.

1.4.14 Circular Foundation with Load Eccentricity, Soil Compressibility, and Elastic Settlement

Develop bearing capacity and elastic settlement design charts for a circular foundation with a load eccentricity of $e_B/B = 0.35$. Include the effect of soil compressibility on bearing capacity calculations. Also, include elastic settlement in your analyses with a maximum foundation elastic settlement of 1 in. Top of rock is at a depth of 12 ft below the base of the circular foundation. Groundwater table is just 6 ft below the foundation base. Use a safety factor of 3. Table 1.30 provides a list of soil properties and other required design parameters.

1.4.14.1 Hand Solution

Follow the steps below to develop bearing capacity and elastic settlement design charts for this circular foundation:

Step 1:
Assume foundation diameter, $B = D = 1$ ft.

Step 2:
Determine bearing capacity factors.
Using Eq. (1.2):

$$N_q = \tan^2\left(45 + \frac{10}{2}\right) e^{\pi \tan(10)} = 2.47$$

Using Eq. (1.3)

$$N_c = (2.47 - 1)\cot(10) = 8.35$$

Using Eq. (1.4):

$$N_\gamma = N_c = (2.47 - 1)\cot(10) = 1.22$$

Table 1.28 Bearing capacity results

B (m)	L (m)	B' (m)	L' (m)	V_{all} (kN)	q_{u} (kN/m^2)	V_{u} (kN)	$M_{\text{u-}B}$ (kN m)	$M_{\text{u-}L}$ (kN m)
1	1.5	0.692344806	1.5	373.4609986	1,078.829495	1,120.382996	112.0382996	112.0382996
2	3	1.384689611	3	1,415.608541	1,022.329141	4,246.825623	424.6825623	424.6825623
3	4.5	2.077034417	4.5	2,993.239961	960.7415704	8,979.719884	897.9719884	897.9719884
4	6	2.769379223	6	5,184.895822	936.1115623	15,554.68747	1,555.468747	1,555.468747
5	7.5	3.461724029	7.5	8,016.180877	926.2645793	24,048.54263	2,404.854263	2,404.854263

Table 1.29 Foundation loads based on elastic settlement analysis

B (m)	L (m)	q_{all} (kN/m^2)	V_{all} (kN)	$M_{all\text{-}B}$ (kN m)	$M_{all\text{-}L}$ (kN m)
1	1.5	619.2022754	464.4017066	46.44017066	69.66025599
2	3	316.8823341	1,267.529336	253.5058673	380.2588009
3	4.5	227.7185821	2,305.650643	691.695193	1,037.542789
4	6	187.7870517	3,605.511392	1,442.204557	2,163.306835
5	7.5	163.1091798	5,097.161869	2,548.580935	3,822.871402

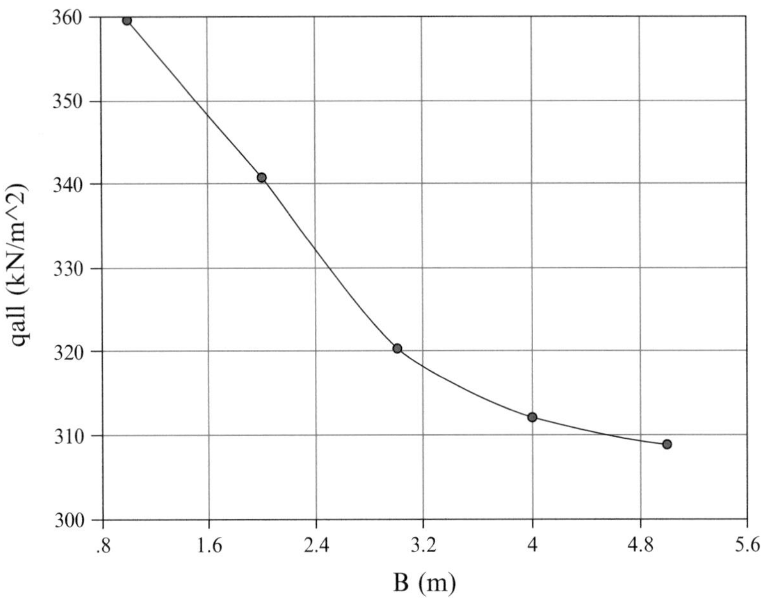

Square/Rectangular footing [BxL]; [L/B = 1.5]

Fig. 1.43 Allowable bearing capacity chart

Step 3: Determine depth factors.

For $\phi' > 0$ and using Eq. (1.8):

$$\eta = \frac{D_f}{B} = \frac{4}{1} = 4 > 1$$

Use Eq. (1.6):

$$F_{qd} = 1 + 2\tan(10)(1 - \sin(10))^2 \tan^{-1}\left(\frac{4}{1}\right) = 1.31$$

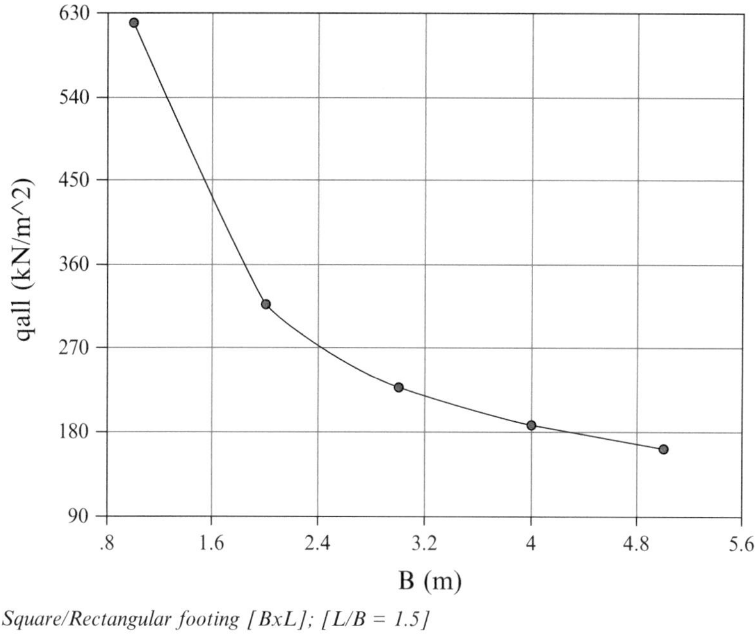

Fig. 1.44 Allowable elastic settlement capacity chart

Table 1.30 Soil properties and other design parameters

Property	Value	Unit
D_f	4	ft
c	1,250	lb/ft^2
ϕ	10	°
γ_1	120	lb/ft^3
γ_2	135	lb/ft^3
E_f	520,000,000	lb/ft^2
t	1.5	ft
E_s	220,000	lb/ft^2
μ_s	0.3	–
k	1,500	lb/ft^2/ft
S_e	1	in.
H	12	ft
GWT depth	10	ft

Use Eq. (1.7):

$$F_{cd} = 1.31 - \frac{1 - 1.31}{8.35 \times \tan(10)} = 1.53$$

$$F_{\gamma d} = 1$$

Step 4:
Determine load inclination factors.
All load inclination factors are set to 1, since the *H*/*V* ratio is 0 for this problem. So,

$$F_{ci} = F_{qi} = F_{\gamma i} = 1$$

Step 5:
Determine soil compressibility factors.
The effective stress, q', at a depth of ($D_f + B/2$) can be calculated as

$$q' = 120 \times (4 + 0.5) = 540\,\text{kN/m}^2$$

I_r can be now calculated using Eq. (1.14) as follows:

$$I_r = \frac{220{,}000}{2 \times (1 + 0.3)(1{,}250 + 540 \times \tan(10))} = 62.90$$

Using Eq. (1.15):

$$I_{r(cr)} = \frac{1}{2}\left\{\exp\left[\left(3.3 - 0.45\left(\frac{1}{1}\right)\right)\cot\left(45 - \frac{10}{2}\right)\right]\right\} = 14.926$$

Since $I_r > I_{r(cr)}$, then

$$F_{cc} = F_{qc} = F_{\gamma c} = 1$$

Step 6:
Determine q and γ to account for the GWT.

Since the groundwater table is too deep and Case III applies, one can determine the unit weight of the foundation soil using Eq. (1.23) as follows:

$$\gamma = \gamma_1 = 120\,\text{lb/ft}^3$$

And the effective stress at the base of the foundation is determined using Eq. (1.21):

$$q = 120 \times 4 = 4{,}80\,\text{lb/ft}^2$$

Step 7:
Determine the effective dimensions to account for the effect of load eccentricity
From Eq. (1.34):

$$f_1 = 43.473\left(\frac{e_D}{D}\right)^4 - 61.224\left(\frac{e_D}{D}\right)^3 + 32.094\left(\frac{e_D}{D}\right)^2 - 8.7505\left(\frac{e_D}{D}\right) + 1.2896$$
$$= 0.185$$

From Eq. (1.35):

$$f_2 = 1.5303\left(\frac{e_D}{D}\right)^2 - 2.438\left(\frac{e_D}{D}\right) + 0.8257 = 0.155$$

From Eq. (1.32):

$$B^{'} = 0.185 \times 1 = 0.185\,\text{ft}$$

From Eq. (1.33):

$$L^{'} = \frac{0.155 \times 1^2}{0.185} = 0.837\,\text{ft}$$

Step 8:
Determine shape factors.
Use Eq. (1.54):

$$F_{cs} = 1 + (0.22)\left(\frac{2.47}{8.35}\right) = 1.06$$

Use Eq. (1.55):

$$F_{qs} = 1 + (0.22)\tan(8) = 1.06$$

Use Eq. (1.56):

$$F_{\gamma s} = 1 - 0.4(0.22) = 0.911$$

Step 9:
The ultimate bearing capacity of the foundation can now be calculated from Eq. (1.1) as follows:

$$q_u = (1,250)(8.35)(1.447)(0.93)(1)(1.273) + (476)(2.47)(2.06)(0.93)(1)(1.23)$$
$$+ \frac{1}{2}(120)(1.22)(0.185)(0.91)(1)(1)(1) = 18,719.64437\,\text{lb/ft}^2$$

Thus,

The ultimate vertical foundation load based on bearing capacity can be computed from Eq. (1.68):

$$V_u = 18,719.64437 \times 0.185 \times 0.837 = 2,992.23\,\text{lb}$$

From Eq. (1.67):

$$q_{\text{all}} = \frac{18,719.64437}{3} = 6,222.385\,\text{lb/ft}^2$$

From Eq. (1.69):

$$V_{\text{all}} = 6,222.385 \times 0.135 \times 0.837 = 996.4\,\text{lb}$$

From the ratio given in the problem statement:

$$M_u = 2,992.23 \times 0.35 = 1,042.77\,\text{lb ft}$$

Step 10:
Several foundation diameters can be selected and the procedure can be repeated to evaluate the impact of the size of the foundation on the baring capacity of the foundation.

To determine the allowable foundation loads based on elastic settlement, the steps below must be followed:

Step 11:
For circular foundations, use Eq. (1.67) to determine the effective dimension:

$$B_e = B = 1\,\text{ft}$$

Given that, $E_s = 220{,}750$ lb/ft^2 at $D_f/2$.

From Eq. (1.64):

$$\beta_1 = \frac{12}{1} = 12$$

From Eq. (1.65):

$$\beta_2 = 2.16$$

Thus, I_G can be computed from Eq. (1.66):

$$\begin{aligned} I_G = &\left(-0.01189e^{-1.26658\beta_1} + 0.012608\right) \times (0.34865 \times 2.16^5 + 1.05867 \times 2.16^4 \\ &- 4.2618 \times 2.16^3 - 7.1333 \times 2.16^2 + 28.92718 \times 2.16 + 51.4275 \\ = &\ 0.94 \end{aligned}$$

From Eq. (1.61):

$$I_F = \frac{\pi}{4} + \frac{1}{4.6 + 10\left(\dfrac{522 \times 10^6}{250,000 + \frac{2}{2} \times 0}\right)\left(\dfrac{2 \times 1.4}{2}\right)^3} = 0.785$$

From Eq. (1.62):

$$I_E = 1 - \left(\frac{1}{3.5\exp(1.22 \times 0.3 - 0.4)\left(\dfrac{2}{4} + 1.6\right)}\right) = 0.84$$

Given that $S_e = 1$ in. and from Eq. (1.57), one can solve for the net allowable bearing capacity:

$$0.0833 = \frac{q_{\text{all(net)}} \times 1 \times 0.94 \times 0.785 \times 0.84}{250,000}\left(1 - 0.3^2\right)$$

Thus,

$$q_{\text{all(net)}} = 31{,}920.8 \text{ lb/ft}^2.$$

From Eq. (1.67):

$$q_{\text{all}} = 31,920.81 + \frac{476}{3.2} = 32,069.55 \,\text{lb/ft}^2$$

Similarly, Eq. (1.75) can be used to solve for the allowable vertical load based on settlement analysis as follows:

$$32,069.55 = V_{\text{all}}\left(\frac{4}{\pi \times 1^2} + \frac{32 \times 0.35}{\pi \times 1^3}\right)$$

Therefore,

$$V_{\text{all}} = 6,624.89 \,\text{lb ft}$$

1.4.14.2 foundationPro Solution

General Information Section

In this section, we enter the factor of safety and select the BS units to be used throughout the analysis.

Groundwater Table and Soil Information Section

Soil and groundwater table information are entered in this section. We click the "GWT exists" button, and enter the depth of the groundwater table. We enter γ_1, γ_2, c', and ϕ. We enter the elastic modulus of soil at the base of foundation, rate of increase in the elastic modulus of soil, and the Poisson's ratio of soil.

Foundation Shape/Depth Section

We enter the depth of the foundation and then we select the circular foundation as the foundation shape. Then, we enter the values we would like to consider for the foundation diameter; we do not enter any value for the *L*/*B* ratio since this is a circular foundation.

Load Conditions Section

We do not consider the horizontal loading for this problem. For the load eccentricity section, we enter the e_D/D values given in the problem.

Elastic Settlement Section

We enter the "S_e," "H," "E_f," and "t" to consider the elastic settlement in this problem.

And now, we can hit the run button to view the results! Various foundation loads based on bearing capacity and settlement analyses are summarized in Tables 1.31 and 1.32. Also, bearing capacity and elastic settlement design charts are shown in Figs. 1.44 and 1.45, respectively (Fig. 1.46).

1.5 Suggested Projects

In this section, you will find some suggested design projects for the reader to practice the concepts learned in the design of shallow foundations on homogeneous soil. These projects will cover a variety of foundation design configurations (e.g., at different depths, different groundwater table depths).

Table 1.31 Bearing capacity results

B (ft)	B' (ft)	q_{all} (lb/ft^2)	V_{all} (lb)	q_u (lb/ft^2)	V_u (lb)	$M_{u\text{-}D}$ (lb ft)
1	0.185827706	6,229.068235	995.7897	18,687.2047	2,987.369247	1,045.579236
1.5	0.278741559	6,049.527072	2,175.984	18,148.88122	6,527.951802	2,284.783131
2	0.371655413	5,884.407111	3,762.766	17,653.22133	11,288.29942	3,950.904798
2.5	0.464569266	5,735.019751	5,730.064	17,205.05925	17,190.19301	6,016.567553
3	0.557483119	5,601.563882	8,059.426	16,804.99165	24,178.27836	8,462.397426
3.5	0.650396972	5,483.578583	10,738.53	16,450.73575	32,215.58172	11,275.4536
4	0.743310825	5,721.798339	14,635.15	17,165.39502	43,905.44139	15,366.90449
4.5	0.836224678	5,546.533707	17,955.57	16,639.90112	53,866.69519	18,853.34332
5	0.929138531	5,406.917675	21,608.98	16,220.75303	64,826.94913	22,689.43219

Table 1.32 Foundation loads based on settlement analysis

B (ft)	q_{all} (lb/ft^2)	V_{all} (lb)	$M_{all\text{-}D}$ (lb ft)
1	32,664.93565	6,751.310651	2,362.958728
1.5	21,598.71592	13,314.44707	6,990.084709
2	16,307.14599	21,346.00419	14,942.20293
2.5	13,070.83712	30,264.77437	26,481.67758
3	10,913.8163	39,902.70079	41,897.83582
3.5	9,391.425548	50,197.8348	61,492.34763
4	8,271.040744	61,139.39021	85,595.1463
4.5	7,419.164236	72,737.69721	114,561.8731
5	6,754.034371	85,009.7146	148,767.0006

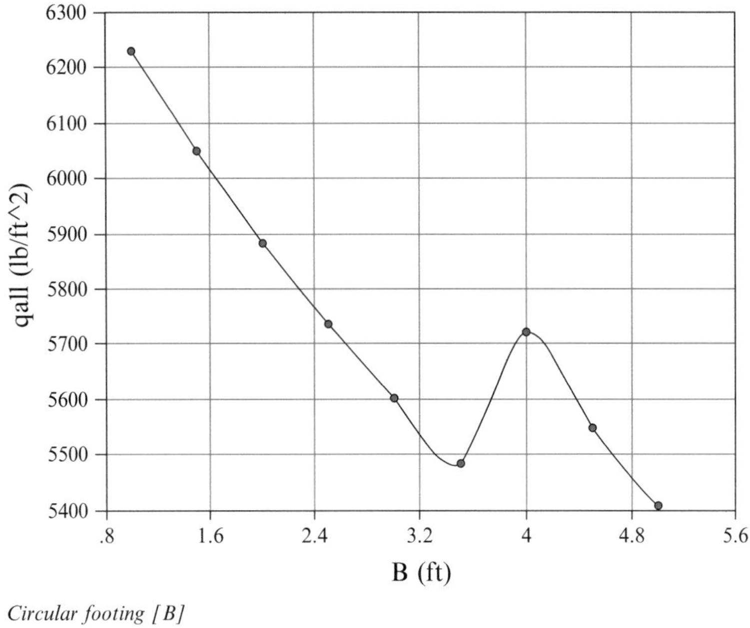

Fig. 1.45 Allowable bearing capacity design chart

1.5.1 Suggested Projects: Rectangular Foundation

Develop a bearing capacity design chart for a rectangular foundation with L/B ratio of 2, with no horizontal loading and no eccentricity. Ignore the effect of soil compressibility on bearing capacity calculations. Factor of safety is 3.5. Soil properties and other design parameters are provided in Table 1.33.

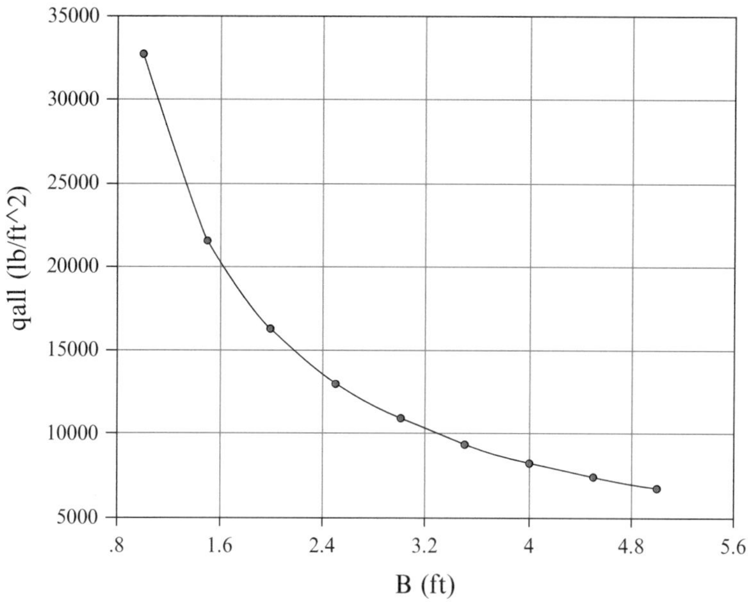

Circular footing [B]

Fig. 1.46 Allowable elastic settlement capacity chart

Table 1.33 Soil properties and other design parameters

Property	Value	Unit
D_f	1.7	m
c	50	kN/m^2
ϕ	10	°
γ_1	20	kN/m^3
γ_2	22	kN/m^3
GWT depth	5	m

1.5.2 Suggested Projects: Circular Foundation with Load Eccentricity

Develop a bearing capacity design chart for a circular foundation with a load eccentricity of 0.5 in the *D* direction. Use a safety factor of 3 in your bearing capacity calculations and do not include the effect of soil compressibility. Other required parameters are provided in Table 1.34.

Table 1.34 Soil properties and design parameters

Property	Value	Unit
D_f	3	ft
c	400	lb/ft^2
ϕ	31	°
γ	120	lb/ft^3
GWT depth	1	ft

Table 1.35 Soil properties and design parameters

Property	Value	Unit
D_f	5	ft
c	200	lb/ft^2
ϕ	30	°
γ_1	135	lb/ft^3
γ_2	145	lb/ft^3
GWT depth	Ground surface	ft

1.5.3 Suggested Projects: Strip/Continuous Foundation with Horizontal Loading

Develop a bearing capacity design chart for a strip/continuous foundation with no load eccentricity, and with a horizontal loading of 50 % of the vertical loading (i.e., $H/V = 0.5$). Factor of safety is 4.0. Following are the soil properties and other design parameters (Table 1.35).

1.5.4 Suggested Projects: Rectangular Foundation with Load Eccentricity

Develop a bearing capacity design chart for a rectangular foundation with L/B ratio of 1.8, and load eccentricity of $e_B/B = 0.3$ and $e_L/L = 0.25$. Groundwater table is 0.5 m below the foundation base. Include the effect of soil compressibility on bearing capacity calculations. Table 1.36 below summarizes the other required information to solve the problem.

1.5.5 Suggested Projects: Circular Foundation with Load Eccentricity and Settlement

Develop bearing capacity and elastic settlement design charts for a circular foundation with a load eccentricity of $e_B/B = 0.5$ with no horizontal loading. Include the effect of soil compressibility on bearing capacity calculations and also elastic

Table 1.36 Soil properties and other design parameters

Property	Value	Unit
D_f	1.5	m
c	60	kN/m^2
ϕ	25	°
γ_1	18	kN/m^3
γ_2	20	kN/m^3
FS	3.5	–
GWT depth	2	m
E_s	1,900	kN/m^2
μ_s	0.30	–

Table 1.37 Soil properties and other design parameters

Property	Value	Unit
D_f	5	ft
c	1,300	lb/ft^2
ϕ	15	°
γ_1	125	lb/ft^3
γ_2	140	lb/ft^3
E_f	520,000,000	lb/ft^2
t	1.5	ft
E_s	220,000	lb/ft^2
μ_s	0.3	–
k	1,750	lb/ft^2/ft
S_e	1.5	in.
H	15	ft
GWT depth	3	ft

settlement analysis with a maximum foundation elastic settlement of 1.5 in. Use a factor of safety of 3 for bearing capacity analysis. Other required information can be found in Table 1.37.

References

Caquot A, Kerisel J. Sur le terme de surface dans le calcul des foundations en milieu pulverulent. Proceedings, Third International Congress on Soil Mechanics and Foundation Engineering, Zurich, vol 1, p 336–337; 1953.

De Beer EE. Experimental determination of the shape factors and bearing capacity factors of sand. Geotechnique. 1970;20(4):387–411.

Hanna AM, Meyerhof GG. Experimental evaluation of bearing capacity of footings subjected to incline loads. Can Geotech J. 1981;18(4):599–603.

Hansen JB. A revised and extended formula for bearing capacity. Danish Geotech Inst Bull. 1970;28.

Highter WH, Anderes JC. Dimensioning footings subjected to eccentric loads. J Geotech Eng. 1985;111(5):659–65.

Mayne PW, Poulos HG. Approximate displacement influence factors for elastic shallow foundation. J Geotech Geoenviron Eng ASCE. 1999;125(6):453–60.
Meyerhof G. The bearing capacity of foundation under eccentric and inclined loads. Proceedings, Third International Congress on Soil Mechanics and Foundation Engineering, Zurich, vol 1, p 440–445; 1953.
Meyerhof G. The bearing capacity of foundation under eccentric and inclined loads. J Soil Mech Found Div ASCE. 1963;97:95–117.
Prandtl L. Uber die Eindringungsfestigkeit (Harte) plastischer Baustoffe Und die Festigkeit von schneiden. Zeitschrift fur angewandte Mathematik und Mechanik. 1921;1(1):15–20.
Reissner H. Zum Erddruckproblem. Proceedings, First International Congress of Applied Mechanics, Delft, Netherlands, p. 295–311; 1924.
Terzaghi K. Theoretical soil mechanics. New York: Wiley; 1943.
Vesic AS. Analysis of ultimate loads of shallow foundations. J Soil Mech Found. 1973;99(1):45–73.

Further Readings

Bowles JE. Foundation analysis and design. New York: McGraw-Hill; 1996.
Budhu M. Soil mechanics and foundations. New York: Wiley; 2000.
Coduto D. Foundation design principle and practices. 2nd ed. Englewood Cliffs, NJ: Prentice-Hall; 2001.
Das B. Principles of foundation engineering. 7th ed. Stamford, CA: Cengage Learning; 2010.
Kumbhojkar AS. Numerical evaluation of Terzaghi's N_{γ}. J Geotech Eng. 1993;119(3):598–607.
Saran S, Agarwal R. Bearing capacity of eccentrically obliquely loaded footing. J Geotech Eng. 1991;117(11):1669–90.

Chapter 2
Axial Capacity of Single Pile Foundations in Soil

Abstract This chapter deals with single pile foundations in homogeneous and nonhomogeneous soils. Calculations of axial load a single pile foundation can withstand are explained in details in this chapter. The calculations were performed to satisfy both bearing capacity and elastic settlement requirements. For the bearing capacity condition, several methods were utilized and the effects of many factors were considered in these methods such as pile shape (square/circular), pile size, pile length, soil type, and groundwater table. Then again, the effects of skin friction and the end bearing components were considered in the elastic settlement analyses. Additionally, a step-by-step procedure is introduced in this chapter to develop bearing capacity and elastic settlement design charts which can be useful in the design process of single piles in soil. A number of design problems are also presented in this chapter and its solution are explained in details. These problems were first hand-solved, and then, resolved using the Pile-1 (in soil) and Pile-2 (in rock) applications of the foundationPro program. Finally, a set of design projects was suggested at the end of this chapter to allow the reader practice the concepts learned.

Keywords Single pile foundation • Bearing capacity • Elastic settlement • Pile-1 • Pile-2 • foundationPro

2.1 Introduction

This chapter deals with single pile foundations in various types of soils. Calculations of allowable and ultimate loads which a single pile foundation can sustain are explained in details in this chapter. Allowable and ultimate axial loads on a single pile foundation are estimated to satisfy both bearing capacity and elastic settlement requirements.

In the bearing capacity analyses, the classical bearing capacity equations for a single pile foundation were utilized. Effect of pile foundation shape (i.e., square or circular) is considered in the bearing capacity calculations. Piles in sand, clay, and rock are dealt with herein. Effect of the groundwater table depth is also considered in the bearing capacity equation. In this chapter, a number of methods are summarized and can be utilized to estimate the axial capacity of a single pile in

M. Yamin, *Problem Solving in Foundation Engineering using foundationPro*,
DOI 10.1007/978-3-319-17650-5_2

homogeneous and nonhomogeneous soils. The following methods are discussed in this chapter: Meyerhof, Vesic, Coyle and Castello, alpha, modified alpha, lambda, and critical depth methods.

The axial capacity of single drilled shaft is also estimated based on elastic settlement (theory of elasticity). The allowable load a single pile foundation can sustain to satisfy elastic settlement requirements and not to exceed the allowable permissible settlement is determined herein. The total elastic settlement of a single pile in soil was estimated as a result of adding the various types of settlements occurred due to end bearing and skin friction loads.

A step-by-step procedure was introduced in this chapter to develop bearing capacity and elastic settlement design charts. These design charts present the relationship between various applied loads on a single pile foundation versus pile dimensions (side or diameter and total length). These charts can be useful in the pile foundation design process to find what will control the final design; the bearing capacity, or the elastic settlement of the foundation.

Ten design problems are presented in this chapter. First, these design problems were hand-solved and their solution was explained in details, and then the foundationPro program was used to resolve the problems to replicate and verify the hand solution. Also, the program was used to investigate a wider and detailed solution and design alternatives for the hand-solved problems. Since the foundationPro includes a set of several applications, the Pile-1 and Pile-2 applications of the foundationPro are the responsible applications to perform bearing capacity and elastic settlement calculations for single pile foundations embedded in homogeneous and nonhomogeneous soils. Therefore, only Pile-1 and Pile-2 applications will be used throughout this chapter to replicate the hand-solved problems. Three design projects were suggested at the end of this chapter to allow the reader to practice and apply the learned concepts.

2.2 Theory

In this section, the procedures to estimate the axial capacity of a single pile in soil are discussed. The governing methodologies to calculate the ultimate and allowable loads that can be applied to a single pile foundation with the given pile configurations in different types of soil based on bearing capacity and elastic settlement are summarized in the following subsections. There will be several different theories that can be used to calculate the bearing capacity of a foundation and these theories will depend on various factors as will be introduced herein. Figure 2.1 defines the main soil properties and design parameters used in the analyses of single pile foundations. As in the figure below, C is the soil cohesion, ϕ is the soil friction angle, γ' is the effective unit weight of the soil, E_s is the elastic modulus of the soil, μ_s is Poisson's ratio of the soil, and B is the pile diameter (circular) or side (square). The thicknesses of each soil layer in the case of multiple soil layers can be defined by providing the depths (Z_0, Z_1, Z_2, etc.) at the boundaries between different soil layers (see Fig. 2.1).

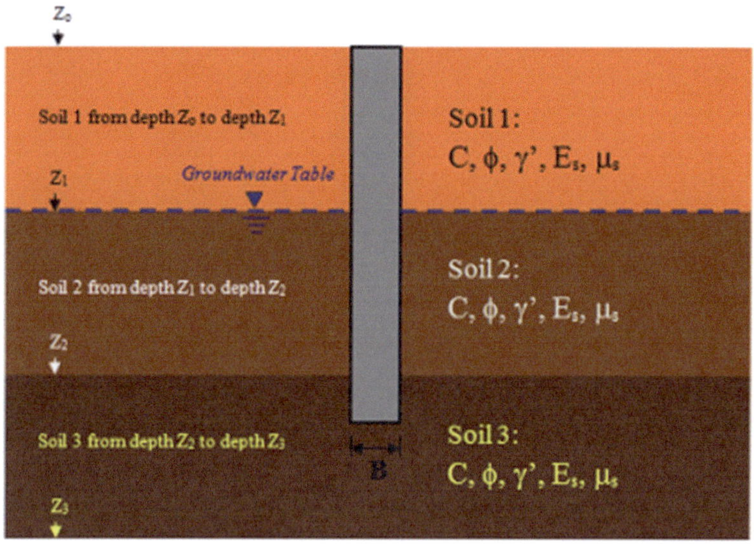

Fig. 2.1 Single pile embedded in multiple soil layers

2.2.1 Axial Capacity of a Single Pile (Bearing Capacity)

The ultimate axial capacity (load), Q_u, a single pile foundation embedded in soil can sustain as shown in Fig. 2.2 is estimated as in Eq. (2.1) by summing the end bearing capacity component (Q_p) and the frictional skin resistance component (Q_s):

$$Q_u = Q_p + Q_s \tag{2.1}$$

where Q_p is the load carrying capacity of the pile point, and Q_s is the frictional resistance (skin friction).

2.2.1.1 End Bearing Capacity (Q_p)

One can estimate the pile tip resistant–end bearing capacity of a single pile depending on the soil which the pile is resting on. Figure 2.3 shows the available methods in Pile-1 application of foundationPro which can be used to estimate Q_p.

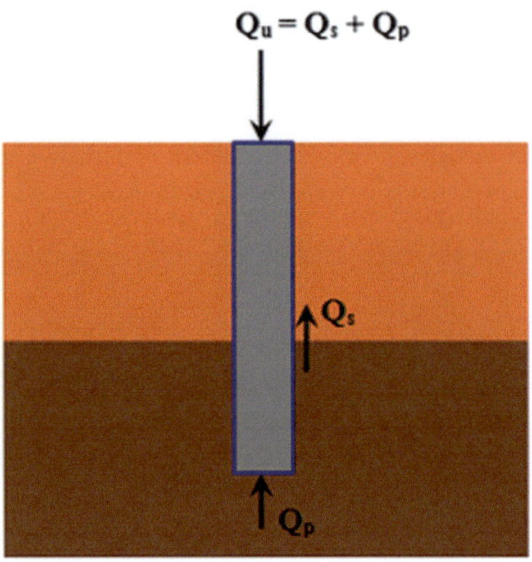

Fig. 2.2 Ultimate load on a single pile foundation

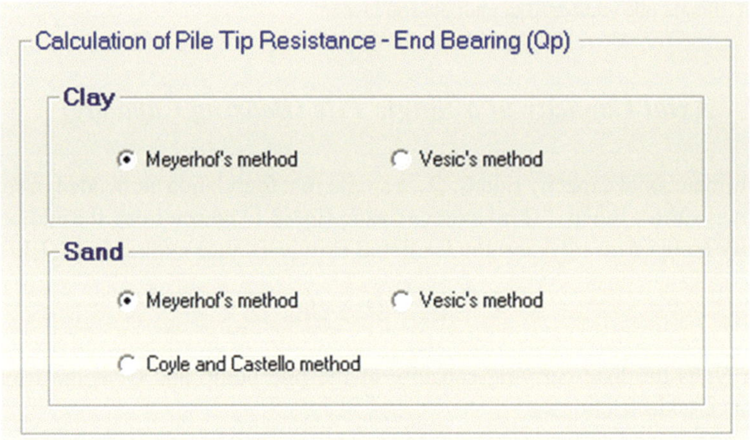

Fig. 2.3 Available methods of analysis in Pile-1 application for pile tip resistance–end bearing component

1. End bearing capacity in sand
   Q_p can be estimated using any of the following methods in the case where the pile tip is resting on sand:

   (a) Meyerhof's method:
   Q_p can be estimated according to Meyerhof's method (Meyerhof 1976) using the following equation:

$$Q_{\mathrm{p}} = A_{\mathrm{p}} q_{\mathrm{p}} = A_p q' N_q^* \tag{2.2}$$

where

$$N_q^* = 0.3147 \times e^{0.1752\phi} \tag{2.3}$$

A_{p} is the cross-sectional area of the pile
q' is the effective stress at the pile tip
One must be aware that the Q_{p} value should be limited to $A_{\mathrm{p}} q_l$. Therefore,

$$Q_{\mathrm{p}} = A_{\mathrm{p}} q' N_q^* \leq A_p q_l \tag{2.4}$$

$$q_{\mathrm{l}} = 0.5 p_a N_q^* \tan\left(\phi'\right) \tag{2.5}$$

where
p_{a} is the atmospheric pressure and can be approximated as 100 kN/m^2 or 2,000 lb/ft^2
ϕ' is the internal friction angle of the soil underneath the pile tip

(b) Vesic's method:
Q_{p} can be estimated according to Vesic's method (Vesic 1977) using the following equation:

$$Q_{\mathrm{p}} = \left(B\gamma' N_\gamma^* + q' N_q^*\right) A_{\mathrm{p}} \tag{2.6}$$

where
B is the pile diameter
N_γ^* and N_q^* are the bearing capacity factors
γ' is the effective unit weight of the soil
The bearing capacity factors can be estimated as in the following equations:

$$N_\gamma^* - 0.6\left(N_q^* - 1\right) \tan\left(\phi'\right) \tag{2.7}$$

$$N_q^* = \left(\frac{1 + 2K_{\mathrm{o}}}{3}\right) N_\sigma \tag{2.8}$$

where
K_{o} is the earth pressure coefficient at rest and can be taken as $\left(1 - \sin\left(\phi'\right)\right)$

$$N_\sigma = \frac{3}{3 - \sin\phi'} e^{\left(\frac{90-\phi'}{180}\right)\pi} \tan^2\left(45 + \frac{\phi'}{2}\right) I_r^{\frac{4\sin\phi'}{3(1+\sin\phi')}} \tag{2.9}$$

where

I_r is the rigidity index and can be computed using Eq. (2.10) below:

$$I_r = \frac{E_s}{2(1 + \mu_s)(q' \tan(\phi))} \tag{2.10}$$

(c) Coyle and Castillo's method:

Q_p can be estimated according to the method suggested by Coyle and Castello (1981) as in the following equation:

$$Q_p = q' N_q^* A_p \tag{2.11}$$

where

$$N_q^* = \left(1.04529\left(\frac{\phi^\circ}{40}\right)^2 - 1.58056\left(\frac{\phi^\circ}{40}\right) + 0.6289\right) \times \left(-0.32039\left(\frac{L}{D}\right)^2 + 14.2496\left(\frac{L}{D}\right) + 935.565\right) \tag{2.12}$$

2. End bearing capacity in clay

Q_p can be estimated using any of the following methods in the case where the pile tip is resting on clay:

(a) Meyerhof's method:

Q_p can be estimated according to Meyerhof's method using the following equation:

$$Q_p = 9 c_u A_p \tag{2.13}$$

where

c_u is the undrained cohesion of the soil below the tip of the pile

(b) Vesic's method:

Q_p can be estimated according to Vesic's method using the following equation:

$$Q_p = A_p c_u N_c^* \tag{2.14}$$

According to Vesic (1977), N_c^* can be calculated using the following equation:

$$N_c^* = \frac{4}{3}(\ln I_r + 1) + \frac{\pi}{2} + 1 \tag{2.15}$$

where

I_r is the rigidity index in the clay and can be calculated based on the suggestion made by O'Neill and Reese (1999) as in the following equation:

$$I_r = 347\left(\frac{c_u}{p_a}\right) - 33 \leq 300 \tag{2.16}$$

3. End bearing capacity in rock:

   Q_p can be estimated based on the suggestion by Goodman (1980) in the case where the pile tip extends to rock:

$$Q_p = q_u(N_\phi + 1) \tag{2.17}$$

where

q_u is the unconfined compression strength of rock

ϕ' is the friction angle

$$N_\phi = \tan^2\left(45 + \frac{\phi'}{2}\right) \tag{2.18}$$

2.2.1.2 Skin Friction Resistance (Q_s)

One can estimate the skin friction resistance of a single pile depending on the type of soil that is surrounding the pile shaft. Figure 2.4 shows the available methods in Pile-1 application of foundationPro which can be used to estimate Q_s.

1. Skin friction resistance in sand

   Q_s can be estimated using any of the following methods in the case where the pile shaft is surrounded by sand:

   (a) Critical depth method:

$$Q_s = \sum pfL \tag{2.19}$$

where

f is the frictional resistance

L is the distance from surface to the bottom of the pile

p is the pile perimeter

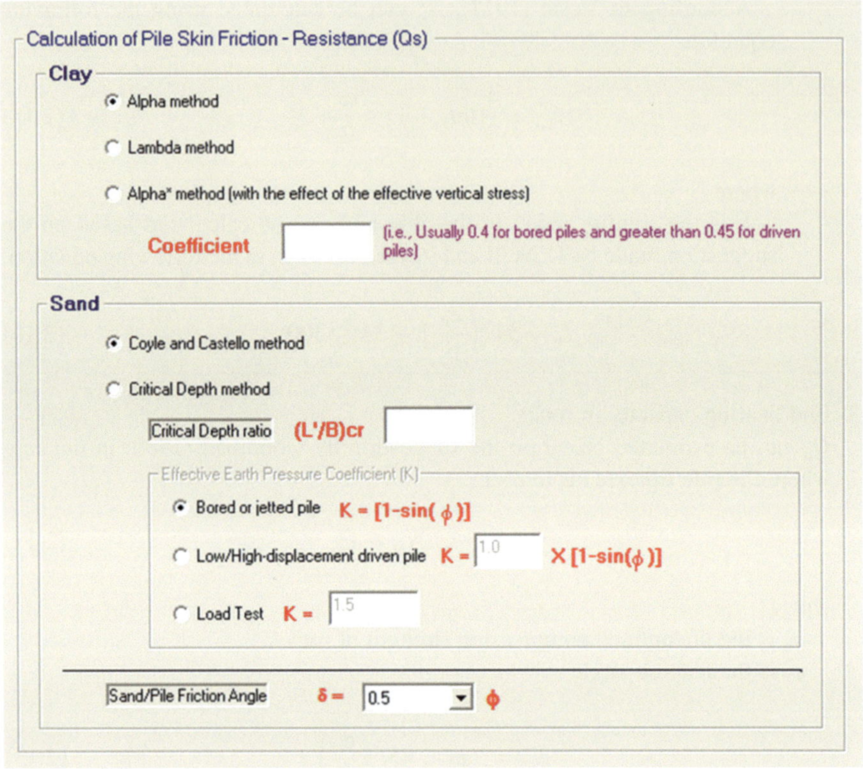

Fig. 2.4 Available methods of analysis in Pile-1 application for skin friction resistance component

In order to calculate (f) which is the unit skin friction, one must assume a critical depth ratio $\left(\frac{L'}{B}\right)_{\text{cr}}$ which usually ranges from 15 to 20. From this ratio, one can determine the critical depth (L'). Therefore,

$$L' = \left(\frac{L'}{B}\right)_{\text{cr}} \times B \tag{2.20}$$

Thus,

For $z = 0$ to L', f can be estimated using the following equation:

$$f = K \times \sigma_o' \tan\left(\delta'\right) \tag{2.21}$$

However, for $z = L'$ to L, f can be estimated using the following equation:

$$f = f_{z=L'} \tag{2.22}$$

where

K is the earth pressure coefficient and can range from $1 - \sin\left(\phi'\right)$ to $1.8\left(1 - \sin\left(\phi'\right)\right)$ depending on the pile type (bored or jetted and low or high displacement driven)

σ_o' is the effective vertical stress at depth under consideration

δ' is the soil-pile friction angle and can be approximated in terms of the soil friction angle as in the following equation:

$$\delta' = 0.8\phi' \tag{2.23}$$

(b) Coyle and Castello method:
Coyle and Castello (1981) suggested the following equation to estimate Q_s:

$$Q_s = \sigma_o' K \tan\left(\delta'\right) pL \tag{2.24}$$

where

$$\delta' = 0.8\phi' \tag{2.25}$$

$$K = \left(-1.32484\left(\frac{\phi^\circ}{36^\circ}\right) + 1.03116\right) \times \left(0.2879\left(\frac{L}{D}\right) - 13.572\right) \tag{2.26}$$

where

D is the width or diameter of pile

ϕ' is the soil-pile friction angle

2. Skin friction resistance in clay
   (a) Alpha (α*) method:
   The unit skin friction f can be estimated as follows:

$$f = \alpha \times c_u \tag{2.27}$$

where

$$\alpha = 0.026 \times \left(\frac{c_u}{p_a}\right)^4 - 0.02233\left(\frac{c_u}{p_a}\right)^3 + 0.7498\left(\frac{c_u}{p_a}\right)^2 - 1.198\left(\frac{c_u}{p_a}\right) + 1.1187 \tag{2.28}$$

where c_u is the undrained cohesion and p_a is the atmospheric pressure as defined earlier

Table 2.1 Typical values of Ψ in Eq. (2.31)

Bored piles	Driven piles
0.4–0.5	≥0.5

Thus,

$$Q_s = \sum f \times p \times \Delta L = \sum \alpha \times c_u p \times \Delta L \tag{2.29}$$

(b) Modified Alpha (α^*) method:
The unit skin friction f can be estimated as follows:

$$f = \alpha^* \times c_u \tag{2.30}$$

where Sladen (1992) suggested the following equation to estimate α^*

$$\alpha^* = \Psi\left(\frac{\overline{\sigma}'_o}{c_u}\right)^{0.45} \tag{2.31}$$

where
$\overline{\sigma}'_o$ is the mean vertical effective stress
Ψ as in Table 2.1 depends on the pile type
Therefore,

$$Q_s = \sum f \times p \times \Delta L = \sum \alpha \times c_u p \times \Delta L \tag{2.32}$$

(c) Lambda (λ) method:
Vijayvergiya and Focht (1972) suggested the following equation to estimate the average unit skin friction:

$$f_{avg} = \lambda\left(\sigma'_o + 2c_u\right) \tag{2.33}$$

where
λ is a factor depending on the pile embedment length and can be obtained from Fig. 2.5.
Thus,

$$Q_s = pLf_{avg} \tag{2.34}$$

3. Skin friction resistance in rock
Skin friction resistance is usually ignored when the pile is surrounded by rock. Therefore,

$$Q_s = 0$$

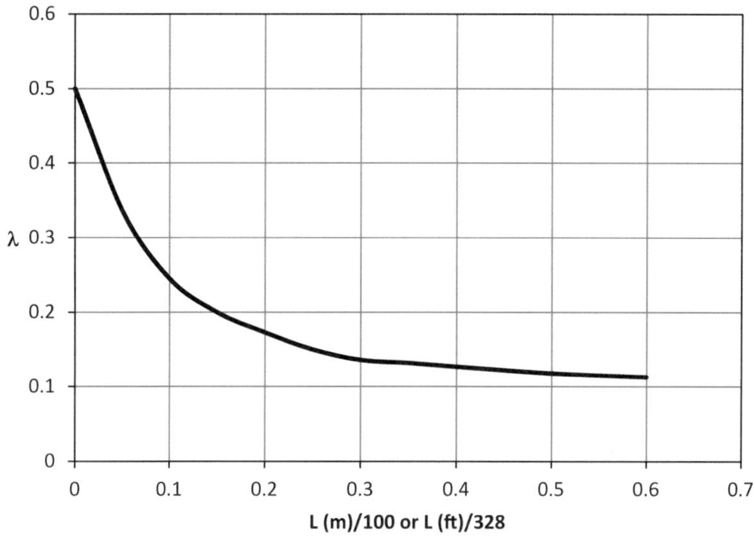

Fig. 2.5 The factor λ in Eq. (2.33)

2.2.2 *Axial Capacity (Elastic Settlement)*

The total elastic settlement (S_e) of a single pile foundation can be estimated using the following equation:

$$S_e = S_{e_1} + S_{e_2} + S_{e_3} \tag{2.35}$$

where

S_{e_1} is the elastic settlement of pile length
S_{e_2} is the settlement of the pile caused by the load at the pile tip
S_{e_3} is the settlement of the pile caused by the load transmitted along the pile shaft
S_{e_1} can be estimated using the following equation:

$$S_{e_1} = \frac{(Q_{wp} + \varepsilon Q_{ws})L}{A_p E_p} \tag{2.36}$$

where

Q_{wp} is the working point bearing
Q_{ws} is the working frictional skin resistance
A_p is the cross-sectional area of the pile
L is the pile length
E_p is the modulus of elasticity of the pile material

ε varies between 0.5 and 0.67

S_{e_2} can be estimated using the following equation:

$$S_{e_2} = \frac{(q_{wp}D)}{E_s}(1 - \mu_s^2)I_{wp} \tag{2.37}$$

where

D is the width or diameter of pile, q_{wp} is the point load per unit area at the pile point (Q_{wp}/A_p), E_s is the modulus of elasticity of soil at or below the pile point, μ_s is Poisson's ratio of the soil, and I_{wp} is the influence factor which can be taken as 0.85.

S_{e_3} can be estimated using the following equation:

$$S_{e_3} = \frac{Q_{ws}}{pL}\left(\frac{D}{E_s}\right)(1 - \mu_s^2)I_{ws} \tag{2.38}$$

where

I_{ws} is the influence factor and can be estimated from the equation suggested by Vesic (1977) as follows:

$$I_{ws} = 2 + 0.35\left(\sqrt{\frac{L}{D}}\right) \tag{2.39}$$

2.2.3 Allowable Loads

Allowable load a single pile foundation can sustain based on bearing capacity analyses can be determined using the following equation:

$$Q_{all} = \frac{Q_u}{FS} \tag{2.40}$$

where

Q_u is the ultimate load a pile can carry and can be determined using Eq. (2.1)

FS is the factor of safety

However, the allowable axial load a single pile foundation can sustain based on total elastic settlement is determined by solving Eq. (2.42) below for Q_{wp}, and then finding Q_w using Eq. (2.43):

Let

$$\eta = \frac{Q_p}{Q_s} \tag{2.41}$$

Then

$$S_{\mathrm{e}} = Q_{\mathrm{wp}}\left(\frac{\left(1+\frac{\varepsilon}{\eta}\right)L}{A_{\mathrm{p}}E_{\mathrm{p}}} + \frac{D\left(1-\mu_{\mathrm{s}}^{2}\right)I_{\mathrm{wp}}}{A_{\mathrm{p}}E_{\mathrm{s}}} + \frac{D\left(1-\mu_{\mathrm{s}}^{2}\right)I_{\mathrm{ws}}}{\eta PLE_{\mathrm{s}}}\right) \tag{2.42}$$

$$Q_{\mathrm{w}} = Q_{\mathrm{wp}}\left(1+\frac{1}{\eta}\right) \tag{2.43}$$

2.3 Step-by-Step Procedure

To determine the axial capacity of a single pile foundation in soil based on bearing capacity and total elastic settlement requirements and establish the associated design charts, the following steps must be followed:

Step 1:
Assume the pile length, L

Step 2:
Determine Q_{p}, end bearing capacity of the pile point depending on the type of soil below the pile tip

For sand
- Meyerhof's method, use Eqs. (2.2)–(2.5)
- Vesic's method, use Eqs. (2.6)–(2.10)
- Coyle and Castello method, use Eqs. (2.11) and (2.12)

For clay
- Meyerhof's method, use Eq. (2.13)
- Vesic's method, use Eqs. (2.14)–(2.16)

For rock
- Use Eqs. (2.17) and (2.18)

Step 3:
Determine Q_{s}, skin frictional resistance of the pile depending on the type of soil surrounding the pile shaft.

For sand
- Critical depth method, use Eqs. (2.19)–(2.23)
- Coyle and Castello method, use Eqs. (2.24) and (2.26)

For clay
- Alpha (α*) method, use Eqs. (2.27)–(2.29)
- Alpha (α*) method, use Eqs. (2.30)–(2.32) and Table 2.1
- Lambda (λ) method, use Eqs. (2.33) and (2.34)

Step 4:
Find Q_u using Eq. (2.1)

Step 5:
Find Q_{all} using Eq. (2.40)

Step 6:
Find Q_{ws} using Eqs. (2.38) and (2.39)

Step 7:
Find Q_{wp} using Eqs. (2.39), (2.41), and (2.42)

Step 8:
Find Q_w using Eq. (2.43),

Step 9:
Steps 1 through 8 can be repeated for several L values to develop the required pile foundation capacity charts

2.4 Design Problems

Many design problems are presented in this section to help the reader reiterate the theory and integrate the Pile-1 and Pile-2 applications of the foundationPro program in the design process of the single pile foundation. These examples were selected to give exposure to a wide variety of challenges and design configurations that can be faced in designing a single pile foundation while helping us iron out the finer details of the theories introduced earlier.

2.4.1 Circular Pile in Sandy Soil (Meyerhof's and Critical Depth Methods)

Develop axial capacity (bearing capacity and settlement) design charts for a circular bored pile with a diameter of 0.75 m in homogeneous sandy soil. Use Meyerhof's method for the end bearing capacity and the critical depth method for the skin frictional resistance. Table 2.2 provides the soil properties and other required design parameters.

2.4.1.1 Hand Solution

To determine the axial capacity of the pile for the given diameter and soil properties, one must follow the steps below:

Step 1:
Assume pile length, $L = 6$ m.

Table 2.2 Soil properties and other design parameters

Property	Value	Unit
FS	3	–
c	0	kN/m^2
ϕ	28	°
γ	18	kN/m^3
E_s	25,000	kN/m^2
μ_s	0.35	–
E_p	21,000,000	kN/m^2
δ	0.5ϕ	–
S_e	0.025	m
ε	0.6	–
$(L'/B)_{cr}$	15	–

Step 2:
Determine Q_p from Eq. (2.2):

$$Q_p = A_p q_p = 0.441 \times 18 \times 6 \times N_q^*$$

From Eq. (2.3):

$$N_q^* = 0.3147 \times e^{0.1752 \times 28} = 42.498$$

Therefore,

$$Q_p = A_p q_p = 0.441 \times 18 \times 6 \times 42.498 = 2,024.109\,\text{kN}$$

From Eq. (2.5):

$$q_l \times A_p = 0.5 \times 100 \times 42.498 \tan(28) \times 0.441 = 498\,\text{kN}$$

We must use the smaller value; therefore,

$$Q_p = 498\,\text{kN}$$

Step 3:
Determine Q_s frictional skin resistance from Eq. (2.20):

$$L' = 15 \times 0.75 = 11.25\,\text{m}$$

$$\sigma_o' = 0 \text{ at } z = 0\,\text{m}$$

The unit skin friction can be then determined from Eq. (2.21):

$$f = 0$$

However,
At $z = 6\,\text{m}$:

$$\sigma_o^{'} = 6 \times 18 = 108\,\text{kN/m}^2$$

The unit skin friction from Eq. (2.21):

$$f = (1 - \sin(28)) \times 108 \tan(14) = 14.28\,\text{kN/m}^2$$

Thus,
The frictional skin resistance, Q_s, can be determined from Eq. (2.19):

$$Q_s = \frac{f_{z=0} + f_{z=6}}{2} \times (3.14 \times 0.75) \times 6$$

$$Q_s = \frac{0 + 14.28}{2} \times (3.14 \times 0.75) \times 6 = 101\,\text{kN}$$

Step 4:
From Eq. (2.1):

$$Q_u = 498 + 101 = 599\,\text{kN}$$

Step 5:
The allowable load based on bearing capacity is determined using Eq. (2.40):

$$Q_{all} = \frac{599}{3} = 199.6\,\text{kN}$$

To determine the axial capacity of the pile foundation based on elastic settlement given that $S_e = 0.025$ m, one must follow Steps 6 through 8 as below:

Steps 6–8:
From Eq. (2.41):

$$\eta = \left(\frac{498}{101}\right) = 4.93$$

Solving for Q_{wp} from Eq. (2.42):

$$0.025 = Q_{wp}\left(\frac{\left(1 + \frac{0.6}{4.93}\right)6}{0.441 \times 21{,}000{,}000} + \frac{0.75(1 - 0.35^2)0.85}{0.441 \times 25{,}000}\right.$$
$$\left. + \frac{0.75(1 - 0.35^2)2.98}{4.93 \times 3.14 \times 0.75 \times 6 \times 25{,}000}\right)$$

Therefore,

$$Q_{\mathrm{wp}} = 475.47kN$$

From Eq. (2.43):

$$Q_{\mathrm{w}} = 475.47\left(1 + \frac{1}{4.93}\right) = 571.91\,\mathrm{kN}$$

Step 9:
Repeat Steps 1 through 8 for another assumed pile length. Another solution is provided below:

Step 1 (second pile length):
Assume $L = 10$ m.

Step 2 (second pile length):
Determine end bearing capacity, Q_{p}, for pile tip in sand from Eq. (2.2):

$$Q_{\mathrm{p}} = A_{\mathrm{p}}q_{\mathrm{p}} = 0.441 \times 18 \times 10 \times N_q^*$$

From Eq. (2.3):

$$N_q^* = 0.3147 \times e^{0.1752 \times 28} = 42.498$$

Therefore,

$$Q_{\mathrm{p}} = A_{\mathrm{p}}q_{\mathrm{p}} = 0.441 \times 18 \times 10 \times 42.498 = 3,373.49\,\mathrm{kN}$$

From Eq. (2.5):

$$q_{\mathrm{l}} \times A_{\mathrm{p}} = 0.5 \times 100 \times 42.498 \tan(28) \times 0.441 = 498\,\mathrm{kN}$$

We use the smaller value:

$$Q_{\mathrm{p}} = 498\,\mathrm{kN}$$

Step 3 (second pile length):
Determine Q_{s} frictional resistance (skin friction) for sand using Eq. (2.20):

$$L' = 15 \times 0.75 = 11.25 \text{ m}$$

At $z = 0$ m:

$$\sigma_o' = 0$$

From Eq. (2.21):

$$f = 0$$

At $z = 10\,\mathrm{m}$:

$$\sigma_o^{'} = 10 \times 18 = 180\,\mathrm{kN/m^2}$$

From Eq. (2.21):

$$f = (1 - \sin(28)) \times 180 \tan(14) = 23.80\,\mathrm{kN/m^2}$$

From Eq. (2.19):

$$Q_\mathrm{s} = \frac{f_{z=0} + f_{z=10}}{2} \times (3.14 \times 0.75) \times 10$$

$$Q_\mathrm{s} = \frac{0 + 23.80}{2} \times (3.14 \times 0.75) \times 10 = 280.358\,\mathrm{kN}$$

Steps 4 and 5 (second pile length):
From Eq. (2.1):

$$Q_\mathrm{u} = 498 + 280.358 = 778.35\,\mathrm{kN}$$

From Eq. (2.40):

$$Q_\mathrm{all} = \frac{778.35}{3} = 259.452\,\mathrm{kN}$$

Steps 6–8 (second pile length):
From Eq. (2.41):

$$\eta = \left(\frac{498}{280.3}\right) = 1.77$$

Solving for Q_wp from Eq. (2.42):

$$0.025 = Q_\mathrm{wp}\left(\frac{\left(1 + \frac{0.6}{1.77}\right)10}{0.441 \times 21{,}000{,}000} + \frac{0.75(1 - 0.35^2)0.85}{0.441 \times 25{,}000} + \frac{0.75(1 - 0.35^2)3.27}{1.77 \times 3.14 \times 0.75 \times 10 \times 25{,}000}\right)$$

Therefore,

$$Q_{\mathrm{wp}} = 463.20\,\mathrm{kN}$$

From Eq. (2.43):

$$Q_{\mathrm{w}} = 463.20\left(1 + \frac{1}{1.77}\right) = 724.98\,\mathrm{kN}$$

Step 9 (second pile length):
Repeat Steps 1 through 8 for another assumed pile length. Another solution is provided below for a third pile length:

Step 1 (third pile length):
Assume $L = 16$ m.

Step 2 (third pile length):
Determine Q_{p} for pile tip in sand from Eq. (2.2):

$$Q_{\mathrm{p}} = A_{\mathrm{p}} q_{\mathrm{p}} = 0.441 \times 18 \times 16 \times N_q^*$$

From Eq. (2.3):

$$N_q^* = 0.3147 \times e^{0.1752 \times 28} = 42.498$$

Therefore,

$$Q_{\mathrm{p}} = A_{\mathrm{p}} q_{\mathrm{p}} = 0.441 \times 18 \times 16 \times 42.498 = 5,397.58\,\mathrm{kN}$$

From Eq. (2.5):

$$q_{\mathrm{l}} \times A_{\mathrm{p}} = 0.5 \times 100 \times 42.498 \tan(28) \times 0.441 = 498\,\mathrm{kN}$$

We use the smaller value:

$$Q_{\mathrm{p}} = 498\,\mathrm{KN}$$

Step 3 (third pile length):
Determine Q_{s} frictional resistance (skin friction) in sand from Eq. (2.20):

$$L' = 16 \times 0.75 = 12\,\mathrm{m}$$

At $z = 0\,\mathrm{m}$

$$\sigma_o' = 0$$

From Eq. (2.21):

$$f = 0$$

At $z = 12\,\text{m}$

$$\sigma_o^{'} = 12 \times 18 = 216\,\text{kN/m}^2$$

From Eq. (2.21):

$$f = (1 - \sin(28)) \times 216 \tan(14) = 28.57\,\text{kN/m}^3$$

From Eq. (2.19):

$$Q_\text{s} = \frac{f_{z=0} + f_{z=12}}{2} \times (3.14 \times 0.75) \times 12 + f_{z=12}(3.14 \times 0.75)(16 - 11.25)$$

$$\begin{aligned} Q_\text{s} &= \frac{0 + 28.57}{2} \times (3.14 \times 0.75) \times 12 + 28.57(3.14 \times 0.75)(16 - 11.25) \\ &= 673.186\,\text{kN} \end{aligned}$$

Steps 4 and 5 (third pile length):
From Eq. (2.1):

$$Q_\text{u} = 498 + 673.186 = 1,171.186\,\text{kN}$$

From Eq. (2.40):

$$Q_\text{all} = \frac{1,171.186}{3} = 390.4\,\text{kN}$$

Steps 6–8 (third pile length):
From Eq. (2.41):

$$\eta = \frac{(498)}{(673.186)} = 0.73$$

Solving for Q_wp from Eq. (2.42):

$$0.025 = Q_\text{wp}\left(\frac{\left(1 + \frac{0.6}{0.73}\right)16}{0.441 \times 21,000,000} + \frac{0.75(1 - 0.35^2)0.85}{0.441 \times 25,000} + \frac{0.75(1 - 0.35^2)3.616}{0.73 \times 3.14 \times 0.75 \times 16 \times 25,000}\right)$$

Therefore,

$$Q_{\mathrm{wp}} = 439.14\,\mathrm{kN}$$

From Eq. (2.43):

$$Q_{\mathrm{w}} = 439.14\left(1 + \frac{1}{0.73}\right) = 1{,}032.76\,\mathrm{kN}$$

2.4.1.2 foundationPro Solution

After launching the Pile-1 application of foundationPro, the five sections (General, Pile Information, Soil Properties, Analysis Methods, and OUTPUT) will appear in the main screen as shown in Fig. 2.6.

General Information Section

In this section, the user can provide some general information such as user name and project name. This information is optional. Also, the safety factor which will be used in the bearing capacity calculations and the unit type are required in this section as shown in Fig. 2.7.

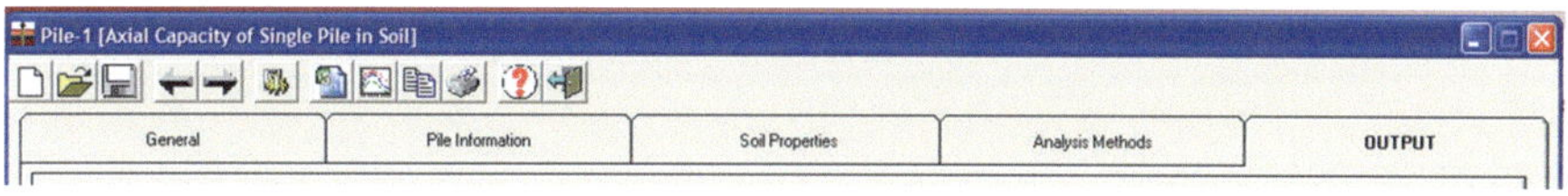

Fig. 2.6 Main sections of Pile-1 application of foundationPro program

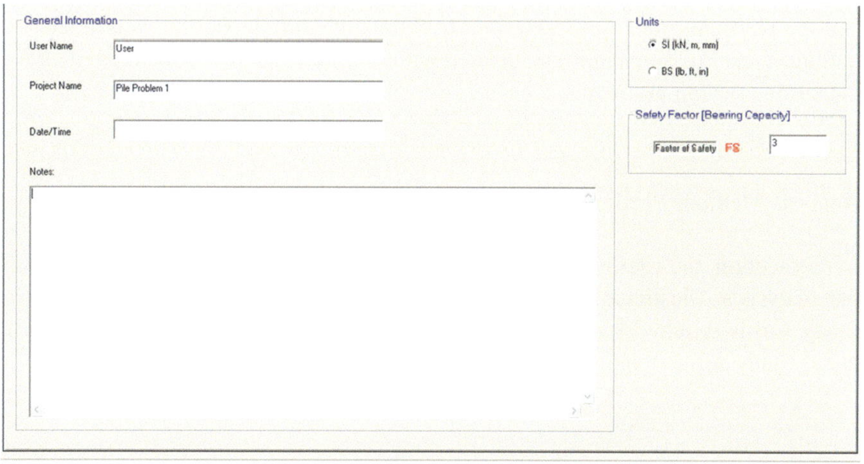

Fig. 2.7 The General Information section for Pile-1 application

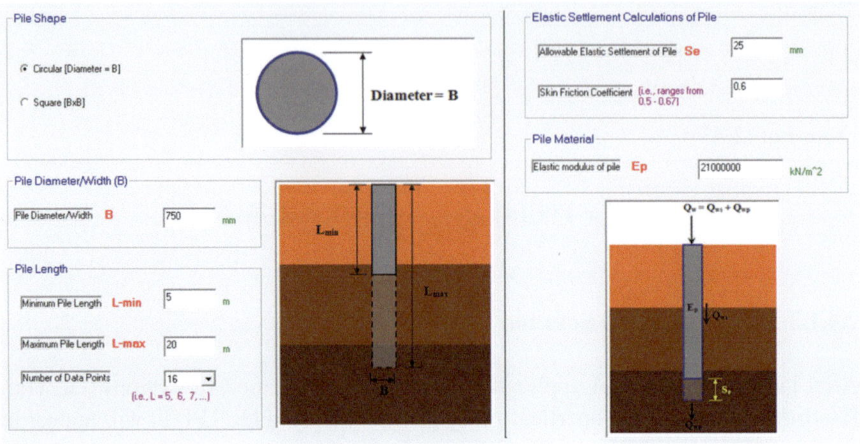

Fig. 2.8 Input data for Pile Information section

Pile Information Section

In this section, the user specifies pile shape (circular or square), provides the pile diameter/width, defines the required range for the pile length to be investigated, and enters elastic settlement of the pile and its allowable elastic settlement. Figure 2.8 shows the data entered for this design problem.

Soil Properties Section

In this section, the user must specify the number of soil layers to describe the soil problem in problem. Also, the user must provide the thickness of each soil layers and the physical properties (cohesion, friction angle, effective unit weight, elastic modulus, and Poisson's ratio) associated with it. Figure 2.9 shows the input data for the current design problem.

Analysis Methods Section

In this section, the user must specify the method of analysis to be used throughout the analysis as illustrated in Fig. 2.10. In this problem, the critical depth method in sandy soil is required for the skin friction resistance and the Meyerhof's method under sandy soil is required for the end bearing capacity.

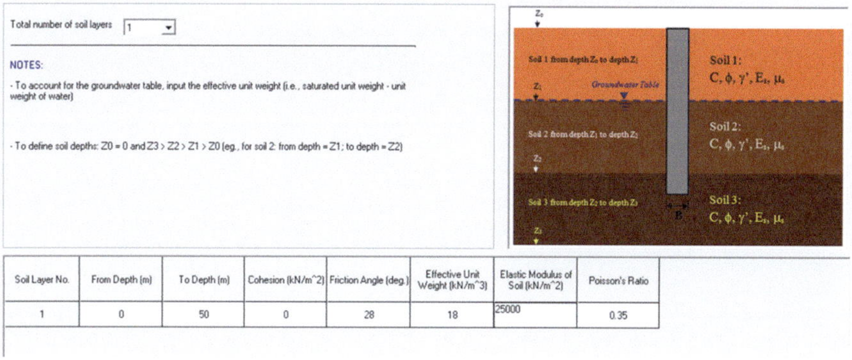

Fig. 2.9 Properties and thicknesses of soil layers

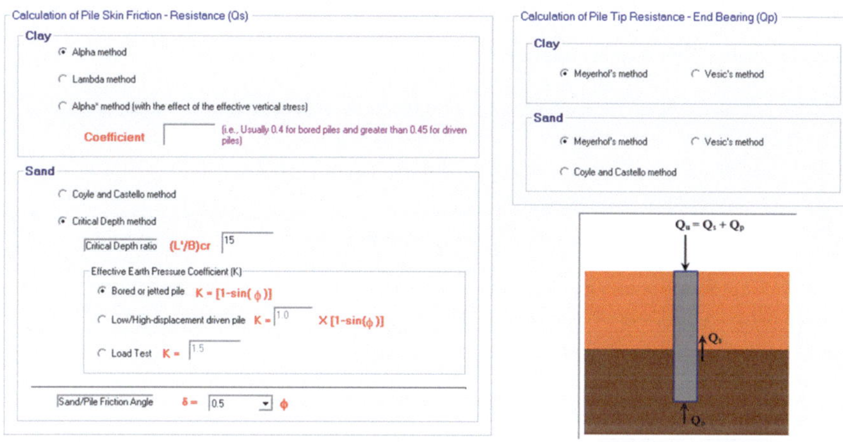

Fig. 2.10 Analysis Methods section for Pile-1 application

Output Section

Now we can hit the RUN button to perform the analyses and view the OUTPUT section and navigate through the results as shown in Figs. 2.11 and 2.12 in table and graph formats, respectively. Of course, the user can specify the required chart and results to view. The various axial capacities based on bearing capacity and elastic settlement analyses at pile lengths from 5 m to 20 m are summarized in Table 2.3. The working load (settlement analysis) and allowable load (bearing capacity analysis) design charts are shown in Figs. 2.13 and 2.14. It is very clear from the two design charts how the bearing capacity is controlling the design for this design problem and not the settlement.

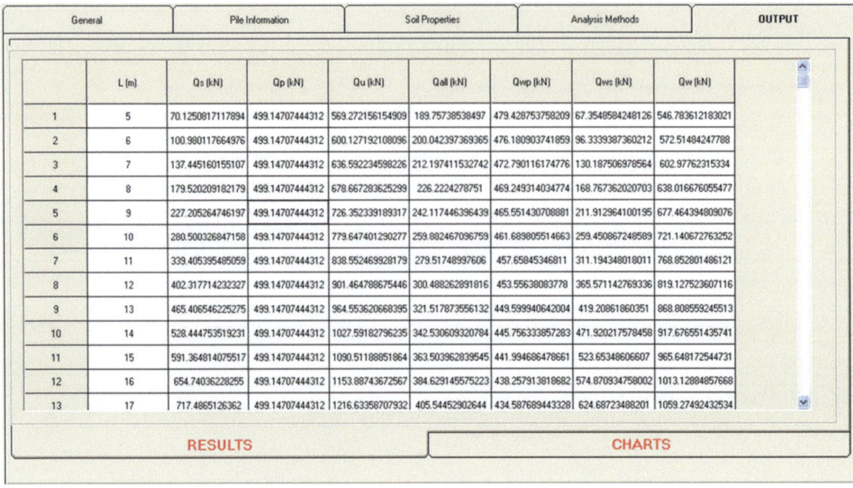

	L (m)	Qs (kN)	Qp (kN)	Qu (kN)	Qall (kN)	Qwp (kN)	Qws (kN)	Qw (kN)
1	5	70.1250817117894	499.14707444312	569.272156154909	189.75738538497	479.428753758209	67.3548584248126	546.783612183021
2	6	100.980117664976	499.14707444312	600.127192108096	200.042397369365	476.180903741859	96.3339387360212	572.51484247788
3	7	137.445160155107	499.14707444312	636.592234598226	212.197411532742	472.790116174776	130.187506978564	602.97762315334
4	8	179.520209182179	499.14707444312	678.667283625299	226.2224278751	469.249314034774	168.767362020703	638.016676055477
5	9	227.205264746197	499.14707444312	726.352339189317	242.117446396439	465.551430708881	211.912964100195	677.464394809076
6	10	280.500326847158	499.14707444312	779.647401290277	259.882467096759	461.689805514663	259.450867248589	721.140672763252
7	11	339.405395485059	499.14707444312	838.552469928179	279.51748997606	457.65845346811	311.194348018011	768.852801486121
8	12	402.317714232327	499.14707444312	901.464788675446	300.488262891816	453.55638083778	365.571142769336	819.127523607116
9	13	465.406546225275	499.14707444312	964.553620668395	321.517873556132	449.599940642004	419.20861860351	868.808559245513
10	14	528.444753519231	499.14707444312	1027.59182796235	342.530609320784	445.756333857283	471.920217578458	917.676551435741
11	15	591.364814075517	499.14707444312	1090.51188851864	363.503962839545	441.994686478661	523.65348606607	965.648172544731
12	16	654.74036228255	499.14707444312	1153.88743672567	384.629145575223	438.257913818682	574.870934758002	1013.12884857668
13	17	717.4865126362	499.14707444312	1216.63358707932	405.54452902644	434.587689443328	624.68723488201	1059.27492432534

Fig. 2.11 Output results (table format)

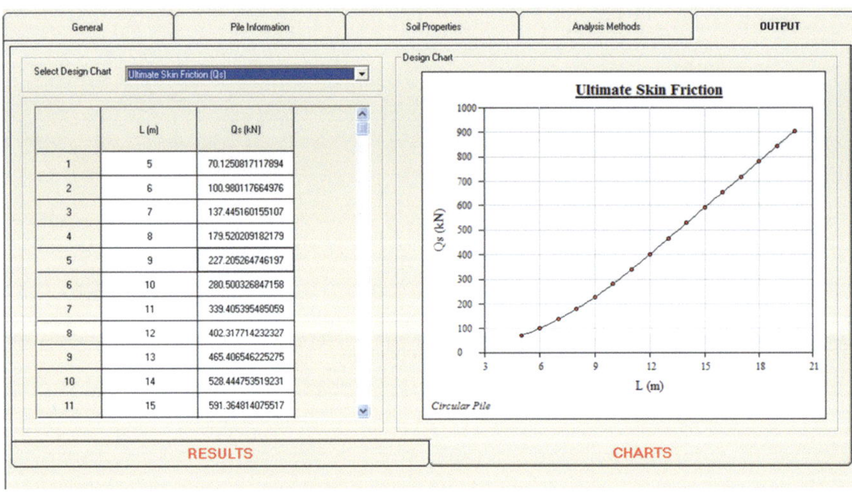

	L (m)	Qs (kN)
1	5	70.1250817117894
2	6	100.980117664976
3	7	137.445160155107
4	8	179.520209182179
5	9	227.205264746197
6	10	280.500326847158
7	11	339.405395485059
8	12	402.317714232327
9	13	465.406546225275
10	14	528.444753519231
11	15	591.364814075517

Fig. 2.12 Output results (chart format)

2.4.2 *Circular Pile in Sandy Soil (Vesic and Critical Depth Methods)*

Develop pile axial capacity design charts (bearing capacity and elastic settlement charts) for a circular bored pile with a diameter of 0.75 m in sandy soil. Use a design factor of safety of 3.0. Use Vesic's method for the end bearing capacity and use the

Table 2.3 Axial pile capacities

L (m)	Q_s (kN)	Q_p (kN)	Q_u (kN)	Q_{all} (kN)	Q_{wp} (kN)	Q_{ws} (kN)	Q_w (kN)
5	70.15973	499.585	569.7447	189.9149	479.6251	67.35664	546.9817
6	101.03	499.585	600.615	200.205	476.3768	96.33666	572.7134
7	137.5131	499.585	637.098	212.366	472.9855	130.1914	603.1769
8	179.6089	499.585	679.1939	226.398	469.4441	168.7728	638.2169
9	227.3175	499.585	726.9025	242.3008	465.7457	211.9202	677.666
10	280.6389	499.585	780.2239	260.0746	461.8835	259.4604	721.3439
11	339.5731	499.585	839.158	279.7193	457.8516	311.2065	769.0581
12	402.5165	499.585	902.1014	300.7005	453.749	365.5863	819.3353
13	465.6365	499.585	965.2214	321.7405	449.7919	419.227	869.019
14	528.7058	499.585	1,028.291	342.7636	445.9477	471.9421	917.8898
15	591.657	499.585	1,091.242	363.7473	442.1854	523.6789	965.8644
16	655.0638	499.585	1,154.649	384.8829	438.4481	574.9001	1,013.348
17	717.841	499.585	1,217.426	405.8086	434.7772	624.7204	1,059.498
18	781.4049	499.585	1,280.99	426.9966	431.0832	674.2607	1,105.344
19	843.634	499.585	1,343.219	447.7397	427.466	721.849	1,149.315
20	906.2167	499.585	1,405.802	468.6006	423.8304	768.8026	1,192.633

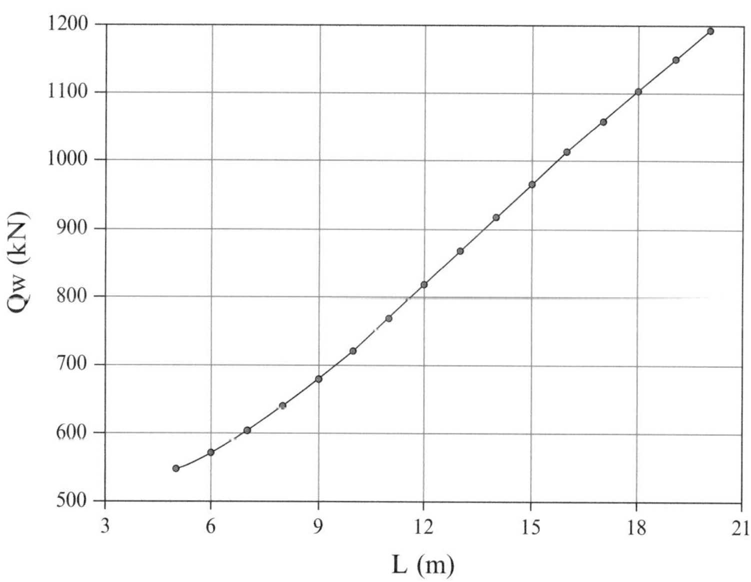

Circular Pile [Allowable Elastic Settlement = 25 mm]

Fig. 2.13 Total working load design chart (settlement analysis)

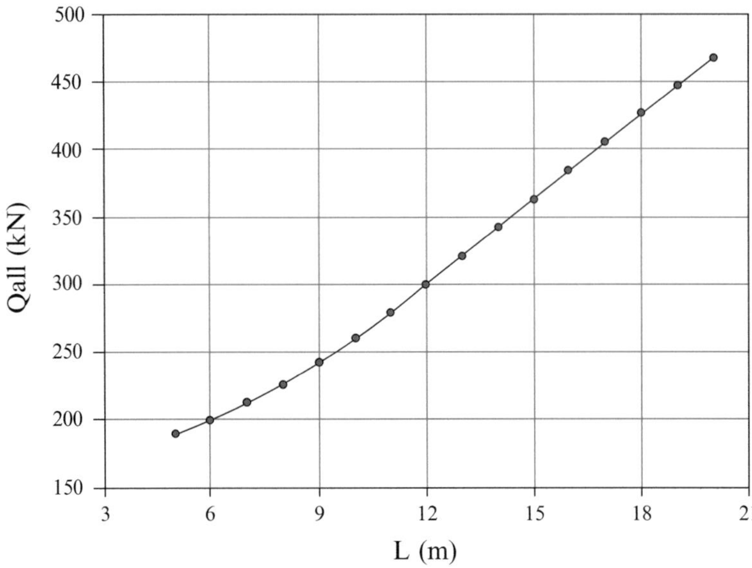

Fig. 2.14 Allowable load carrying capacity (bearing capacity analysis)

Table 2.4 Soil properties and other design parameters

Property	Value	Unit
c	0	kN/m^2
ϕ	28	°
γ	18	kN/m^3
E_s	25,000	kN/m^2
μ_s	0.35	–
E_p	21,000,000	kN/m^2
δ	0.5ϕ	–
S_e	0.025	m
ε	0.6	–
$(L'/B)_{cr}$	15	–

critical depth method for the frictional skin resistance. Table 2.4 provides a list of soil properties and other required design parameters to perform the analyses.

2.4.2.1 Hand Solution

To develop the required design capacity charts, the following steps must be considered:

Step 1:
Assume pile length, $L = 6$ m.

Step 2:
Determine Q_p from Eq. (2.6):

$$Q_p = \left(0.75 \times 18 \times N_\gamma^* + 6 \times 18 \times N_q^*\right)0.441$$

Equation (2.10):

$$I_r = \frac{25,000}{2(1+0.35)(108\tan(28))} = 161.24$$

Equation (2.9):

$$N_\sigma = \frac{3}{3-\sin(28)} e^{\left(\frac{90-28}{180}\right)\pi} \tan^2\left(45+\frac{28}{2}\right)(161.24)^{\frac{4\sin(28)}{3(1+\sin 28)}} = 84.37$$

Equation (2.8):

$$N_q^* = \left(\frac{1+2(1-\sin(28))}{3}\right)84.37 = 57.96$$

Equation (2.7):

$$N_\gamma^* = 0.6(57.96-1)\tan\left(\phi'\right) = 18.17$$

Equation (2.14):

$$Q_p = (0.75 \times 18 \times 18.17 + 6 \times 18 \times 57.96)0.441 = 2,868.69\,\text{kN}$$

Step 3:
Determine Q_s frictional resistance (skin friction).
For sand

$$L' = 15 \times 0.75 = 11.25\,\text{m}$$

At $z = 0\,\text{m}$

$$\sigma_o' = 0$$

From Eq. (2.21):

$$f = 0$$

At $z = 6\,\text{m}$

$$\sigma_o^{'} = 6 \times 18 = 108\,\text{kN/m}^2$$

From Eq. (2.21):

$$f = (1 - \sin(28)) \times 108 \tan(14) = 14.28\,\text{kN/m}^3$$

From Eq. (2.19):

$$Q_\text{s} = \frac{f_{z=0} + f_{z=6}}{2} \times (3.14 \times 0.75) \times 6$$

$$Q_\text{s} = \frac{0 + 14.28}{2} \times (3.14 \times 0.75) \times 6 = 101\,\text{kN}$$

Steps 4 and 5:
From Eq. (2.1):

$$Q_\text{u} = 2,868.69 + 101 = 2,969.69\,\text{KN}$$

From Eq. (2.40):

$$Q_\text{all} = \frac{2,969.69}{3} = 989.89\,\text{kN}$$

Steps 6–8:
From Eq. (2.41):

$$\eta = \left(\frac{2,868.69}{101}\right) = 28.4$$

Solving for Q_wp from Eq. (2.42):

$$0.025 = Q_\text{wp}\left(\frac{\left(1 + \frac{0.5}{28.4}\right)6}{0.441 \times 21,000,000} + \frac{0.75\left(1 - 0.35^2\right)0.85}{0.441 \times 25,000} + \frac{0.75\left(1 - 0.35^2\right)2.98}{28.4 \times 3.14 \times 0.75 \times 6 \times 25,000}\right)$$

Therefore,

$$Q_{\mathrm{wp}} = 484.54\,\mathrm{kN}$$

From Eq. (2.43):

$$Q_{\mathrm{w}} = 484.54\left(1 + \frac{1}{28.4}\right) = 501.60\,\mathrm{kN}$$

Step 9:
Repeat Steps 1 through 8 with another assumed pile length. The solution below is for a second assumed pile length.

Step 1 (second pile length):
Assume pile length, $L = 10$ m.

Step 2 (second pile length):
Determine end bearing capacity, Q_{p}, using Eq. (2.6):

$$Q_{\mathrm{p}} = \left(0.75 \times 18 \times N_{\gamma}^{*} + 10 \times 18 \times N_{q}^{*}\right)0.441$$

Equation (2.10):

$$I_{\mathrm{r}} = \frac{25,000}{2(1 + 0.35)(180\tan(28))} = 96.74$$

Equation (2.9):

$$N_{\sigma} = \frac{3}{3 - \sin(28)}\, e^{\left(\frac{90-28}{180}\right)\pi} \tan^{2}\left(45 + \frac{28}{2}\right)(96.74)^{\frac{4\sin(28)}{3(1+\sin 28)}} = 67.90$$

Equation (2.8):

$$N_{q}^{*} = \left(\frac{1 + 2(1 - \sin(28))}{3}\right)67.90 = 46.64$$

Equation (2.7):

$$N_{\gamma}^{*} = 0.6(46.64 - 1)\tan(28) = 14.56$$

Equation (2.14):

$$Q_{\mathrm{p}} = (0.75 \times 18 \times 14.56 + 10 \times 18 \times 46.64)0.441 = 3,788.96\,\mathrm{kN}$$

Step 3 (second pile length):
Determine Q_s frictional resistance (skin friction) for sand.

$$L' = 15 \times 0.75 = 11.25\,\text{m}$$

At $z = 0\,\text{m}$

$$\sigma_o' = 0$$

From Eq. (2.21):

$$f = 0$$

At $z = 10\,\text{m}$

$$\sigma_o' = 10 \times 18 = 180\,\text{kN/m}^2$$

From Eq. (2.21):

$$f = (1 - \sin(28)) \times 180 \tan(14) = 23.80\,\text{kN/m}^3$$

From Eq. (2.19):

$$Q_s = \frac{f_{z=0} + f_{z=10}}{2} \times (3.14 \times 0.75) \times 10$$

$$Q_s = \frac{0 + 23.80}{2} \times (3.14 \times 0.75) \times 10 = 280.358\,\text{kN}$$

Steps 4 and 5 (second pile length):
From Eq. (2.1):

$$Q_u = 3{,}788.96 + 280.358 = 4{,}069.32\,\text{kN}$$

From Eq. (2.40):

$$Q_{all} = \frac{4{,}069.32}{3} = 1{,}356.44\,\text{kN}$$

Steps 6–8 (second pile length):
From Eq. (2.41):

$$\eta = \left(\frac{2{,}308.05}{280.358}\right) = 13.51$$

Solving for Q_{wp} from Eq. (2.42):

$$0.025 = Q_{wp}\left(\frac{\left(1+\frac{0.5}{13.51}\right)10}{0.441 \times 21{,}000{,}000} + \frac{0.75\left(1-0.35^2\right)0.85}{0.441 \times 25{,}000} + \frac{0.75\left(1-0.35^2\right)3.27}{13.51 \times 3.14 \times 0.75 \times 10 \times 25{,}000}\right)$$

Therefore,

$$Q_{wp} = 479.56\,\text{kN}$$

From Eq. (2.43):

$$Q_w = 479.56\left(1+\frac{1}{13.51}\right) = 515.06\,\text{kN}$$

Steps 9:
Repeat Steps 1 through 8 with another assumed pile length. The solution below is for a third assumed pile length.

Step 1 (third pile length):
Assume pile length, $L = 15$ m.

Step 2 (third pile length):
Determine end bearing capacity, Q_p, from Eq. (2.6):

$$Q_p = \left(0.75 \times 18 \times N_\gamma^* + 16 \times 18 \times N_q^*\right)0.441$$

Equation (2.10):

$$I_r - \frac{25{,}000}{2(1+0.35)(288\tan(28))} = 60.46$$

Equation (2.9):

$$N_\sigma = \frac{3}{3-\sin(28)} e^{\left(\frac{90-28}{180}\right)\pi} \tan^2\left(45+\frac{28}{2}\right)(60.46)^{\frac{4\sin(28)}{3(1+\sin 28)}} = 55.58$$

Equation (2.8):

$$N_q^* = \left(\frac{1 + 2(1 - \sin(28))}{3}\right) 55.58 = 38.18$$

Equation (2.7):

$$N_\gamma^* = 0.6(38.18 - 1)\tan(28) = 11.86$$

Equation (2.14):

$$Q_\mathrm{p} = (0.75 \times 18 \times 11.86 + 16 \times 18 \times 38.18)0.441 = 4{,}919.77\,\mathrm{kN}$$

Step 3:
Determine Q_s frictional resistance (skin friction) in sand:

$$L' = 15 \times 0.75 = 11.25\,\mathrm{m}$$

At $z = 0\,\mathrm{m}$

$$\sigma_o' = 0$$

From Eq. (2.21):

$$f = 0$$

At $z = 11.25\,\mathrm{m}$

$$\sigma_o' = 11.25 \times 18 = 202.5\,\mathrm{kN/m^2}$$

From Eq. (2.21):

$$f = (1 - \sin(28)) \times 202.5\tan(14) = 26.78\,\mathrm{kN/m^3}$$

From Eq. (2.19):

$$Q_\mathrm{s} = \frac{f_{z=0} + f_{z=11.5}}{2} \times (3.14 \times 0.75) \times 11.25 + f_{z=11.5}(3.14 \times 0.75)(16 - 11.25)$$

$$Q_\mathrm{s} = \frac{0 + 26.78}{2} \times (3.14 \times 0.75) \times 11.25 + 26.78(3.14 \times 0.75)(16 - 11.25)$$
$$= 654.39\,\mathrm{kN}$$

Steps 4 and 5 (third pile length):
From Eq. (2.1):

$$Q_u = 4{,}919.77 + 654.39 = 5{,}574.16\,\text{kN}$$

From Eq. (2.40):

$$Q_{all} = \frac{5{,}574.16}{3} = 1{,}858.05\,\text{kN}$$

Q_{all} is for assumed value $L = 16$ m.

Steps 6–8 (third pile length):
Let

$$\eta = \frac{(4{,}919.77)}{(654.39)} = 7.51$$

Solving for Q_{wp} from Eq. (2.42):

$$0.025 = Q_{wp}\left(\frac{\left(1+\frac{0.5}{7.51}\right)16}{0.441 \times 21{,}000{,}000} + \frac{0.75\left(1-0.35^2\right)0.85}{0.441 \times 25{,}000} + \frac{0.75\left(1-0.35^2\right)3.616}{7.51 \times 3.14 \times 0.75 \times 16 \times 25{,}000}\right)$$

Therefore,

$$Q_{wp} = 472.41\,\text{kN}$$

From Eq. (2.43):

$$Q_w = 472.41\left(1 + \frac{1}{0.76}\right) = 535.326\,\text{kN}$$

2.4.2.2 foundationPro Solution

General Information Section

In this section, we select the SI units and we enter a safety factor of 3.0 in the provided textbox. These units and this safety factor will be used throughout the entire analyses.

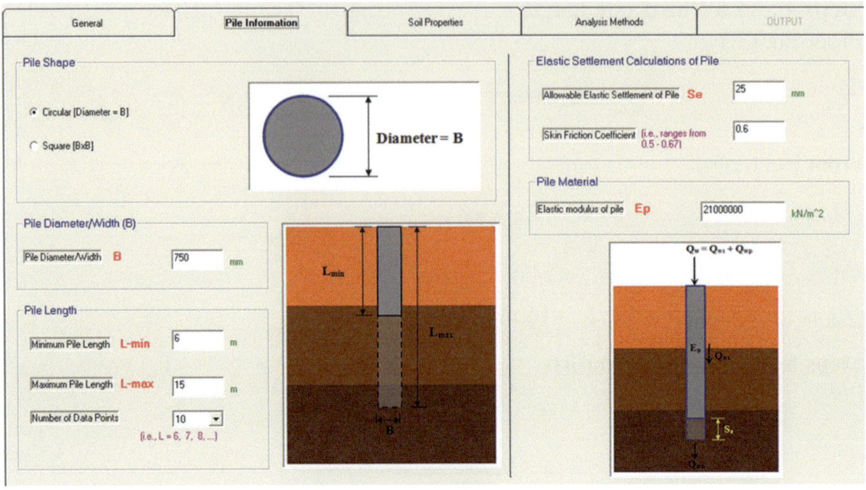

Fig. 2.15 Pile information

Pile Information Section

In this section, the pile information is entered as shown in Fig. 2.15. Number of data points was 10 points (i.e., 10 pile lengths will be considered including the minimum and the maximum lengths).

Soil Properties Section

The thicknesses and physical properties of the soil layers must be entered in this section. As provided in the problem statement, we have a 20 m homogeneous sand layer with properties as shown in Fig. 2.16.

Analysis Methods Section

The methods of analysis to be used to perform end bearing capacity and skin friction analyses are specified in this section as shown in Figs. 2.17 and 2.18.

Now, we can hit the RUN button to view the output results. Summary of results for this design problem is provided in Table 2.5. An example of the available design charts that can be obtained for this design problem is shown in Fig. 2.19.

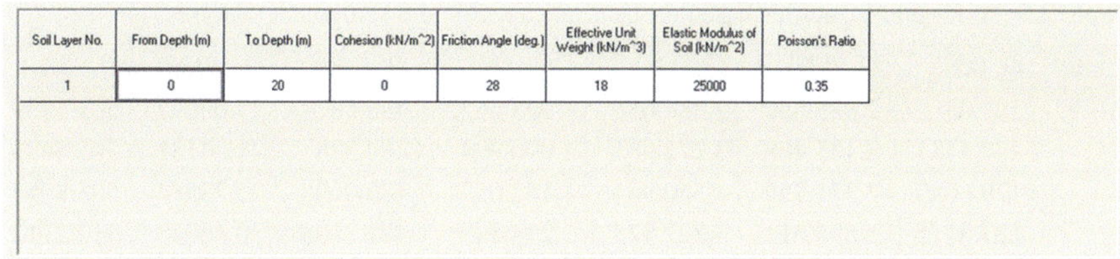

Soil Layer No.	From Depth (m)	To Depth (m)	Cohesion (kN/m^2)	Friction Angle (deg.)	Effective Unit Weight (kN/m^3)	Elastic Modulus of Soil (kN/m^2)	Poisson's Ratio
1	0	20	0	28	18	25000	0.35

Fig. 2.16 Pile information

Fig. 2.17 Specifying Vesic's method for end bearing capacity analysis

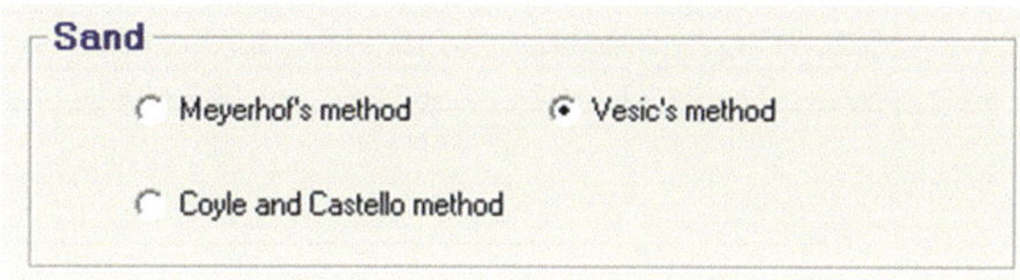

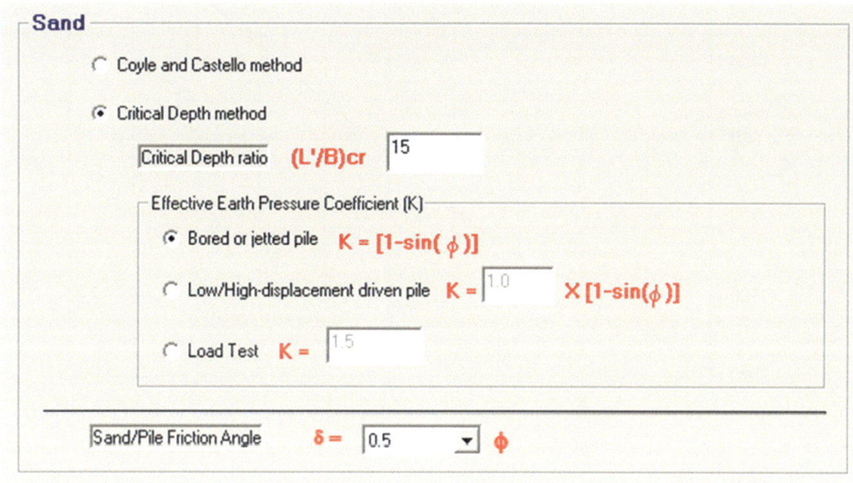

Fig. 2.18 Specifying critical depth method for skin friction analysis

2.4.3 *Square Pile in Clay Soil (Meyerhof's and α Methods)*

Develop pile capacity design charts (bearing capacity and settlement) for a square pile (bored) in clay soil with a side of 0.6 m. Use a factor of safety of 3.0. Use Meyerhof's method for the end bearing capacity and α method for the frictional skin resistance. Other required information is provided in Table 2.6.

Table 2.5 Axial pile capacities based on bearing capacity and settlement analyses

L (m)	Q_s (kN)	Q_p (kN)	Q_u (kN)	Q_{all} (kN)	Q_{wp} (kN)	Q_{ws} (kN)	Q_w (kN)
6	101.03	2,885.656	2,986.686	995.5622	485.5879	17.00097	502.5889
7	137.5131	3,135.468	3,272.981	1,090.994	484.328	21.24131	505.5693
8	179.6089	3,371.355	3,550.964	1,183.655	483.0698	25.73554	508.8053
9	227.3175	3,595.611	3,822.928	1,274.309	481.8108	30.46048	512.2712
10	280.6389	3,809.955	4,090.594	1,363.531	480.5492	35.39695	515.9461
11	339.5731	4,015.715	4,355.288	1,451.763	479.2835	40.52872	519.8122
12	402.5165	4,213.941	4,616.458	1,538.819	478.0263	45.66116	523.6875
13	465.6365	4,405.484	4,871.12	1,623.707	476.8035	50.39562	527.1991
14	528.7058	4,591.042	5,119.747	1,706.582	475.6072	54.77108	530.3783
15	591.657	4,771.197	5,362.854	1,787.618	474.4313	58.83231	533.2636

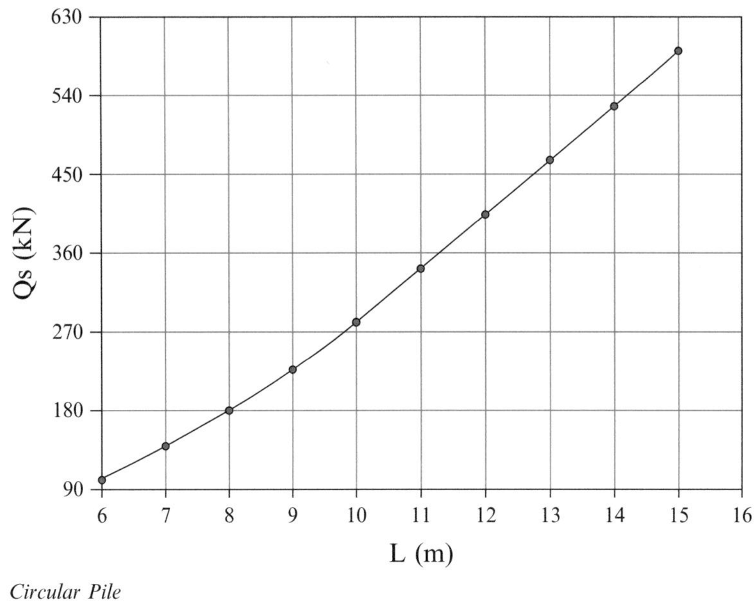

Fig. 2.19 Ultimate skin friction design chart (bearing capacity)

2.4.3.1 Hand Solution

The following steps must be followed to determine the required design charts:

Step 1:
Assume pile length, $L = 6$ m.

Step 2:
Determine Q_p using Eq. (2.13):

$$Q_p = 9 \times 40 \times 0.6 \times 0.6 = 129.6\,\text{kN}$$

Table 2.6 Soil properties and other design parameters

Property	Value	Unit
c	4	kN/m^2
ϕ	0	°
γ	19	kN/m^3
E_s	9,000	kN/m^2
μ_s	0.35	–
E_p	21,000,000	kN/m^2
δ	0.5ϕ	–
S_e	0.025	m
ε	0.6	–

Step 3:
Determine Q_s frictional using Eq. (2.28):

$$\alpha = 0.026 \times \left(\frac{40}{100}\right)^4 - 0.02233\left(\frac{40}{100}\right)^3 + 0.7498\left(\frac{40}{100}\right)^2 - 1.198\left(\frac{40}{100}\right) + 1.1187$$

$$= 0.758$$

From Eq. (2.27):

$$f = 0.758 \times 40 = 30.32$$

From Eq. (2.29):

$$Q_s = 0.758 \times 40 \times 4 \times 0.6 \times 6 = 436.608\,\text{kN}$$

Steps 4 and 5:
The ultimate load can be determined from Eq. (2.1):

$$Q_u = 129.6 + 436.608 = 566.20\,\text{kN}$$

The allowable load can also be determined from Eq. (2.40):

$$Q_{all} = \frac{566.20}{3} = 188.736\,\text{kN}$$

Steps 5–8:
Calculating Q_w for given settlement from Eq. (2.41):

$$\eta = \left(\frac{129.6}{436.608}\right) = 0.296$$

Solving for Q_{wp} from Eq. (2.42):

$$0.025 = Q_{wp}\left(\frac{\left(1+\frac{0.6}{0.296}\right)6}{0.36\times 21{,}000{,}000}+\frac{0.6\left(1-0.35^2\right)0.85}{0.36\times 9{,}000}+\frac{0.6\left(1-0.35^2\right)2.94}{0.296\times 4\times 0.6\times 6\times 9{,}000}\right)$$

Therefore,

$$Q_{wp} = 135\,\text{kN}$$

From Eq. (2.43):

$$Q_{w} = 135\left(1+\frac{1}{0.296}\right) = 591.08\,\text{kN}$$

Step 9:
The above steps can be repeated for several pile lengths to evaluate the effect of pile length on the axial capacity of this pile in sand.

2.4.3.2 foundationPro Solution

General information Section

As described in previous design examples, the user must enter the factor of safety and specify the units to be used in the analysis.

Pile Information Section

In this design problem, pile properties are entered as shown in Fig. 2.20.

Soil Properties Section

In this section, we enter the number of soil layers we will be dealing with and the given depth parameters for the soil layer. If we are dealing with a single layer of soil and the depth parameters are not given, we select a reasonable depth to consider. We also enter the soil cohesion, friction angle, effective unit weight, elastic modulus, and the Poisson's ratio for the soil. Notice that the cohesion is equal to 40 kN/m^2 in this problem.

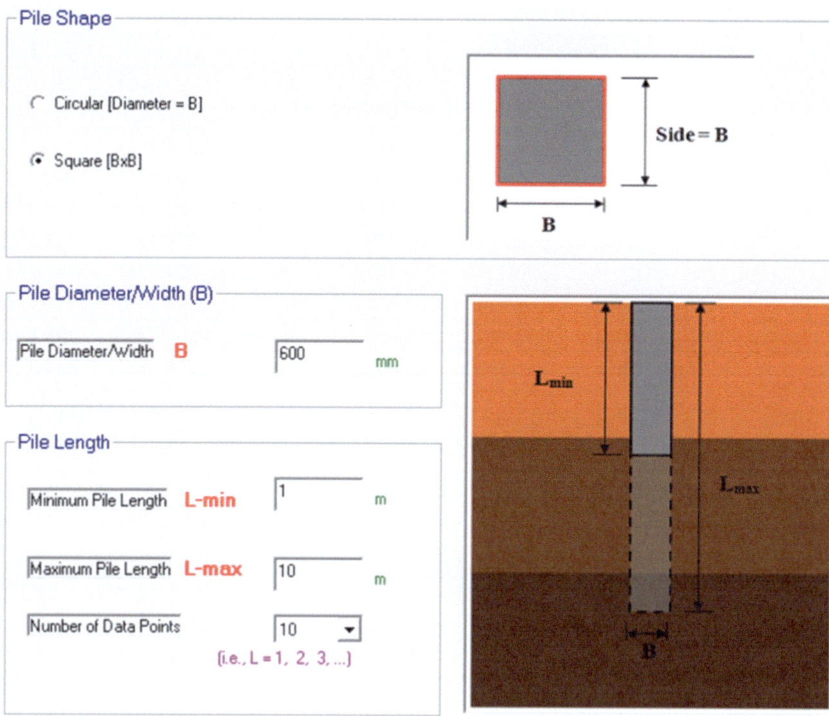

Fig. 2.20 Pile Information section

Analysis Method Section

We specify the methods as provided in the problem statement in this section. Alpha (α*) method for skin friction and Meyerhof's method for end bearing were selected for this problem.

After saving the input data into a file and performing the analysis, one can view the output results for this design problem. Summary of the axial capacity results is provided in Table 2.7. A snapshot for one of the design charts is also shown in Fig. 2.21.

2.4.4 Circular Pile in Clay Soil (Vesic's and α Methods)*

Develop design capacity charts (bearing capacity and elastic settlement) for a circular bored pile in clay soil (undrained cohesion = 900 lb/ft^2) that has a diameter of 3 ft. Use Vesic's method for the end bearing capacity and α* method for the frictional skin resistance. The total elastic settlement must not exceed 1.2 in. Soil properties and other design parameters are provided in Table 2.8.

Table 2.7 Pile capacity based on bearing capacity and settlement requirements at various lengths

L (m)	Q_s (kN)	Q_p (kN)	Q_u (kN)	Q_{all} (kN)	Q_{wp} (kN)	Q_{ws} (kN)	Q_w (kN)
1	71.58935	129.6	201.1894	67.06312	145.931	80.61037	226.5414
2	143.1787	129.6	272.7787	90.92623	143.5976	158.6428	302.2404
3	214.7681	129.6	344.3681	114.7894	141.7371	234.8813	376.6184
4	286.3574	129.6	415.9574	138.6525	140.0936	309.5434	449.637
5	357.9468	129.6	487.5468	162.5156	138.5692	382.7192	521.2885
6	429.5361	129.6	559.1361	186.3787	137.1154	454.4446	591.56
7	501.1255	129.6	630.7255	210.2418	135.7041	524.7281	660.4322
8	572.7148	129.6	702.3148	234.1049	134.3179	593.5635	727.8814
9	644.3042	129.6	773.9042	257.9681	132.9455	660.9361	793.8815
10	715.8935	129.6	845.4935	281.8312	131.5793	726.8267	858.4059

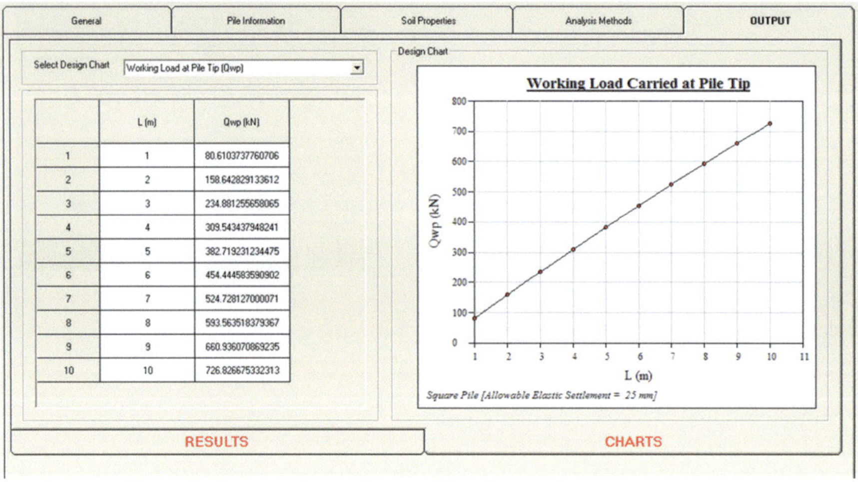

Fig. 2.21 A snapshot of one of the design charts from Pile-1 application

Table 2.8 Soil properties and design parameters

Property	Value	Unit
FS	4.0	–
c_u	900	lb/ft^2
ϕ	0	°
γ	65	lb/ft^3
E_s	94,000	lb/ft^2
μ_s	0.25	–
E_p	460,000,000	lb/ft^2
δ	0.5ϕ	–
S_e	1.2	in.
ε	0.6	–

2.4.4.1 Hand Solution

The steps below must be followed to develop the required design charts for this design problem:

Step 1:
Assume pile length, $L = 15$ ft.

Step 2:
Determine end bearing capacity, Q_p, for clay using Eq. (2.14):

$$Q_p = 7.06 \times 900 \times N_c^*$$

From Eq. (2.16):

$$I_r = 347\left(\frac{900}{2,000}\right) - 33 = 123.15 \leq 300$$

From Eq. (2.15):

$$N_c^* = \frac{4}{3}(\ln(123.15) + 1) + \frac{\pi}{2} + 1 = 10.32$$

$$Q_p = 7.06 \times 900 \times 10.32 = 65,573.28\,\text{lb}$$

Step 3:
Determine Q_s, the frictional resistance (skin friction) for clay from Eq. (2.30):

$$f = \alpha^* \times 900$$

From Eq. (2.31):

$$\alpha^* = 0.5\left(\frac{\frac{1}{2} \times 65 \times 15}{900}\right)^{0.45} = 0.379$$

$$f = 0.379 \times 900 = 341.5$$

From Eq. (2.32):

$$Q_s = 0.379 \times 9.42 \times 15 \times 900 = 48,297.430\,\text{lb}$$

Steps 4 and 5:
From Eq. (2.1):

$$Q_u = 65,573.28 + 48,297.43 = 113,870.71\,\text{lb}$$

From Eq. (2.40):

$$Q_{\text{all}} = \frac{113,870.71}{4} = 28,467.67\,\text{lb}$$

Steps 6–8:
From Eq. (2.41):

$$\eta = \left(\frac{65,573.28}{48,297.43}\right) = 1.35$$

Solving for Q_{wp} from Eq. (2.42):

$$1 = Q_{\text{wp}}\left(\frac{\left(1+\frac{0.6}{1.35}\right)15}{7.06 \times 4.6 \times 10^8} + \frac{3\left(1-0.25^2\right)0.85}{7.06 \times 94,000} + \frac{3\left(1-0.25^2\right)2.67}{1.35 \times 3.14 \times 3 \times 15 \times 94,000}\right)$$

Therefore,

$$Q_{\text{wp}} = 24,811.907\,\text{lb}$$

From Eq. (2.43):

$$Q_{\text{w}} = 24,811.907\left(1 + \frac{1}{1.36}\right) = 43,055.95\,\text{lb}$$

Step 9:
The above steps must be repeated for other pile lengths to allow us to develop the required design charts and have enough points on the charts.

2.4.4.2 foundationPro Solution

General Information Section

For this design problem, we select the BS units to be used and enter a value of 4.0 for the factor of safety as shown in Fig. 2.22.

Pile Information Section

In this section, we select the circular shape as shown in Fig. 2.23. Also, enter the pile diameter, specify the pile lengths to be considered (10, 11, 12, 13, . . ., 20 ft with a total of 11 points), and enter elastic settlement properties.

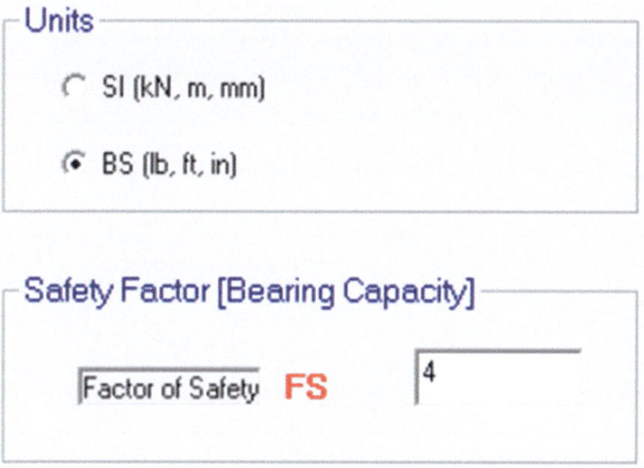

Fig. 2.22 Units and safety factor selection

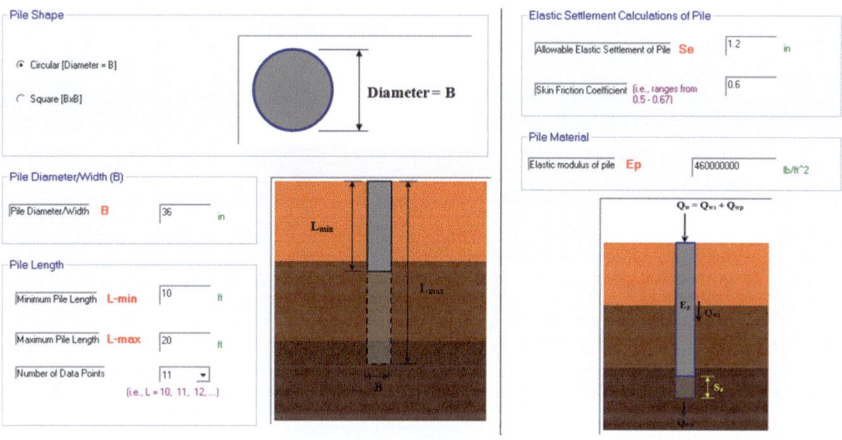

Fig. 2.23 Pile Information section for pile capacity

Soil Properties Section

The soil properties for the homogeneous clay layer are shown in Fig. 2.24. The thickness of the clay soil layer can be assumed 25 ft; any value would work as long as it is larger than the maximum pile length specified in the Pile Information section.

Analysis Methods Section

In this section, we select the method of analysis that we wish to use under the soil type that is given to us. In this problem, we use the Alpha (α*) method for clay soil

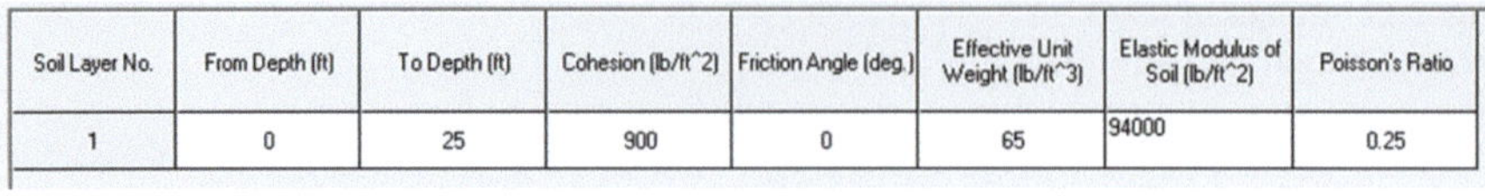

Soil Layer No.	From Depth (ft)	To Depth (ft)	Cohesion (lb/ft^2)	Friction Angle (deg.)	Effective Unit Weight (lb/ft^3)	Elastic Modulus of Soil (lb/ft^2)	Poisson's Ratio
1	0	25	900	0	65	94000	0.25

Fig. 2.24 Soil properties and thickness of clay layer

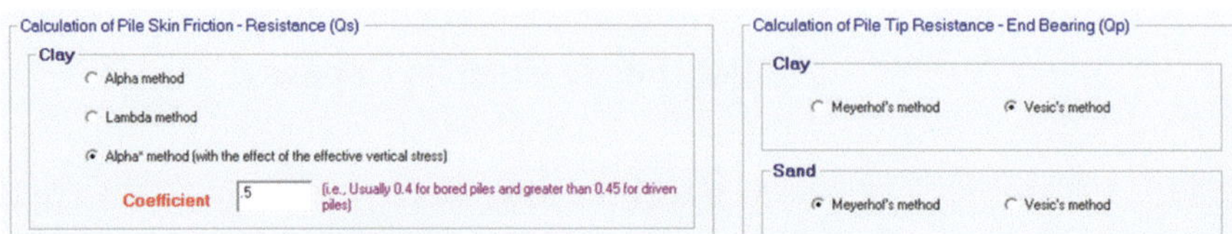

Fig. 2.25 Specifying the methods of analyses in Pile-1 application

Table 2.9 Axial capacity of pile

L (ft)	Q_s (lb)	Q_p (lb)	Q_u (lb)	Q_{all} (lb)	Q_{wp} (lb)	Q_{ws} (lb)	Q_w (lb)
10	26,817.73	65,665.72	92,483.46	23,120.86	25,355.62	10,355.17	35,710.80
11	30,792.24	65,665.72	96,457.97	24,114.49	25,229.47	11,830.71	37,060.18
12	34,932.91	65,665.72	100,598.64	25,149.66	25,108.11	13,357.03	38,465.15
13	39,231.94	65,665.72	104,897.67	26,224.41	24,990.96	14,930.83	39,921.80
14	43,682.51	65,665.72	109,348.24	27,337.06	24,877.55	16,549.18	41,426.74
15	48,278.56	65,665.72	113,944.29	28,486.07	24,767.49	18,209.48	42,976.97
16	53,014.66	65,665.72	118,680.39	29,670.09	24,660.44	19,909.39	44,569.83
17	57,885.92	65,665.72	123,551.65	30,887.91	24,556.12	21,646.81	46,202.94
18	62,887.90	65,665.72	128,553.63	32,138.40	24,454.31	23,419.83	47,874.15
19	68,016.57	65,665.72	133,682.30	33,420.57	24,354.79	25,226.69	49,581.48
20	73,268.19	65,665.72	138,933.92	34,733.48	24,257.38	27,065.78	51,323.16

for the calculation of the skin friction resistance component. We enter 0.5 for the coefficient to stay consistent with the hand solution. This coefficient can be obtained from Table 2.1. Next, we select the Vesic's method for clay soil for the calculation of the end bearing capacity component (see Fig. 2.25).

Now we can hit the RUN button to perform the analysis and then view the results. Axial capacities at pile lengths from 10 to 20 ft are summarized in Table 2.9. The allowable load bearing capacity design chart is also shown in Fig. 2.26.

2.4.5 Circular Pile in Clay Soil (Meyerhof's and λ Methods)

Develop pile capacity design charts for a circular bored pile in clay soil with a diameter of 3 ft. Use Meyerhof's method for the end bearing capacity and use λ method for the frictional skin resistance. Table 2.10 provides a list of required other parameters.

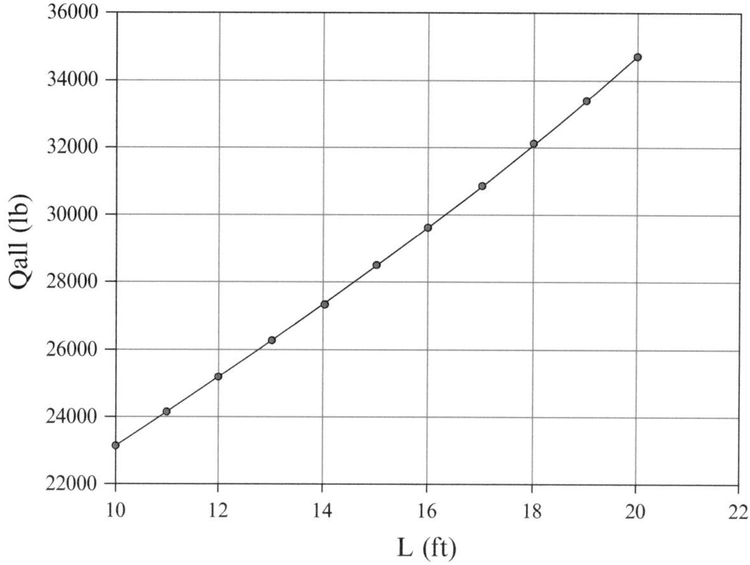

Circular Pile

Fig. 2.26 Allowable load bearing capacity design chart

Table 2.10 Soil properties and other information

Property	Value	Unit
FS	4.0	–
c	1,000	lb/ft^2
ϕ	0	°
γ	110	lb/ft^3
E_s	100,000	lb/ft^2
μ_s	0.25	–
E_p	460,000,000	lb/ft^2
δ	0.5ϕ	–
S_e	1.2	in.
ε	0.6	–

2.4.5.1 Hand Solution

Below are the steps to follow to develop the required pile capacity design charts:

Step 1:
Assume $L = 20$ ft.

Step 2:
Determine Q_p from Eq. (2.13):

$$Q_p = 9 \times 1{,}000 \times 7.06 = 63{,}617.25\,\text{lb}$$

Step 3:
Determine Q_s.

λ can be acquired from Fig. 2.5. Therefore,

$\lambda = 0.31$.

From Eq. (2.33):

$$f_{avg} = 0.31(0.5 \times 110 \times 20 + 2 \times 1,000) = 961$$

From Eq. (2.34):

$$Q_s = \pi \times 3 \times 20 \times 961 = 181,144.23\,\text{lb}$$

Steps 4 and 5:
From Eq. (2.1):

$$Q_u = 63,617.25 + 181,144.23 = 244,761.48\,\text{lb}$$

From Eq. (2.40):

$$Q_{all} = \frac{244,761.48}{4} = 61,190.058\,\text{lb}$$

Steps 6–8:
From Eq. (2.41):

$$\eta = \left(\frac{63,617}{181,144.23}\right) = 0.35$$

Solving for Q_{wp} from Eq. (2.42):

$$0.1 = Q_{wp}\left(\frac{\left(1 + \frac{0.6}{0.35}\right)20}{7.06 \times 4.6 \times 10^8} + \frac{3(1 - 0.25^2)0.85}{7.06 \times 100,000} + \frac{3(1 - 0.25^2)2.904}{0.35 \times 3.14 \times 3 \times 20 \times 100,000}\right)$$

Therefore,

$$Q_{wp} = 21,547.69\,\text{lb}$$

From Eq. (2.43):

$$Q_w = 21,547.69\left(1 + \frac{1}{0.35}\right) = 83,112.528\,\text{lb}$$

Step 9:
One must repeat Steps 1 through 8 for several pile lengths.

2.4.5.2 foundationPro Solution

General information Section

Select the BS units and enter a safety factor of 4 in the provided textbox.

Pile Information Section

In this section, we enter required pile information. We enter a minimum pile length of 15 ft and a maximum pile length of 25 with 11 points (i.e., a step of 1 ft).

Soil Properties Section

We enter soil properties as provided in the problem statement. The thickness of the single homogeneous clay layer can be assumed 30 ft which is larger than the maximum pile length (25 ft).

Analysis Method Section

Lambda method is selected for skin friction calculations and Meyerhof's method is also selected for end bearing calculations.

Now, we can save our progress and hit run to get the results. Axial capacities at various pile lengths are shown in Fig. 2.27. Also, skin friction resistance results based on elastic settlement analyses are shown in Fig. 2.28.

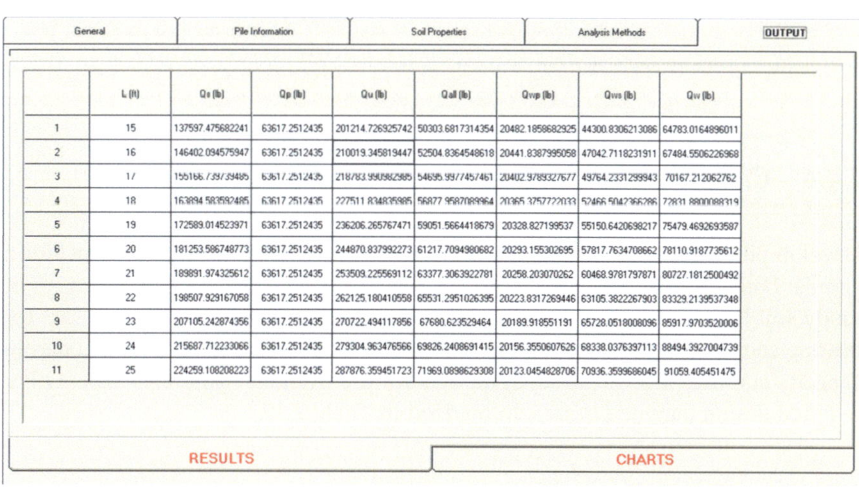

	L (ft)	Qs (lb)	Qp (lb)	Qu (lb)	Qall (lb)	Qwp (lb)	Qws (lb)	Qw (lb)
1	15	137597.475682241	63617.2512435	201214.726925742	50303.6817314354	20482.1858682925	44300.8306213086	64783.0164896011
2	16	146402.094575947	63617.2512435	210019.345819447	52504.8364548618	20441.8387995058	47042.7118231911	67484.5506226968
3	17	155166.739739485	63617.2512435	218783.990982985	54695.9977457461	20402.9789327677	49764.2331299943	70167.212062762
4	18	163894.583592485	63617.2512435	227511.834835985	56877.9587089964	20365.3757722033	52466.5042366286	72831.8800088319
5	19	172589.014523971	63617.2512435	236206.265767471	59051.5664418679	20328.827199537	55150.6420698217	75479.4692693587
6	20	181253.586748773	63617.2512435	244870.837992273	61217.7094980682	20293.155302695	57817.7634708662	78110.9187735612
7	21	189891.974325612	63617.2512435	253509.225569112	63377.3063922781	20258.203070262	60468.9781797871	80727.1812500492
8	22	198507.929167058	63617.2512435	262125.180410558	65531.2951026395	20223.8317269446	63105.3822267903	83329.2139537348
9	23	207105.242874356	63617.2512435	270722.494117856	67680.623529464	20189.918551191	65728.0518008096	85917.9703520006
10	24	215687.712233066	63617.2512435	279304.963476566	69826.2408691415	20156.3550607626	68338.0376397113	88494.3927004739
11	25	224259.108208223	63617.2512435	287876.359451723	71969.0898629308	20123.0454828706	70936.3599686045	91059.405451475

Fig. 2.27 Axial capacities at various pile lengths

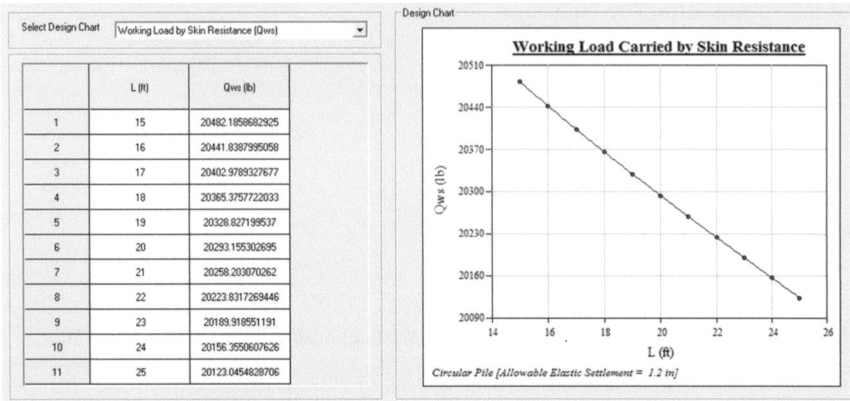

Fig. 2.28 Skin friction resistance results based on settlement analyses

Table 2.11 Thicknesses and physical properties of soil layers

Soil layer	Thickness (m)	Cohesion (kN/m^2)	Friction angle (°)	Effective unit weight (kN/m^3)	Elastic modulus of soil (kN/m^2)	Poisson's ratio
1	10	0	30	19	9,000	0.35
2	9	0	32	10	10,000	0.35
3	11	0	35	9.9	11,000	0.35

Table 2.12 Other design parameters

Property	Value	Unit
E_p	21,000,000	kN/m^2
δ	0.5ϕ	–
S_e	0.025	m
ε	0.6	–
$(L'/B)_{cr}$	18	–

2.4.6 *Circular Pile in Sandy Soil (Three Sand Layers)*

Develop pile capacity design charts (bearing capacity and elastic settlement) for a circular bored pile (diameter = 0.8 m) in a sandy soil. Thicknesses and properties of sandy soil layers are listed in Table 2.11. Use a design factor of safety of 3.0 for bearing capacity calculations. Use Coyle and Castillo's method for the end bearing capacity and use the critical depth method for the frictional skin resistance. Other required design parameters are summarized in Table 2.12.

2.4.6.1 Hand Solution

Design capacity charts for this problem can be developed by following the steps below:

Step 1:
Assume $L_1 = 7$ m.

Step 2:
Determine Q_p from Eq. (2.12):

$$N_q^* = \left(1.04529\left(\frac{30}{40}\right)^2 - 1.58056\left(\frac{30}{40}\right) + 0.6289\right) \times \left(-0.32039\left(\frac{7}{0.8}\right)^2 + 14.2496\left(\frac{7}{0.8}\right) + 935.565\right) = 32.57$$

From Eq. (2.11):

$$Q_p = 7 \times 18.75 \times 32.57 \times 0.502 = 2,157.81\,\text{kN}$$

Step 3:
Determine Q_s frictional resistance (skin friction) from Eq. (2.20):

$$L' = 18 \times 0.8 = 14.4\,\text{m}$$

At $z = 0\,\text{m}$:

$$\sigma_o' = 0$$

From Eq. (2.21):

$$f = 0$$

At $z = 7\,\text{m}$:

$$\sigma_o' = 7 \times 18.75 = 131.25\,\text{kN/m}^2$$

From Eq. (2.21):

$$f = (1 - \sin(30)) \times 131.25 \tan(15) = 17.58\,\text{kN/m}^2$$

From Eq. (2.19):

$$Q_s = \frac{f_{z=0} + f_{z=7}}{2} \times (3.141 \times 0.8) \times 7$$

$$Q_s = \frac{0 + 17.58}{2} \times (3.141 \times 0.8) \times 7 = 154.64\,\text{kN}$$

Steps 4 and 5:
From Eq. (2.1):

$$Q_u = 2,157.8 + 154.64 = 2,312.44\,\text{kN}$$

From Eq. (2.40):

$$Q_{all} = \frac{2,312.44}{3} = 770.816\,\text{kN}$$

Steps 6–8:
From Eq. (2.41):

$$\eta = \left(\frac{2,157.8}{154.64}\right) = 13.95$$

Solving for Q_{wp} from Eq. (2.42):

$$0.025 = Q_{wp}\left(\frac{\left(1 + \frac{0.6}{13.95}\right)7}{0.502 \times 21,000,000} + \frac{0.8(1 - 0.35^2)0.85}{0.502 \times 9,000} + \frac{0.8(1 - 0.35^2)3.03}{13.95 \times \pi \times 0.8 \times 7 \times 9,000}\right)$$

Therefore,

$$Q_{wp} = 188.11\,\text{kN}$$

From Eq. (2.43):

$$Q_w = 188.11\left(1 + \frac{1}{13.95}\right) = 201.59\,\text{kN}$$

Step 9:
Repeat the above steps for another assumed pile length.

Step 1 (second pile length):
Assume $L_2 = 13$ m.

Step 2 (second pile length):
Determine Q_p from Eq. (2.12):

$$N_q^* = \left(1.04529\left(\frac{32}{40}\right)^2 - 1.58056\left(\frac{32}{40}\right) + 0.6289\right)$$
$$\times\left(-0.32039\left(\frac{13}{0.8}\right)^2 + 14.2496\left(\frac{13}{0.8}\right) + 935.565\right)$$
$$= 36.196$$

From Eq. (2.11):

$$Q_p = (190 + 3 \times 10.1) \times 36.196 \times 0.502 = 4,002.9373\,\text{kN}$$

Step 3 (second pile length):
Determine Q_s frictional resistance (skin friction) from Eq. (2.20):

$$L' = 18 \times 0.8 = 14.4\,\text{m}$$

At $z = 0\,\text{m}$:

$$\sigma_o' = 0$$

From Eq. (2.21):

$$f = 0$$

At $z = 10\,\text{m}$:

$$\sigma_o' = 10 \times 18.75 = 187.5\,\text{kN/m}^2$$

From Eq. (2.21):

$$f = (1 - \sin(30)) \times 187.5 \tan(15) = 25.12\,\text{kN/m}^2$$

From Eq. (2.19) with $z = 0$ to $z = 10$:

$$Q_s = \frac{f_{z=0} + f_{z=10} +}{2} \times (3.141 \times 0.8) \times 10$$

$$Q_s = \frac{0 + 25.12}{2} \times (3.141 \times 0.8) \times 10 = 315.60\,\text{kN}$$

At

$$z = 13\,\text{m}$$

$$\sigma_o^{'} = 187.5 + 3 \times 10.1 = 217.8\,\text{kN/m}^2$$

From Eq. (2.21):

$$f = (1 - \sin(32)) \times 217.8 \tan(16) = 29.35\,\text{KN/m}^2$$

From Eq. (2.19) with $z = 10$ to $z = 13$:

$$Q_s = \frac{f_{z=10} + f_{z=13}}{2} \times (3.141 \times 0.8) \times 3$$

$$Q_s = \frac{25.12 + 29.35}{2} \times (3.141 \times 0.8) \times 3 = 205.24\,\text{kN}$$

$$Q_s = 205.24 + 315.60 = 520.8\,\text{kN}$$

Steps 4 and 5 (second pile length):
From Eq. (2.1):

$$Q_u = 4{,}002.93 + 520.8 = 4{,}523.73\,\text{kN}$$

From Eq. (2.40):

$$Q_{all} = \frac{4{,}523.73}{3} = 1{,}507.912\,\text{kN}$$

Steps 6–8 (second pile length):
From Eq. (2.41):

$$\eta = \left(\frac{4{,}002.93}{520.8}\right) = 7.686$$

Solving for Q_{wp} from Eq. (2.42):

$$0.025 = Q_{\mathrm{wp}}\left(\frac{\left(1+\dfrac{0.6}{7.686}\right)13}{0.502\times 21{,}000{,}000}+\frac{0.8\left(1-0.35^2\right)0.85}{0.502\times 10{,}000}\right.$$
$$\left.+\frac{0.8\left(1-0.35^2\right)3.41}{7.686\times \pi\times 0.8\times 13\times 10{,}000}\right)$$

Therefore,

$$Q_{\mathrm{wp}} = 209.86\,\mathrm{kN}$$

From Eq. (2.43):

$$Q_{\mathrm{w}} = 209.86\left(1+\frac{1}{7.686}\right) = 237.16\,\mathrm{kN}$$

Step 9 (second pile length):
Repeat Steps 1 through 8 for another assumed pile length.

Step 1 (third pile length):
Assume $L_2 = 19$ m.

Step 2 (third pile length):
Determine Q_{p} using Eq. (2.12):

$$N_q^{\ *} = \left(1.04529\left(\frac{32}{40}\right)^2 - 1.58056\left(\frac{32}{40}\right) + 0.6289\right)$$
$$\times\left(-0.32039\left(\frac{19}{0.8}\right)^2 + 14.2496\left(\frac{19}{0.8}\right) + 935.565\right)$$
$$= 36.58$$

From Eq. (2.11):

$$Q_{\mathrm{p}} = (187.5 + 9\times 10.1)\times 36.58\times 0.502 = 5{,}112.303\,\mathrm{kN}$$

Step 3 (third pile length):
Determine Q_{s}, the frictional resistance (skin friction) from Eq. (2.20):

$$L^{'} = 18\times 0.8 = 14.4\,\mathrm{m}$$

At $z = 0\,\mathrm{m}$:

$$\sigma_o^{'} = 0$$

From Eq. (2.21):

$$f = 0$$

At $z = 10\,\mathrm{m}$:

$$\sigma_o^{'} = 10 \times 18.75 = 187.5\,\mathrm{kN/m^2}$$

From Eq. (2.21):

$$f = (1 - \sin(30)) \times 187.5 \tan(16) = 25.12$$

From Eq. (2.19) with $z = 0$ to $z = 10$:

$$Q_s = \frac{f_{z=0} + f_{z=10}+}{2} \times (3.141 \times 0.8) \times 10$$

$$Q_s = \frac{0 + 25.12}{2} \times (3.141 \times 0.8) \times 10 = 315.60\,\mathrm{kN}$$

At $z = 10\,\mathrm{m}$

$$\sigma_o^{'} = 10 \times 18.75 = 187.5\,\mathrm{kN/m^2}$$

$$f = (1 - \sin(30)) \times 187.5 \tan(16) = 25.12$$

$$z = 14.4\,\mathrm{m}$$

$$\sigma_o^{'} = 187.5 + 4.4 \times 10.1 = 231.94\,\mathrm{kN/m^2}$$

So

$$f = (1 - \sin(32)) \times 231.94 \tan(16) = 31.26\,\mathrm{kN/m^2}$$

$$Q_s = \frac{f_{z=10} + f_{z=14.4}+}{2} \times (\pi \times 0.8) \times 4.4$$

$$Q_s = \frac{25.12 + 31.26}{2} \times (\pi \times 0.8) \times 4.4 = 311.73\,\mathrm{KN}$$

At $z = 19\,\mathrm{m}$

$$\sigma_o^{'} = 187.5 + 4.4 \times 10.1 = 231.94\,\text{kN/m}^2$$

$$f = (1 - \sin(32)) \times 231.94 \tan(16) = 31.26\,\text{kN/m}^2$$

$$Q_\text{s} = \frac{f_{z=14.4} + f_{z=19}}{2} \times (3.14 \times 0.8) \times (19 - 14.4)$$

$$Q_\text{s} = \frac{31.26 + 31.26}{2} \times (3.14 \times 0.8) \times (19 - 14.4) = 361.39\,\text{kN}$$

Steps 4 and 5 (third pile length):
From Eq. (2.1):

$$Q_\text{u} = 5,112.03 + 988.728 = 6,100.75\,\text{kN}$$

From Eq. (2.40):

$$Q_\text{all} = \frac{6,100.75}{3} = 2,033.58\,\text{kN}$$

Q_all is for assumed value $L_3 = 19$ m.

Steps 6–8 (third pile length):
From Eq. (2.41):

$$\eta = \left(\frac{5,112.03}{988.728}\right) = 5.170$$

Solving for Q_wp from Eq. (2.42):

$$0.025 = Q_\text{wp}\left(\frac{\left(1 + \frac{0.6}{5.170}\right)19}{0.502 \times 21,000,000} + \frac{0.8(1 - 0.35^2)0.85}{0.502 \times 10,000} + \frac{0.8(1 - 0.35^2)3.70}{5.170 \times \pi \times 0.8 \times 19 \times 10,000}\right)$$

Therefore,

$$Q_\text{wp} = 209.6\,\text{kN}$$

From Eq. (2.43):

$$Q_{\mathrm{w}} = 209.6\left(1 + \frac{1}{5.17}\right) = 250.14\,\mathrm{kN}$$

Step 9:
Repeat Steps 1 through 8 for as many pile lengths as needed.

2.4.6.2 foundationPro Solution

General Information and Pile Information Sections

Information is entered as provided in the design problem statement. Pile lengths of 5, 6, 7, 8, …, 20 m are considered for the development of design charts.

Soil Properties Section

In this section, we must specify the number of soil layers which is three in this problem. Then, thicknesses and physical soil properties are entered in the provided table as shown in Fig. 2.29.

Analysis Method Section

The required methods of analysis are selected in this section according to the provided problem statement.

Total number of soil layers 3

NOTES:

- To account for the groundwater table, input the effective unit weight (i.e., saturated unit weight - unit weight of water)

- To define soil depths: Z0 = 0 and Z3 > Z2 > Z1 > Z0 (eg., for soil 2: from depth = Z1; to depth = Z2)

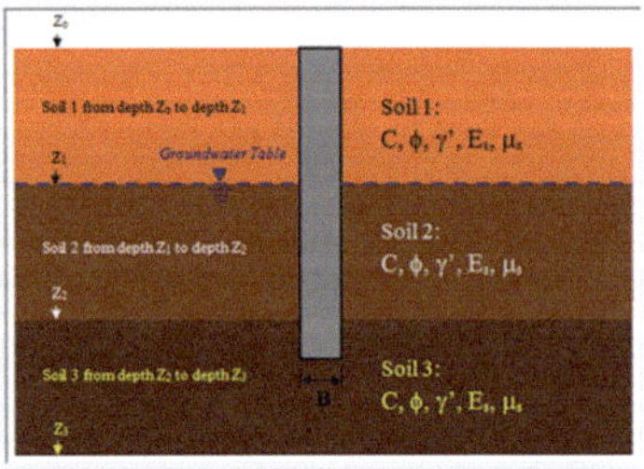

Soil Layer No.	From Depth (m)	To Depth (m)	Cohesion (kN/m^2)	Friction Angle (deg.)	Effective Unit Weight (kN/m^3)	Elastic Modulus of Soil (kN/m^2)	Poisson's Ratio
1	0	10	0	30	19	9000	0.35
2	10	19	0	32	10	10000	0.35
3	19	30	0	35	9.9	11000	0.35

Fig. 2.29 Thicknesses and physical properties of soil layers

Now we can save our progress and hit run to get the results. Axial capacity results at various pile lengths are summarized in Table 2.13. Also, ultimate skin friction design chart based on bearing capacity analyses is shown in Fig. 2.30.

Table 2.13 Axial capacity results at various pile lengths

L (m)	Q_s (kN)	Q_p (kN)	Q_u (kN)	Q_{all} (kN)	Q_{wp} (kN)	Q_{ws} (kN)	Q_w (kN)
5	80.00648	1,520.051	1,600.058	400.0144	187.5836	9.873289	197.4569
6	115.2093	1,846.238	1,961.448	490.3619	187.4195	11.69539	199.1149
7	156.8127	2,177.713	2,334.525	583.6313	187.2564	13.48396	200.7404
8	204.8166	2,513.572	2,718.388	679.5971	187.0937	15.24519	202.3389
9	259.221	2,852.914	3,112.135	778.0337	186.9307	16.98487	203.9156
10	320.0259	3,194.836	3,514.862	878.7156	186.7671	18.70842	205.4756
11	386.1409	3,601.148	3,987.289	996.8223	207.1797	22.2153	229.395
12	455.6378	3,803.035	4,258.673	1,064.668	206.9273	24.79175	231.7191
13	528.5465	4,003.987	4,532.534	1,133.133	206.6841	27.28335	233.9675
14	604.6678	4,200.221	4,804.888	1,201.222	206.4465	29.72023	236.1667
15	683.5709	4,395.163	5,078.734	1,269.684	206.215	32.07221	238.2873
16	763.2559	4,592.29	5,355.546	1,338.887	205.9949	34.23713	240.2321
17	842.2705	4,778.215	5,620.486	1,405.121	205.7814	36.2737	242.0551
18	921.6413	4,965.323	5,886.964	1,471.741	205.575	38.15792	243.7329
19	1,000.548	5,143.515	6,144.062	1,536.016	205.3718	39.95017	245.3219
20	1,079.695	7,350.498	8,430.194	2,107.548	225.9695	33.19207	259.1616

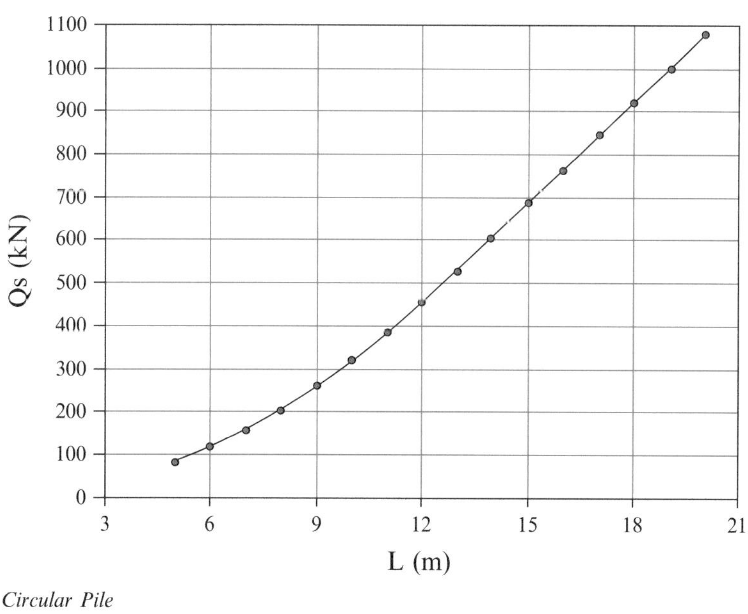

Fig. 2.30 Ultimate skin friction design chart (bearing capacity analysis)

Table 2.14 Useful design parameters

Property	Value	Unit
FS	3.0	–
E_p	22,500,000	kN/m^2
δ	0.8ϕ	–
S_e	0.025	m
ε	0.6	–

Table 2.15 Thicknesses and physical properties of sand layers

Soil layer	Thickness (m)	Cohesion (kN/m^2)	Friction angle (°)	Effective unit weight (kN/m^3)	Elastic modulus of soil (kN/m^2)	Poisson's ratio
1	12	0	30	19	12,500	0.30
2	30	0	34	18.1	19,000	0.35

2.4.7 *Square Pile in Sandy Soil (Two Sand Layers)*

Develop pile capacity design charts for a 0.6 m × 0.6 m square bored pile in cohesionless sandy soil. The sandy soil is not homogeneous. Use Meyerhof's method for the end bearing capacity and use Coyle and Castillo's method for the frictional skin resistance. Many useful design parameters are provided in Table 2.14. Thicknesses and physical properties of the two sand layers are listed in Table 2.15.

2.4.7.1 Hand Solution

To develop axial pile capacity design charts, one must follow the steps below:

Step 1:
Assume $L_1 = 8$ m.

Step 2:
Determine end bearing capacity, Q_p, from Eq. (2.3):

$$N_q^* = 0.3147 \times e^{0.1752(30)} = 60.33$$

From Eq. (2.2):

$$Q_p = 0.6 \times 0.6 \times 19 \times 8 \times 60.33 = 3,301.256$$

From Eq. (2.4):

$$q_1 \times A_p = 0.5 \times 100 \times 60.33 \tan(30) \times 0.36 = 626.96\,\text{kN}$$

$$Q_{\mathrm{p}} = 626.96\,\mathrm{KN}$$

Step 3:
Determine Q_{s} using Eq. (2.26):

$$K = \left(-1.32484\left(\frac{30^\circ}{36^\circ}\right) + 1.03116\right) \times \left(0.2879\left(\frac{8}{0.6}\right) - 13.572\right) = 0.709$$

From Eq. (2.24):

$$Q_{\mathrm{s}} = \left(\frac{(19 \times 8)}{2}\right)0.709\tan(0.8 \times 30) \times 4 \times 0.6 \times 8 = 460.62$$

Steps 4 and 5:
From Eq. (2.1):

$$Q_{\mathrm{u}} = 626.96 + 460.62 = 1{,}087.58\,\mathrm{kN}$$

From Eq. (2.40):

$$Q_{\mathrm{all}} = \frac{1{,}087.58}{3} = 362.52\,\mathrm{kN}$$

Steps 6–8:
From Eq. (2.41):

$$\eta = \left(\frac{626.96}{460.62}\right) = 1.36$$

Solving for Q_{wp} from Eq. (2.42):

$$0.025 = Q_{\mathrm{wp}}\left(\frac{\left(1 + \frac{0.6}{1.36}\right)8}{0.36 \times 22{,}500{,}000} + \frac{0.6(1 - 0.3^2)0.85}{0.36 \times 12{,}500} + \frac{0.6(1 - 0.3^2)3.27}{1.36 \times 4 \times 0.6 \times 8 \times 12{,}500}\right)$$

Therefore

$$Q_{\mathrm{wp}} = 227.27\,\mathrm{kN}$$

From Eq. (2.43):

$$Q_{\mathrm{w}} = 227.27\left(1 + \frac{1}{1.36}\right) = 394.38\,\mathrm{kN}$$

Step 9:
Repeat Steps 1 through 8 for other pile lengths. Below is another solution for a second assumed pile length.

Step 1 (second pile length):
Assume $L_2 = 16$ m.

Step 2 (second pile length):
Determine Q_{p}, the end bearing capacity from Eq. (2.3):

$$N_q^{\ *} = 0.3147 \times e^{0.1752(34)} = 121.59$$

From Eq. (2.2):

$$Q_{\mathrm{p}} = 0.6 \times 0.6 \times (12 \times 19 + 18.1 \times 4) \times 121.59 = 13,149.22$$

From Eq. (2.4):

$$q_1 \times A_{\mathrm{p}} = 0.5 \times 100 \times 121.59 \tan(34) \times 0.36 = 1,476.24\,\mathrm{kN}$$

$$Q_{\mathrm{p}} = 1,476.24\,\mathrm{kN}$$

Step 3 (second pile length):
Determine Q_{s}, the frictional resistance (skin friction) from Eq. (2.26) with $z = 0$–12:

$$K = \left(-1.32484\left(\frac{30^\circ}{36^\circ}\right) + 1.03116\right) \times \left(0.2879\left(\frac{16}{0.6}\right) - 13.572\right) = 0.4295$$

From Eq. (2.24):

$$Q_{\mathrm{s}} = \left(\frac{(19 \times 12)}{2}\right) 0.4295 \tan(0.8 \times 30) \times 4 \times 0.6 \times 12 = 627.8\,\mathrm{kN}$$

From Eq. (2.26) with $z = 12$–16:

$$K = \left(-1.32484\left(\frac{34^\circ}{36^\circ}\right) + 1.03116\right) \times \left(0.2879\left(\frac{4}{0.6}\right) - 13.572\right) = 1.3$$

$$Q_s = \left(\frac{((19 \times 12) + (4 \times 18.1) + (12 \times 19))}{2}\right) 1.3 \tan(0.8 \times 34) \times 4 \times 0.6 \times 4$$
$$= 1,694.51\,\text{kN}$$

Steps 4 and 5 (second pile length):
From Eq. (2.1):

$$Q_u = 1,476.24 + 2,322.33 = 3,798.5788\,\text{kN}$$

From Eq. (2.40):

$$Q_{all} = \frac{3,798.5788}{3} = 1,266.19\,\text{kN}$$

Steps 6–8 (second pile length):
From Eq. (2.41):

$$\eta = \left(\frac{1,476.24}{2,322.33}\right) = 0.63$$

Solving for Q_{wp} from Eq. (2.42):

$$0.025 = Q_{wp}\left(\frac{\left(1 + \frac{0.6}{0.63}\right)16}{0.36 \times 22,500,000} + \frac{0.6(1 - 0.35^2)0.85}{0.36 \times 19,000} + \frac{0.6(1 - 0.35^2)3.807}{0.63 \times 4 \times 0.6 \times 16 \times 19,000}\right)$$

Therefore,

$$Q_{wp} = 348.46\,\text{kN}$$

From Eq. (2.43):

$$Q_w = 348.46\left(1 + \frac{1}{0.63}\right) = 901.575\,\text{kN}$$

Step 9:
Repeat the above steps for as many pile lengths as needed to develop required design charts. More solutions are obtained from foundationPro.

2.4.7.2 foundationPro Solution

General Information Section

The SI unit is selected in this section, and then the safety factor of 3.0 is entered in the provided textbox.

Pile Information Section

Pile Information is entered in this section. Pile lengths that will be considered in the analysis are from 5 m to 20 m with 16 points.

Soil Properties Section

The thicknesses and physical properties of sand layers are entered in this section as illustrated in Fig. 2.31.

Analysis Methods Section

The required methods are selected in this section as provided in the problem statement. Coyle and Castello's method is selected for skin friction in sand, and then Meyerhof's method for end bearing capacity in sand is selected as well.

After saving our progress and performing the analyses, one can view the Output section and navigate through the output results. Axial capacities at various pile lengths are summarized in Table 2.16. Also, Figs. 2.32 and 2.33

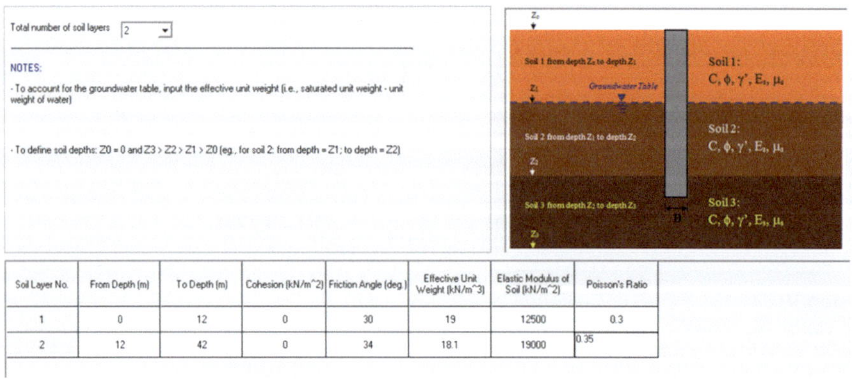

Soil Layer No.	From Depth (m)	To Depth (m)	Cohesion (kN/m^2)	Friction Angle (deg.)	Effective Unit Weight (kN/m^3)	Elastic Modulus of Soil (kN/m^2)	Poisson's Ratio
1	0	12	0	30	19	12500	0.3
2	12	42	0	34	18.1	19000	0.35

Fig. 2.31 Properties of sand layers

Table 2.16 Axial pile capacities at various pile lengths

L (m)	Q_s (kN)	Q_p (kN)	Q_u (kN)	Q_{all} (kN)	Q_{wp} (kN)	Q_{ws} (kN)	Q_w (kN)
5	206.6284	626.9897	833.6181	277.8727	232.5931	76.65249	309.2456
6	284.7664	626.9897	911.7561	303.9187	230.7207	104.7888	335.5095
7	370.2057	626.9897	997.1954	332.3985	228.9183	135.1647	364.0829
8	460.8167	626.9897	1,087.806	362.6021	227.1952	166.981	394.1762
9	554.4695	626.9897	1,181.459	393.8197	225.5609	199.4716	425.0325
10	649.0345	626.9897	1,276.024	425.3414	224.0251	231.9017	455.9268
11	742.3818	626.9897	1,369.371	456.4572	222.598	263.5652	486.1632
12	832.3817	626.9897	1,459.371	486.4571	221.2908	293.7822	515.0729
13	1,249.338	1,476.257	2,725.595	908.5317	354.2315	299.7816	654.013
14	1,642.357	1,476.257	3,118.614	1,039.538	348.6947	387.9277	736.6224
15	2,003.548	1,476.257	3,479.805	1,159.935	343.8554	466.6738	810.5292
16	2,318.926	1,476.257	3,795.183	1,265.061	339.764	533.7061	873.47
17	2,584.546	1,476.257	4,060.804	1,353.601	336.3632	588.8853	925.2485
18	2,797.257	1,476.257	4,273.514	1,424.505	333.6128	632.1395	965.7522
19	2,939.5	1,476.257	4,415.757	1,471.919	331.6576	660.3913	992.0489
20	3,016.784	1,476.257	4,493.041	1,497.68	330.3923	675.1682	1,005.56

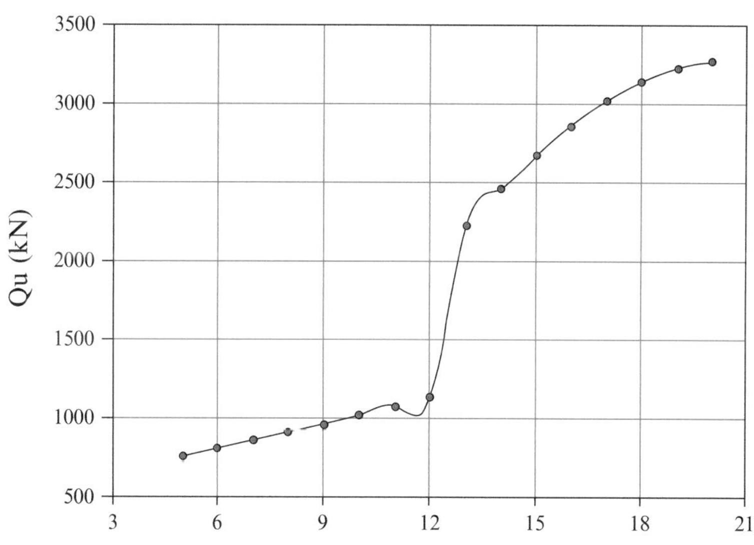

Square Pile

Fig. 2.32 Ultimate load carrying capacity design chart (bearing capacity analysis)

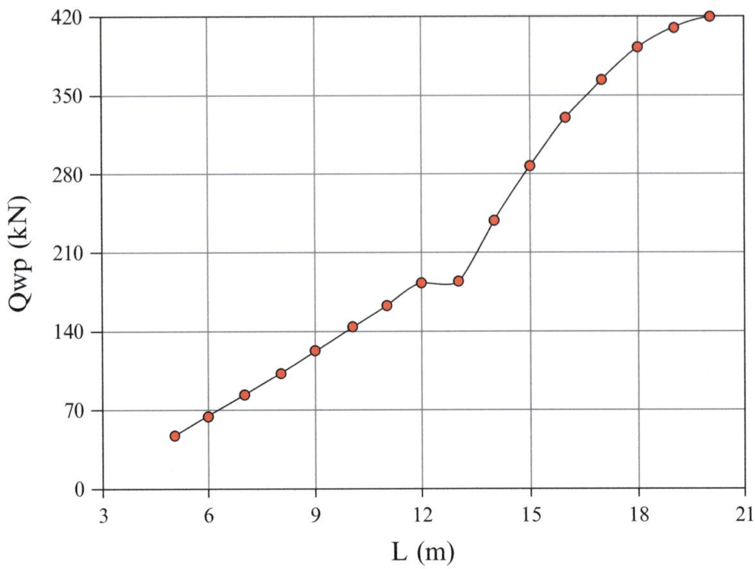

Fig. 2.33 Working load carried by pile tip design chart (elastic settlement analysis)

Table 2.17 Thicknesses and physical properties of clay layers

Soil layer	Thickness (ft)	Cohesion (lb/ft^2)	Friction angle (°)	Effective unit weight (lb/ft^3)	Elastic modulus of soil (lb/ft^2)	Poisson's ratio
1	40	835	0	47.6	135,000	0.3
2	60	1,350	0	53	166,000	0.3

show some obtained design charts based on bearing capacity and elastic settlement analyses, respectively.

2.4.8 Circular Pile in Clayey Soil (Two Clay Layers)

Develop pile capacity design charts (bearing capacity and elastic settlement) for a circular bored pile in nonhomogeneous clayey soil that has a diameter of 2.5 ft. Soil properties and thicknesses of the clay layers are provided in Table 2.17. Use Vesic's method for the end bearing capacity calculations and use α method for the skin frictional resistance calculations. Other required design parameters are provided in Table 2.18.

Table 2.18 Other useful design parameters

Property	Value	Unit
FS	4.0	–
E_p	470,000,000	lb/ft^2
δ	0.5ϕ	–
S_e	1	in.
ε	0.6	–

2.4.8.1 Hand Solution

One must follow the steps below to develop the required capacity design charts:

Step 1:
Assume $L_1 = 20$ ft.

Step 2:
Determine Q_p in clay from Eq. (2.16):

$$I_r = 347\left(\frac{831}{2,000}\right) - 33 = 111.17$$

From Eq. (2.15):

$$N_c^* = \frac{4}{3}(\ln(111.17) + 1) + \frac{\pi}{2} + 1 = 10.18$$

From Eq. (2.14):

$$Q_p = 4.90 \times 831 \times 10.18 = 41,548.89\,\mathrm{lb}$$

Step 3:
Determine Q_s using Eq. (2.28):

$$\begin{aligned}\alpha = 0.026 \times \left(\frac{831}{2,000}\right)^4 - 0.02233\left(\frac{831}{2,000}\right)^3 + 0.7498\left(\frac{831}{2,000}\right)^2 \\ - 1.198\left(\frac{831}{2,000}\right) + 1.1187 \\ = 0.756\end{aligned}$$

From Eq. (2.27):

$$f = 0.756 \times 831 = 628.67$$

From Eq. (2.29):

$$Q_s = 628.67 \times \pi \times 2.5 \times 20 = 98,683.08\,\text{lb}$$

Steps 4 and 5:
From Eq. (2.1):

$$Q_u = 41,548.89 + 98,683.08 = 140,231.97\,\text{lb}$$

From Eq. (2.40):

$$Q_{all} = \frac{140,231.97}{4} = 35,057.99\,\text{lb}$$

Steps 6–8:
From Eq. (2.41):

$$\eta = \left(\frac{41,548.89}{98,683.08}\right) = 0.42$$

Given that $S_e = 0.083$ ft and solving for Q_{wp} from Eq. (2.42):

$$0.083 = Q_{wp}\left(\frac{\left(1 + \frac{0.6}{0.42}\right)20}{4.90 \times 4,700,000,000} + \frac{2.5(1 - 0.3^2)0.85}{4.90 \times 135,000} + \frac{2.5(1 - 0.3^2)2.98}{0.42 \times \pi \times 2.5 \times 20 \times 135,000}\right)$$

Therefore,

$$Q_{wp} = 22,519.11\,\text{lb}$$

From Eq. (2.43):

$$Q_w = 22,519.11\left(1 + \frac{1}{0.42}\right) = 76,136.06\,\text{kN}$$

Step 9:
Repeat the above steps by varying the assumed pile length to obtain it new axial capacities. Below is a solution for a second assumed pile length.

Step 1 (second pile length):
Assume $L_2 = 45$ ft.

Step 2 (second pile length):
Determine end bearing capacity.
From Eq. (2.16):

$$I_r = 347\left(\frac{1,350}{2,000}\right) - 33 = 201.225$$

From Eq. (2.15):

$$N_c^* = \frac{4}{3}(\ln(201.225) + 1) + \frac{\pi}{2} + 1 = 10.97$$

From Eq. (2.14):

$$Q_p = 4.90 \times 1,350 \times 10.97 = 72,610.83\,\text{lb}$$

Step 3 (second pile length):
Determine Q_s, the frictional resistance (skin friction) from Eq. (2.28):
From $z = 0$–40 m:

$$\alpha = 0.026 \times \left(\frac{831}{2,000}\right)^4 - 0.02233\left(\frac{831}{2,000}\right)^3 + 0.7498\left(\frac{831}{2,000}\right)^2 - 1.198\left(\frac{831}{2,000}\right) + 1.1187 = 0.749$$

From Eq. (2.27):

$$f = 0.749 \times 831 = 622.87$$

From Eq. (2.29):

$$Q_s = 622.87 \times \pi \times 2.5 \times 40 = 195,682.231\,\text{lb}$$

From Eq. (2.28):
From $z = 40$–45 ft:

$$\alpha = 0.026 \times \left(\frac{1,350}{2,000}\right)^4 - 0.02233\left(\frac{1,350}{2,000}\right)^3 + 0.7498\left(\frac{1,350}{2,000}\right)^2 - 1.198\left(\frac{1,350}{2,000}\right) + 1.1187 = 0.650$$

From Eq. (2.27):

$$f = 0.650 \times 1,350 = 877.78$$

From Eq. (2.29):

$$Q_s = 877.78 \times \pi \times 2.5 \times 5 = 34,470.34\,\text{lb}$$

Steps 4 and 5 (second pile length):
From Eq. (2.1):

$$Q_u = 72,610.83 + 230,152.57 = 302,763.408\,\text{lb}$$

From Eq. (2.40):

$$Q_{all} = \frac{290,638.644}{4} = 75,690.85\,\text{lb}$$

Steps 6–8 (second pile length):
From Eq. (2.41):

$$\eta = \left(\frac{72,610.83}{230,152.57}\right) = 0.315$$

Solving for Q_{wp} from Eq. (2.42):

$$0.083 = Q_{wp}\left(\frac{\left(1+\frac{0.6}{0.315}\right)45}{4.90 \times 4,700,000,000} + \frac{2.5(1-0.3^2)0.85}{4.90 \times 166,000} + \frac{2.5(1-0.3^2)3.48}{0.315 \times \pi \times 2.5 \times 45 \times 166,000}\right)$$

Therefore,

$$Q_{wp} = 29,542.390\,\text{lb}$$

From Eq. (2.43):

$$Q_w = 29,542.390\left(1 + \frac{1}{0.315}\right) = 123,327.755\,\text{lb}$$

Step 9:
Repeat Steps 1 through 8 for other pile lengths to obtain enough data points for the development of the design charts.

2.4.8.2 foundationPro Solution

General Information Section

Select the BS units to be used throughout the analyses, and then enter the safety factor of 4 in the provided textbox.

Pile Information Section

Pile information is entered in this section. Various pile lengths are considered in this problem as illustrated in Fig. 2.34.

Soil Properties Section

Number of soil layers, thicknesses, and soil properties are entered in this section as shown in Fig. 2.35.

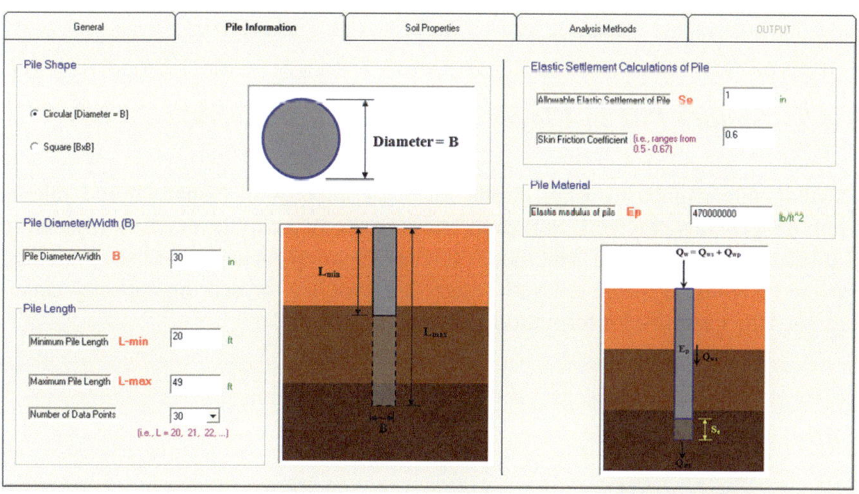

Fig. 2.34 Pile Information section

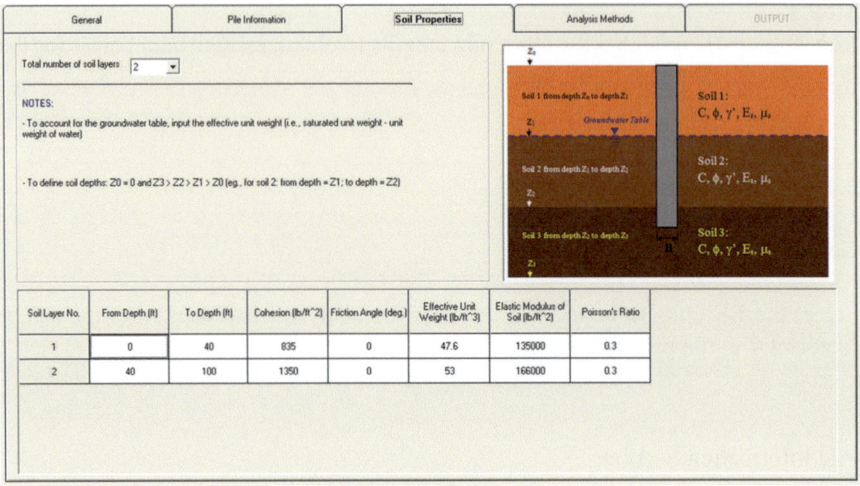

Fig. 2.35 Soil properties used for this design problem

Analysis Methods Section

The user must specify the methods of analysis to be used in this section. In this problem, we use Vesic's method in clay for skin friction computations and we use α method in clay for end bearing capacity computations.

Now we can save our progress and hit run to perform the analyses and get the results. Summary of the axial pile capacity results at various pile lengths (20, 21, 22, 23, . . ., 49 ft) is shown in Table 2.19. A view of the Output section in the Pile-1 application of foundationPro is also shown in Fig. 2.36.

2.4.9 *Square Pile in Clayey Soil (Two Clay Layers)*

Develop pile capacity design charts for a 24 in. × 24 in. square bored pile in nonhomogeneous clay soil. Soil properties for the two clay layers are provided in Table 2.20. Use Meyerhof's method for the end bearing capacity and use α^* method for the frictional resistance. Use a safety factor of 4.0 for bearing capacity calculations. Other required information is listed in Table 2.21.

2.4.9.1 Hand Solution

The steps below describe the procedure the user must follow to determine the axial capacity of the square pile in the two clay layers:

Table 2.19 Axial capacities at various pile lengths

L (ft)	Q_s (lb)	Q_p (lb)	Q_u (lb)	Q_{all} (lb)	Q_{wp} (lb)	Q_{ws} (lb)	Q_w (lb)
20	96,264.62	41,802.31	138,066.9	34,516.73	22,670.02	52,205.75	74,875.77
21	101,077.8	41,802.31	142,880.2	35,720.04	22,622.65	54,701.49	77,324.14
22	105,891.1	41,802.31	147,693.4	36,923.35	22,575.97	57,188.09	79,764.06
23	110,704.3	41,802.31	152,506.6	38,126.66	22,529.93	59,665.63	82,195.56
24	115,517.5	41,802.31	157,319.9	39,329.96	22,484.47	62,134.15	84,618.62
25	120,330.8	41,802.31	162,133.1	40,533.27	22,439.53	64,593.71	87,033.25
26	125,144	41,802.31	166,946.3	41,736.58	22,395.07	67,044.36	89,439.43
27	129,957.2	41,802.31	171,759.5	42,939.89	22,351.05	69,486.12	91,837.16
28	134,770.5	41,802.31	176,572.8	44,143.19	22,307.42	71,919.02	94,226.44
29	139,583.7	41,802.31	181,386	45,346.5	22,264.15	74,343.08	96,607.23
30	144,396.9	41,802.31	186,199.2	46,549.81	22,221.22	76,758.33	98,979.55
31	149,210.2	41,802.31	191,012.5	47,753.12	22,178.58	79,164.77	101,343.4
32	154,023.4	41,802.31	195,825.7	48,956.42	22,136.23	81,562.41	103,698.6
33	158,836.6	41,802.31	200,638.9	50,159.73	22,094.13	83,951.27	106,045.4
34	163,649.9	41,802.31	205,452.2	51,363.04	22,052.26	86,331.33	108,383.6
35	168,463.1	41,802.31	210,265.4	52,566.35	22,010.6	88,702.6	110,713.2
36	173,276.3	41,802.31	215,078.6	53,769.66	21,969.13	91,065.08	113,034.2
37	178,089.5	41,802.31	219,891.9	54,972.96	21,927.84	93,418.76	115,346.6
38	182,902.8	41,802.31	224,705.1	56,176.27	21,886.71	95,763.62	117,650.3
39	187,716	41,802.31	229,518.3	57,379.58	21,845.73	98,099.67	119,945.4
40	192,529.2	41,802.31	234,331.5	58,582.89	21,804.87	100,426.9	122,231.8
41	198,745.5	72,773.8	271,519.3	67,879.83	29,591.76	80,815.22	110,407
42	205,030.2	72,773.8	277,804	69,451.01	29,519.44	83,167	112,686.4
43	211,260.8	72,773.8	284,034.6	71,008.64	29,449.5	85,491.28	114,940.8
44	217,428.6	72,773.8	290,202.4	72,550.59	29,381.97	87,785.44	117,167.4
45	223,781.7	72,773.8	296,555.5	74,138.88	29,311.52	90,133.86	119,445.4
46	229,935.3	72,773.8	302,709.1	75,677.27	29,246.16	92,405.86	121,652
47	236,140.2	72,773.8	308,913.9	77,228.49	29,180.57	94,686.62	123,867.2
48	242,396.4	72,773.8	315,170.2	78,792.54	29,114.74	96,975.94	126,090.7
49	248,703.9	72,773.8	321,477.7	80,369.42	29,048.67	99,273.6	128,322.3

Step 1:
Assume $L_1 = 10$ ft.

Step 2:
Determine Q_p using Eq. (2.13):

$$Q_p = 9c_u A_p$$

$$Q_p = 9 \times 900 \times 2^2 = 32,400\,\text{lb}$$

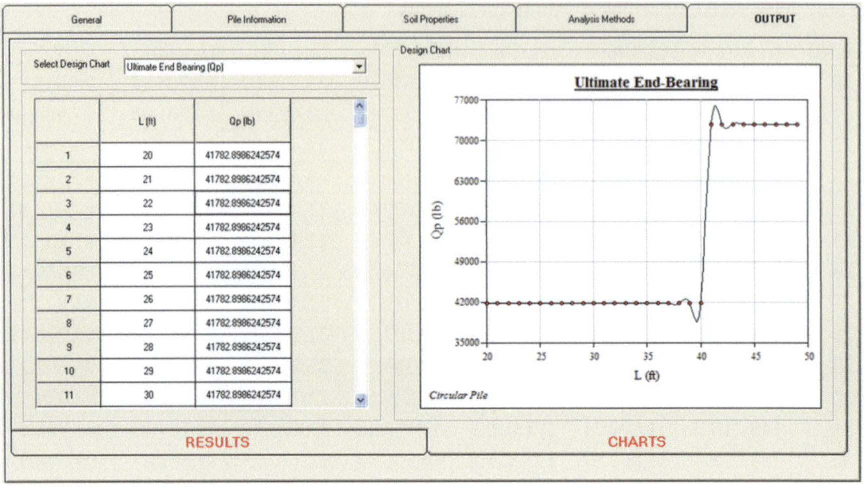

Fig. 2.36 Output section for the solved problem

Table 2.20 Soil properties and thicknesses of clay layers

Soil layer	Thickness (ft)	Cohesion (lb/ft^2)	Friction angle (°)	Effective unit weight (lb/ft^3)	Elastic modulus of soil (lb/ft^2)	Poisson's ratio
1	20	900	0	50	135,000	0.3
2	40	1,400	0	55	166,000	0.3

Table 2.21 Useful design parameters

Property	Value	Unit
E_p	470,000,000	lb/ft^2
δ	0.5ϕ	–
S_e	1	in.
ε	0.6	–

Step 3:
Determine Q_s frictional resistance (skin friction) from Eq. (2.31):

$$\alpha^* = 0.5\left(\frac{\left(\frac{10 \times 50}{2}\right)}{900}\right)^{0.45} = 0.28$$

From Eq. (2.30):

$$f = 0.28 \times 900 = 252.85$$

From Eq. (2.32):

$$Q_s = 0.28 \times 900 \times 4 \times 2 \times 10 = 20,160\,\text{lb}$$

Steps 4 and 5:
From Eq. (2.1):

$$Q_u = 32,400 + 20,160 = 52,560\,\text{lb}$$

From Eq. (2.40):

$$Q_{all} = \frac{52,560}{4} = 13,140\,\text{lb}$$

Steps 6–8:
From Eq. (2.41):

$$\eta = \left(\frac{32,400}{20,160}\right) = 1.6$$

Given that $S_e = 0.083$ ft and solving for Q_{wp} from Eq. (2.42):

$$0.083 = Q_{wp}\left(\frac{\left(1 + \frac{0.6}{1.6}\right)10}{4 \times 4,700,000,000} + \frac{2(1 - 0.3^2)0.85}{4 \times 135,000} + \frac{2(1 - 0.3^2)2.78}{1.6 \times 4 \times 2 \times 10 \times 135,000}\right)$$

Therefore,

$$Q_{wp} = 26,378.55\,\text{lb}$$

From Eq. (2.43):

$$Q_w = 26,378.55\left(1 + \frac{1}{1.6}\right) = 424,865.14\,\text{lb}$$

Step 9:
Repeat the above steps with a different pile length. The solution below is for another assumed pile length.

Step 1 (second pile length):
Assume $L_2 = 30$ ft.

Step 2 (second pile length):
From Eq. (2.13):

$$Q_{\mathrm{p}} = 9c_{\mathrm{u}}A_{\mathrm{p}}$$

$$Q_{\mathrm{p}} = 9 \times 1{,}400 \times 2^2 = 50{,}400\,\mathrm{lb}$$

Step 3 (second pile length):
Determine Q_{s}, the frictional resistance (skin friction) using Eq. (2.31) with $z = 0$–20 ft:

$$\alpha^* = 0.5\left(\frac{20 \times 50}{900}\right)^{0.45} = 0.52$$

From Eq. (2.30):

$$f = 0.52 \times 900 = 471.84$$

From Eq. (2.32):

$$Q_{\mathrm{s}} = 0.52 \times 900 \times 4 \times 2 \times 20 = 74{,}880\,\mathrm{lb}$$

From Eq. (2.31) with $z = 20$–30:

$$\alpha^* = 0.5\left(\frac{\left(\frac{10 \times 55 + 20 \times 50 + 20 \times 50}{2}\right)}{1{,}400}\right)^{0.45} = 0.47$$

From Eq. (2.30):

$$f = 0.47 \times 1{,}400 = 658$$

From Eq. (2.32):

$$Q_{\mathrm{s}} = 0.53 \times 1{,}400 \times 4 \times 2 \times 10 = 52{,}640\,\mathrm{lb}$$

Steps 4 and 5 (second pile length):
From Eq. (2.1):

$$Q_{\mathrm{u}} = 50{,}400 + 127{,}520 = 177{,}920\,\mathrm{lb}$$

From Eq. (2.40):

$$Q_{\text{all}} = \frac{177,920}{4} = 44,480 \text{ lb}$$

Steps 6–8 (second pile length):
From Eq. (2.41):

$$\eta = \left(\frac{50,400}{127,520}\right) = 0.39$$

Solving for Q_{wp} from Eq. (2.42):

$$0.083 = Q_{\text{wp}}\left(\frac{\left(1+\frac{0.6}{0.39}\right)30}{4\times 4,700,000,000} + \frac{2(1-0.3^2)0.85}{4\times 166,000} + \frac{2(1-0.3^2)3.35}{0.39\times 4\times 2\times 30\times 166,000}\right)$$

Therefore,

$$Q_{\text{wp}} = 30,571.83\,\text{lb}$$

From Eq. (2.43):

$$Q_{\text{w}} = 30,571.83\left(1+\frac{1}{0.39}\right) = 108,961.144\,\text{lb}$$

Step 9:
Repeat the above procedure as many times as needed to develop additional data point for the design charts.

2.4.9.2 foundationPro Solution

General Information Section

Enter a safety factor of 4 in the provided textbox, and then select the BS units to be used in the analyses of this problem.

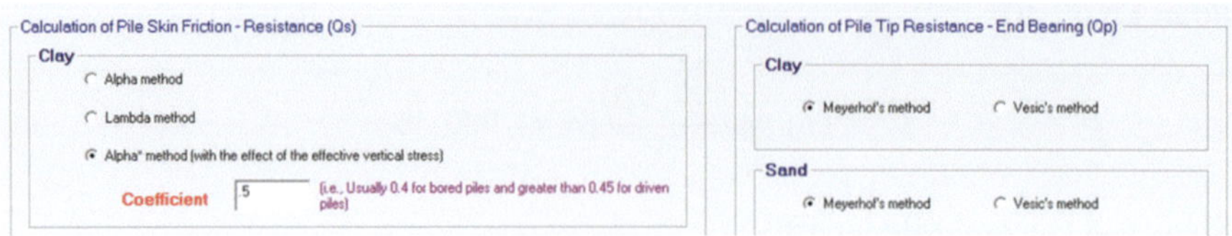

Fig. 2.37 Methods of analysis used for this design problem

Pile Information Section

Pile information is entered in this section as provided in the problem statement. Pile lengths from 10 ft to 34 ft will be considered with 34 data points including the minimum and the maximum pile lengths.

Soil Properties Section

The number of soil layers which is two, the thicknesses, and the physical properties of the clay layers must be entered in the provided table in this section.

Analysis Method Section

In this problem, we use the Alpha (α*) method in clay soil for the skin friction resistance. We enter 0.5 for the coefficient (see Fig. 2.37) to stay consistent with the hand solution. This coefficient can be obtained from Table 2.1. Next, we select the Meyerhof's method under clay soil for the end bearing capacity.

Now we can save our progress and hit run to get the results. Summary of the pile capacity results based on both bearing capacity and elastic settlement analyses is shown in Table 2.22. Total working load design chart based on elastic settlement analyses is shown in Fig. 2.38.

2.4.10 Square Pile in Rock

Determine the design capacity of a square Pile in rock, with a width of 1 m. Consider a factor of safety of 3.0. The rock has unconfined compression strength of 40,000 kN/m^2 and friction angle of 40°.

Table 2.22 Pile capacity results (bearing capacity and elastic settlement analyses)

L (ft)	Q_s (lb)	Q_p (lb)	Q_u (lb)	Q_{all} (lb)	Q_{wp} (lb)	Q_{ws} (lb)	Q_w (lb)
10	20,228.63	32,400	52,628.63	13,157.16	26,330.53	16,439.21	42,769.73
11	23,226.61	32,400	55,626.61	13,906.65	26,180.95	18,768.36	44,949.31
12	26,349.91	32,400	58,749.91	14,687.48	26,036.42	21,174.61	47,211.04
13	29,592.68	32,400	61,992.68	15,498.17	25,896.29	23,652.48	49,548.77
14	32,949.74	32,400	65,349.74	16,337.44	25,760.02	26,197.1	51,957.13
15	36,416.54	32,400	68,816.54	17,204.14	25,627.19	28,804.13	54,431.32
16	39,988.98	32,400	72,388.98	18,097.25	25,497.43	31,469.64	56,967.07
17	43,663.38	32,400	76,063.38	19,015.84	25,370.43	34,190.08	59,560.51
18	47,436.38	32,400	79,836.38	19,959.09	25,245.93	36,962.2	62,208.12
19	51,304.93	32,400	83,704.93	20,926.23	25,123.68	39,782.99	64,906.68
20	55,266.23	32,400	87,666.23	21,916.56	25,003.51	42,649.69	67,653.19
21	60,188.79	50,400	110,588.8	27,647.2	32,134.64	38,375.89	70,510.54
22	65,206.61	50,400	115,606.6	28,901.65	31,985.42	41,382.16	73,367.58
23	70,381.48	50,400	120,781.5	30,195.37	31,837.18	44,459.27	76,296.45
24	75,662.65	50,400	126,062.6	31,515.66	31,692.02	47,577.42	79,269.44
25	81,082.32	50,400	131,482.3	32,870.58	31,548.1	50,753.84	82,301.94
26	86,582.83	50,400	136,982.8	34,245.71	31,407.65	53,955.61	85,363.26
27	92,254.51	50,400	142,654.5	35,663.63	31,266.52	57,231.7	88,498.22
28	97,985.22	50,400	148,385.2	37,096.3	31,129.15	60,519.77	91,648.93
29	103,816.9	50,400	154,216.9	38,554.23	30,993.49	63,842.23	94,835.73
30	109,803.8	50,400	160,203.8	40,050.95	30,857.32	67,227.2	98,084.52
31	115,813.7	50,400	166,213.7	41,553.43	30,725.35	70,603.5	101,328.9
32	121,966.7	50,400	172,366.7	43,091.68	30,592.98	74,034.24	104,627.2
33	128,263.3	50,400	178,663.3	44,665.82	30,460.17	77,518.28	107,978.5
34	134,627	50,400	185,027	46,256.76	30,329.41	81,015.05	111,344.5

2.4.10.1 Hand Solution

Step 1:
First, we determine the force that the soil under the pile tip can carry using Eqs. (2.17) and (2.18):

$$N_\phi = \tan^2\left(45 + \frac{40}{2}\right) = 4.5989$$

$$Q_p = 40,000(4.5989 + 1) = 223,956.40\,\text{kN}$$

Step 2:
Since the skin friction around the pile is ignored, Q_s is considered equal to zero. Therefore,

$$Q_u = Q_p.$$
$$Q_u = 223,956.40\ \text{kN}.$$

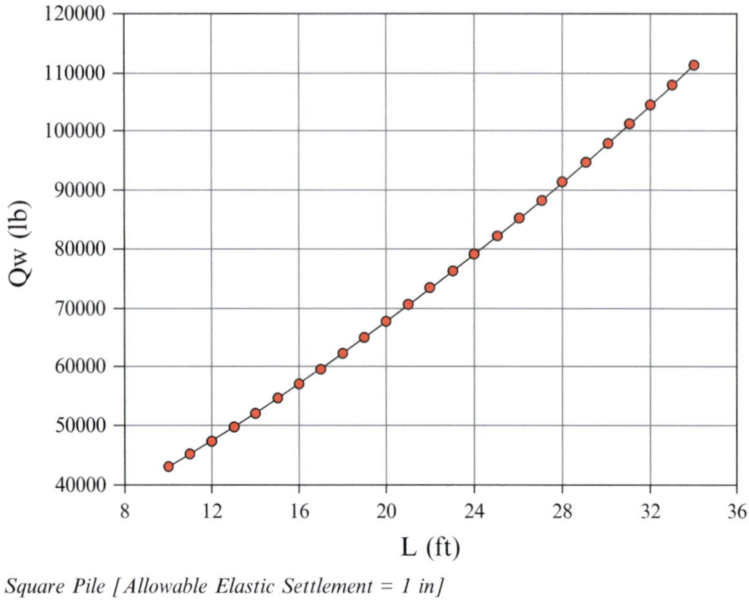

Fig. 2.38 Total working load (elastic settlement) design chart

Step 3:
Using the factor of safety given in the problem we can find Q_{all}:

$$Q_{all} = 223,956.40/3 = 74,652.13\,\text{kN}$$

2.4.10.2 foundationPro Solution

Using Pile-2 application of foundationPro, one can determine the axial capacity of the square pile extending to rock. In this problem, one must enter the design parameters in the Input section as in Fig. 2.39. After that, we hit run to perform the analysis and view the results.

As shown in Fig. 2.40, one can see that the program will calculate soil resistance at the pile tip (Q_p) and consider this value equal to the ultimate resistance of the soil (Q_u). In addition, the program will calculate the allowable (Q_{all}) force that the foundation can resist based on the safety factor given in the problem statement.

2.4.11 Suggested Projects

In this section, you will find some suggested design problems to allow the reader practice the concepts and the ideas discussed in the previous sections. These

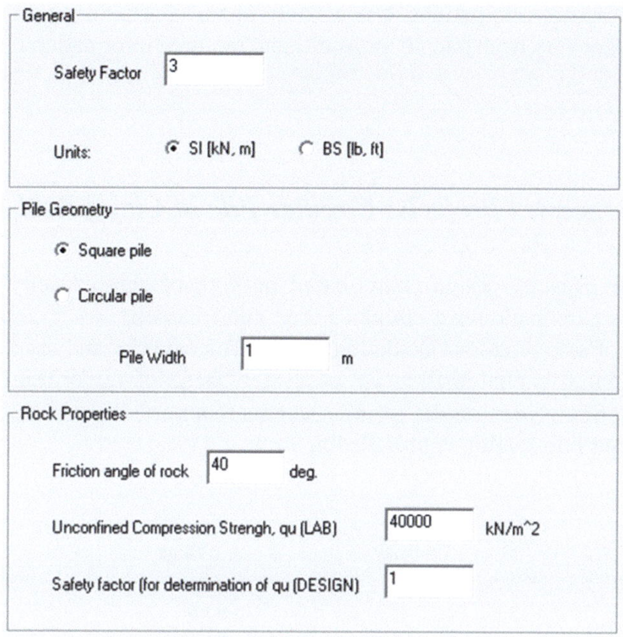

Fig. 2.39 Input section in Pile-2 application

Fig. 2.40 Allowable load of a square pile extending to rock

suggested problems will cover a variety of different pile shapes in homogeneous and nonhomogeneous soils.

2.4.12 Suggested Projects: Circular Pile in Sandy Soil

Develop pile capacity design charts (bearing capacity and elastic settlement) for a 750-mm-diameter circular bored pile in sandy soil. Use Meyerhof's method for the end bearing capacity calculations and use the critical depth method for the skin frictional resistance calculations. Use a safety factor of 3.5. Show your results for pile lengths that vary from 6 to 20 m. Additional required information is provided in Table 2.23.

2.4.13 *Suggested Projects: Circular Pile in Clayey Soil*

Develop pile capacity design charts based on both bearing capacity and elastic settlement requirements for a circular bored pile (diameter = 4 ft) in clayey soil. Use Meyerhof's method for calculating end bearing capacity and use λ method for calculating skin frictional resistance. Use a safety factor of 4.0 for bearing capacity calculations. Show your results for pile lengths that vary from 15 to 60 ft. Additional required information is provided in Table 2.24.

Table 2.23 Soil properties and other required design parameters

Property	Value	Unit
c	0	kN/m^2
ϕ	30	°
γ	20	kN/m^3
E_s	25,000	kN/m^2
μ_s	0.35	–
E_p	21,000,000	kN/m^2
δ	0.5ϕ	–
S_e	25	mm
ε	0.6	–
$(L'/B)_{cr}$	15	–

Table 2.24 Soil properties and other design parameters

Property	Value	Unit
c	1,500	lb/ft^2
ϕ	2	°
γ	110	lb/ft^3
E_s	100,000	lb/ft^2
μ_s	0.25	–
E_p	460,000,000	lb/ft^2
δ	0.5ϕ	–
S_e	1.2	in.
ε	0.6	–

Table 2.25 Properties of sand layers

Soil layer	Thickness (m)	Cohesion (kN/m^2)	Friction angle (°)	Effective unit weight (kN/m^3)	Elastic modulus of soil (kN/m^2)	Poisson's ratio
1	10	0	30	19	9,000	0.35
2	5	0	35	12	11,000	0.35
3	10	0	37	9	13,000	0.35

Table 2.26 Additional design parameters

Property	Value	Unit
FS	3.0	–
E_p	21,000,000	kN/m^2
δ	0.5ϕ	–
S_e	0.025	m
ε	0.6	–
(L'/B)	18	–

2.4.14 Suggested Projects: Square Pile in Sandy Soil (Three Sand Layers)

Develop pile capacity design charts (bearing capacity and elastic settlement) for a 450-mm × 450-mm-square bored pile in nonhomogeneous sandy soil. Thicknesses and physical properties of sand layers are provided in Table 2.25. Use Coyle and Castillo's method for the end bearing capacity and use the critical depth method for the skin frictional resistance. Show your results for pile lengths that vary from 5 to 18 m. Additional required information is provided in Table 2.26.

References

Coyle HM, Castello RR. New design corrections for piles in sand. Geotech. Engrng. Journal, ASCE. 1981;107(GT7):965–86.

Goodman RE. Introduction rock mechanics. New York: Wiley; 1980.

Meyerhof GG. Bearing capacity and settlement of pile foundations. Journal of Geotechnical Engineering Division, American Society of Civil Engineering. 1976;102(GT3):197–228.

O'Neill MW, Reese LC. Drilled shafts: construction procedure and design methods. FHWA report no. IF-99-025. 1999.

Sladen JA. The adhesion factor: applications and limitations. Canadian Geotechnical Journal. 1992;29(2):323–6.

Vesic S. Design of pile foundations. National cooperative highway research program synthesis of practice no. 42. Washington, DC: Transportation Research Board; 1977.

Vijayvergiya VN, Focht Jr JA. A new way to predict capacity of piles in clay. Offshore technology conference, paper 1718, fourth offshore technology conference, Houston; 1972.

Further Readings

American Society of Civil Engineers. Design of pile foundations (Technical Engineering and Design Guides as Adapted from the U.S. Army Corps of Engineers, No. 1). New York: American Society of Civil Engineers; 1993.

Bowles JE. Foundation analysis and design. New York: McGraw-Hill; 1996.

Budhu M. Soil mechanics and foundations. New York: Wiley; 2000.

Coduto D. Foundation design principle and practices. New Jersey: Prentice-Hall Inc.; 2001.

Das B. Principles of foundation engineering. Stamford, CT: Cengage Learning; 2010.

Vesic AS. Bearing capacity of deep foundation in sand. Highway research record no. 39. Washington: National Academy Sciences; 1963. p. 112–53.

Vesic AS. Tests on instrumental piles—Ogeechee river site. Journal of the Soil Mechanics and Foundations Division, American Society of Civil Engineers. 1970;96(SM2):561–84.

Chapter 3
Axial Capacity of Single Drilled Shaft Foundations in Soil

Abstract This chapter deals with single drilled shaft foundations in homogeneous and nonhomogeneous soils. Calculations of axial load a single drilled shaft foundation can withstand are explained in details in this chapter. The calculations were performed to satisfy both bearing capacity and elastic settlement requirements. For the bearing capacity condition, the general bearing capacity equation was utilized and the effects of many factors were considered in the analyses such as shaft type (straight/belled), shaft size, shaft length, soil type, and groundwater table. Then again, the effects of skin friction and the end bearing components were considered in the elastic settlement analyses. Additionally, a step-by-step procedure was introduced in this chapter to develop bearing capacity and elastic settlement design charts which can be useful in the design process of single drilled shafts in soil. A number of design problems were also presented in this chapter and its solution were explained in details. These problems were first hand-solved, and then, resolved using the Shaft-1 (in soil) and Shaft-2 (in rock) applications of the foundationPro program. Finally, a set of design projects was suggested at the end of this chapter to allow the reader practice the concepts learned.

Keywords Single drilled shaft foundation • Bearing capacity • Elastic settlement • Shaft-1 • Shaft-2 • foundationPro

3.1 Introduction

This chapter deals with single drilled shaft foundations in different types of soils. Calculation procedures of allowable and ultimate loads in which a single drilled shaft foundation can sustain are discussed in details in this chapter. Allowable and ultimate axial loads on a single drilled shaft foundation are estimated to satisfy both bearing capacity and elastic settlement requirements.

In the bearing capacity analyses, the classical bearing capacity equations for a single drilled shaft foundation were utilized. Effect of drilled shaft foundation type (i.e., straight or belled) is considered in the bearing capacity calculations. Drilled shafts in sand, clay, and rock are dealt with herein. Effect of the groundwater table depth is also considered in the bearing capacity equation.

M. Yamin, *Problem Solving in Foundation Engineering using foundationPro*,
DOI 10.1007/978-3-319-17650-5_3

The axial capacity of single drilled shaft is also estimated based on elastic settlement (theory of elasticity). The allowable load a single drilled shaft foundation can sustain to satisfy elastic settlement requirements and not to exceed the allowable permissible settlement is determined herein. The total elastic settlement of a single drilled shaft in soil was estimated as a result of adding the various types of settlements occurred due to end bearing and skin friction loads.

A step-by-step procedure was introduced in this chapter to develop bearing capacity and elastic settlement design charts. These design charts present the relationship between various applied loads on a single drilled shaft foundation versus shaft length (for straight and belled shafts). These charts can be useful in the drilled shaft foundation design process to find what will control the final design: the bearing capacity or the elastic settlement of the foundation.

Five design problems were presented in this chapter. First, these design problems were hand-solved and their solution was explained in details, and then the foundationPro program was used to resolve the problems to replicate and verify the hand solution. Also, the program was used to investigate a wider and detailed solution and design alternatives for the hand-solved problems. Since the foundationPro includes a set of several applications, the Shaft-1 and Shaft-2 applications of the foundationPro are the responsible applications to perform bearing capacity and elastic settlement calculations for single drilled shaft foundations embedded in homogeneous and nonhomogeneous soils. Therefore, only Shaft-1 and Shaft-2 applications will be used throughout this chapter to replicate the hand-solved problems. Three design projects were suggested at the end of this chapter to allow the reader to practice and apply the learned concepts.

3.2 Theory

In this section, the procedures to estimate the axial capacity of a single drilled shaft in soil are discussed. The governing methodologies to calculate the ultimate and allowable loads that can be applied to a single drilled shaft foundation with the given shaft design configurations in different types of soil based on bearing capacity and elastic settlement are summarized in the following subsections. Figure 3.1 defines the main soil properties and design parameters used in the analyses of single drilled shaft (straight shaft as in Fig. 3.1a and belled shaft as in Fig. 3.1b) foundations. As can be seen in the figure, C is the soil cohesion, ϕ is the soil friction angle, γ' is the effective unit weight of the soil, E_s is the elastic modulus of the soil, μ_s is Poisson's ratio of the soil, D_s is the shaft diameter, and D_b is bell diameter. The thicknesses of each soil layer in the case of multiple soil layers can be defined by providing the depths (Z_0, Z_1, Z_2, etc.) at the boundaries between different soil layers (see Fig. 3.1).

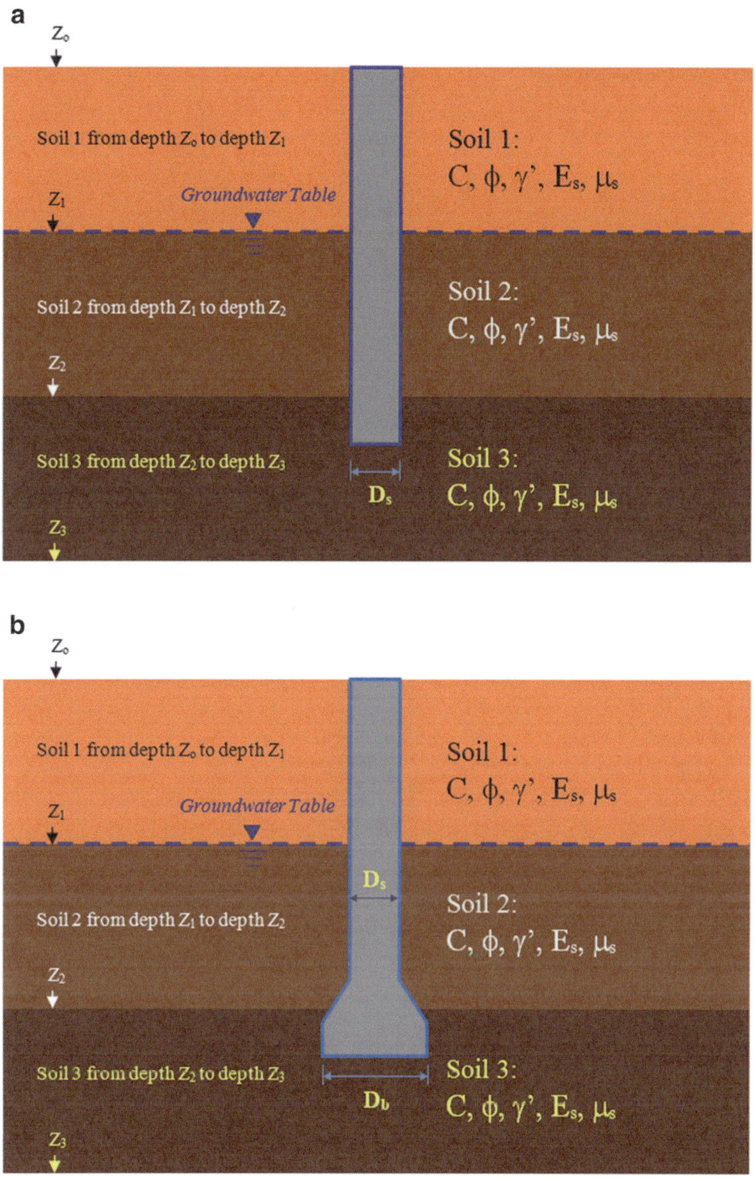

Fig. 3.1 Single drilled shaft embedded in multiple soil layers: (**a**) straight shaft and (**b**) belled shaft

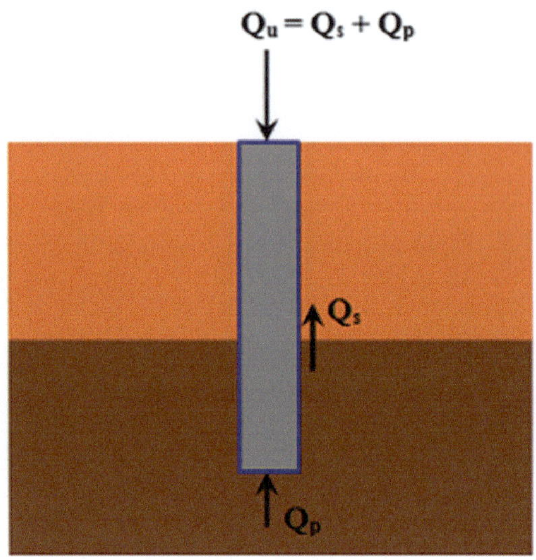

Fig. 3.2 Ultimate load on a single drilled shaft (straight) foundation

3.2.1 *Axial Capacity of a Single Drilled Shaft (Bearing Capacity)*

The ultimate axial capacity (load), Q_u, a single drilled shaft foundation embedded in soil can sustain as shown in Fig. 3.2 can be estimated as in Eq. (3.1) by summing the end bearing capacity component (Q_p) and the frictional skin resistance component (Q_s):

$$Q_u = Q_p + Q_s \tag{3.1}$$

where

Q_p is the load carrying capacity of the pile point
Q_s is the frictional skin resistance

3.2.1.1 End Bearing Capacity (Q_p)

One can estimate the shaft tip resistant–end bearing capacity of a single drilled shaft depending on the soil which the drilled shaft is resting on. Below is a discussion of the procedures to be followed to estimate Q_p in sand, clay, and rock.

1. End bearing capacity in sand:
   From the general bearing capacity equation as suggested by Meyerhof (1963), the ultimate end bearing capacity (at the shaft tip), Q_p, in sandy soil is given as

$$Q_{\mathrm{p}} = A_{\mathrm{p}}\left[q'(N_q - 1)F_{qs}F_{qd}F_{qc}\right] \tag{3.2}$$

where

N_q is the bearing capacity factor
F_{qs} is the shape factor
F_{qd} is the depth factor
F_{qc} is the soil compressibility factor
q' is the effective vertical stress at the base of the shaft

$$A_{\mathrm{p}} = \frac{\pi}{4}D_{\mathrm{s}}^2 \text{ for straight shaft and } = \frac{\pi}{4}D_{\mathrm{b}}^2$$

Shape and depth factors can be determined using the following equations:

$$F_{qs} = 1 + \tan\phi' \tag{3.3}$$

$$F_{qd} = 1 + 2\tan\phi'\left(1 - \sin\phi'\right)^2 \times \tan^{-1}\underbrace{\left[\frac{L}{D_{\mathrm{b}}}\right]}_{\text{radians}} \tag{3.4}$$

To determine the compressibility factor (F_{qc}), Chen and Kulhawy (1994) suggest the following procedure:

$$F_{qc} = 1 \text{ if } \mathrm{I}_{\mathrm{rr}} \geq \mathrm{I}_{\mathrm{cr}} \tag{3.5}$$

However, if $I_{\mathrm{rr}} > I_{\mathrm{cr}}$,

$$F_{qc} = \exp\left\{\left(-3.8\tan\phi'\right) + \left[\frac{(3.07\sin\phi')(\log_{10}2I_{\mathrm{rr}})}{1 + \sin\phi'}\right]\right\} \tag{3.6}$$

where

I_{cr} is the critical rigidity index and can be found from the following equation:

$$I_{\mathrm{cr}} = 0.5\exp\left[2.85\cot\left(45 \quad \frac{\phi'}{2}\right)\right] \tag{3.7}$$

and I_{rr} is the reduced rigidity index which can be determined using the equation below:

$$I_{\mathrm{rr}} = \frac{I_{\mathrm{r}}}{1 + I_{\mathrm{r}}\Delta} \tag{3.8}$$

$$I_r \text{ is the soil rigidity index } = \frac{E_s}{2(1+\mu_s) \times q' \times \tan\phi'} \quad (3.9)$$

$$\Delta = n\left(\frac{q'}{p_a}\right) \quad (3.10)$$

p_a is the atmospheric pressure

$$n = 0.005\left(1 - \frac{\phi' - 25}{20}\right) \quad (3.11)$$

2. End bearing capacity in clay:
From the general bearing capacity equation as suggested by Meyerhof (1963) and the suggestion made by Das (2010) the shaft length is greater than three times the shaft diameter. The net ultimate end bearing capacity (at the shaft tip), Q_p, in clayey soil is given as

$$Q_{p(net)} = A_p c_u N_c^* \quad (3.12)$$

where
c_u is the undrained cohesion:

$$N_c^* = 1.33 \times \left[\left(\ln\left(\frac{E_s}{3c_u}\right)\right) + 1\right] \quad (3.13)$$

In the case that the elastic modulus of the soil is not available, one can review the suggested relationship by O'Neill and Reese (1999) to obtain the elastic modulus of the soil.

3. End bearing capacity in rock:
Based on the full-scale drilled shaft test results by Zhang and Einstein (1998), the end bearing capacity of a drilled shaft extending to and resting on rock can be estimated using the following equation:

$$Q_p = \omega(q_u)^{0.51} A_p \quad (3.14)$$

where
q_u is the unconfined compression
$\omega = 4.83$ when SI units are used in Eq. (3.14) as follows: q_u in MN/m^2, A_p is in m^2, and Q_p will come out in MN, or
$\omega = 8{,}299.29$ when BS units are used in Eq. (3.14) as follows: when q_u in lb/ft^2, A_p is in ft^2, and Q_p will come out in lb.

3.2.1.2 Frictional Skin Resistance (Q_s)

One can estimate the friction skin resistance of a single drilled shaft depending on the type of soil that is surrounding the drilled shaft along its length. Below is a discussion of the procedures to be followed to estimate Q_s when a shaft is surrounded by sand, clay, or rock.

1. Frictional skin resistance in sand:
   The frictional resistance at ultimate load, Q_s, developed in a drilled shaft may be calculated as in the following equation:

$$Q_s = \sum p \times f \times \Delta L \tag{3.15}$$

where
p is the shaft perimeter
f can be determined depending on the depth z as follows:
For

$$z = 0 \text{ to } L'$$
$$f = K \times \sigma_o' \tan\left(\delta'\right) \tag{3.16}$$

For

$$z = L' \text{to } L_s$$
$$f = f_{z=L'} \tag{3.17}$$

L_s is the length of the straight portion of the drilled shaft

$$L' = D_s \times \left(\frac{L'}{D_s}\right)_{cr} \tag{3.18}$$

Usually, the critical depth ration $\left(\frac{L'}{D_s}\right)_{cr}$ varies from 15 to 20.
In Eq. (3.16),
K is the effective earth pressure coefficient
σ_o' is the effective vertical stress at depth under consideration which will increase until L', and will remain constant thereafter
δ' is the soil/shaft friction angle and can be taken as (0.7–0.8)ϕ' for poor construction or ϕ' for good construction.

2. Frictional skin resistance in clay:
   The expression for skin resistance of drilled shafts in clay is given as

$$Q_s = \sum_{L=0}^{L=L_s} \alpha^* c_u p \Delta L \tag{3.19}$$

Table 3.1 Required parameters for Eq. (3.21)

	ξ in Eq. (3.21)		Units used in Eq. (3.21)			
Units	Smooth socket	Rough socket	D_s	L	q_u	Q_s
SI	1.2566	2.5133	m	m	MN/m^2	MN
BS	1,954.793	3,908.644	ft	ft	lb/ft^2	lb

α*, which must not exceed the value of 1.0, can be calculated according to Kulhawy and Jackson (1989) as in the following equation:

$$\alpha^* = 0.21 + 0.25\left(\frac{p_a}{c_u}\right) \tag{3.20}$$

3. Frictional skin resistance in rock:
   Q_s can be estimated again from the work done by Zhang and Einstein (1998) when the shaft is surrounded by rock and is embedded in the rock a length (L) as in the following equation:

$$Q_s = \xi \times D_s \times L \times \sqrt{q_u} \tag{3.21}$$

Required parameters and more details about Eq. (3.21) can be found in Table 3.1.

3.2.2 *Axial Capacity of a Single Drilled Shaft (Elastic Settlement)*

The total elastic settlement (S_e) of a single drilled shaft foundation can be estimated using Eq. (3.22). Definition of the various working loads under total elastic settlement condition is illustrated in Fig. 3.3:

$$S_e = S_{e_1} + S_{e_2} + S_{e_3} \tag{3.22}$$

where

S_{e_1} is the elastic settlement of shaft length

S_{e_2} is the settlement of the pile caused by the load at the drilled shaft tip

S_{e_3} is the settlement of the pile caused by the load transmitted along the drilled shaft length

S_{e_1} can be estimated using the following equation:

$$S_{e_1} = \frac{(Q_{wp} + \varepsilon Q_{ws})L}{A_p E_p} \tag{3.23}$$

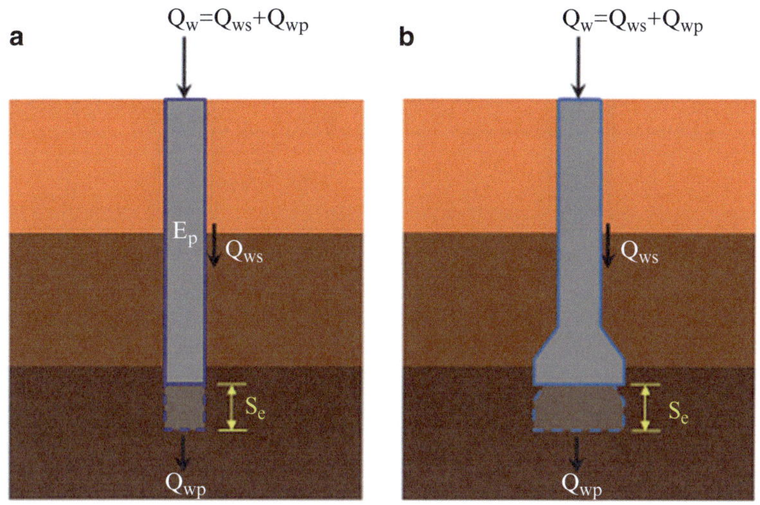

Fig. 3.3 Definition of working loads due to total settlement for (**a**) straight shaft and (**b**) belled shaft

where

Q_{wp} is the load carried at the shaft base under working load

Q_{ws} is the load carried by frictional skin resistance under working load

L is the length of shaft

E_p is the modulus of elasticity of the shaft material

In Eq. (3.23), the parameter ε varies between 0.5 and 0.67.

S_{e_2} can be estimated using the following equation:

$$S_{e_2} = 0.85 \times \frac{(Q_{wp} D_s)}{A_p E_s} (1 - \mu_s^2) \tag{3.24}$$

where

E_s is the modulus of elasticity of soil at or below the shaft point

S_{e_3} can be estimated using the following equation:

$$S_{e_3} = (1 - {\mu_s}^2) \left(\frac{Q_{ws}}{pL} \right) \left(\frac{D_s}{E_s} \right) \left(2 + 0.35 \left(\sqrt{\frac{L}{D_s}} \right) \right) \tag{3.25}$$

where

p is the perimeter of shaft

L is the embedded length of shaft

3.2.3 Allowable Loads

Allowable load single drilled shaft foundation can sustain based on bearing capacity analyses and can be determined using the following equation:

$$Q_{\text{all}} = \frac{Q_{\text{p(net)}} + Q_{\text{s}}}{\text{FS}} \tag{3.26}$$

where

FS is the factor of safety for bearing capacity

However,

To determine the allowable (working) load based on elastic settlement analyses:

First,

Let

$$\eta = \frac{Q_{\text{p}}}{Q_{\text{s}}} \tag{3.27}$$

Then,

One can solve for Q_{wp} at the total allowable settlement (S_{e}) using the following equation:

$$S_{\text{e}} = Q_{\text{wp}} \left(\frac{\left(1 + \frac{\varepsilon}{\eta}\right) L}{A_{\text{p}} E_{\text{p}}} + \frac{0.85 \times D_{\text{s}} (1 - \mu_{\text{s}}^2)}{A_p E_s} + \frac{D_{\text{s}} (1 - \mu_{\text{s}}^2)}{\eta p L E_{\text{s}}} \left(2 + 0.35 \left(\sqrt{\frac{L}{D_{\text{s}}}} \right) \right) \right) \tag{3.28}$$

Thus,

The total working load based on settlement analyses can be found using the equation below:

$$Q_{\text{w}} = Q_{\text{wp}} \left(1 + \frac{1}{\eta} \right) \tag{3.29}$$

3.3 Step-by-Step Procedure

To determine the axial capacity of a single drilled shaft foundation in soil based on bearing capacity and total elastic settlement requirements and establish the associated design charts, the following steps must be followed:

Step 1:
Assume drilled shaft length L

Step 2:
Determine the end bearing capacity, Q_p, for the given soil type as follows:

- For sandy soil use Eqs. (3.2)–(3.11).
- For clayey soil use Eqs. (3.12) and (3.13).
- For rock soil use Eq. (3.14).

Step 3:
Determine the frictional skin resistance, Q_s, for the given soil type as follows:

- For sandy soil use Eqs. (3.15)–(3.18).
- For clayey soil use Eqs. (3.19) and (3.20).
- For rock soil use Eq. (3.21).

Step 4:
Determine the ultimate load based on bearing capacity analyses, Q_u, from Eq. (3.1). Then, determine the allowable load (bearing capacity analyses), Q_{all}, using Eq. (3.26).

Step 5:
In order to find the total working load, Q_w, based on the elastic settlement analyses, we use Eq. (3.29).

Step 6:
Repeat steps 1 through 5 for as many shaft lengths as needed to have enough data points for the capacity design charts.

3.4 Design Problems

Several design problems are presented in this section to help the reader reiterate the theory and integrate the Shaft-1 and Shaft-2 applications of the foundationPro program in the design process of the single drilled shaft foundations. These examples were selected to give exposure to a wide variety of challenges and design configurations that can be faced in designing a single drilled shaft foundation while helping us iron out the finer details of the theories introduced earlier.

3.4.1 Straight Shaft in Homogeneous Sandy Soil

Develop axial capacity design charts based on bearing capacity and elastic settlement for a straight drilled shaft (diameter = 1.2 m) in homogeneous cohesionless sandy soil. Use a factor of safety of 3.0 for bearing capacity calculations. The total allowable settlement of the drilled shaft must not exceed 25 mm. Soil properties and other useful information are provided in Table 3.2.

Table 3.2 Soil properties and other design parameters

Property	Value	Unit
c	0	kN/m^2
ϕ	28	°
γ	18.5	kN/m^3
E_s	25,000	kN/m^2
μ_s	0.35	–
E_p	23,000,000	kN/m^2
δ	0.75ϕ	–
S_e	25	mm
$(L_s/D_s)_{cr}$	16.5	–
ε	0.6	–

3.4.1.1 Hand Solution

To develop axial capacity design charts, one must follow the step-by-step procedure explained in Sect. 3.3.

Step 1:
Assume $L = 15$ m

Step 2:
Determine Q_p in sand.
Calculate the critical rigidity index (I_{cr}) using Eq. (3.7):

$$I_{cr} = 0.5\exp\left[2.85\cot\left(45 - \frac{28}{2}\right)\right] = 57.4$$

Also, calculate the soil rigidity index from Eq. (3.9):

$$I_r = \frac{25{,}000}{2(1 + 0.35) \times (18.5 \times 15) \times \tan(28)} = 62.75$$

In order to calculate I_{rr} we will need to first calculate Δ and n; these can be calculated as follows:

From Eq. (3.11):

$$n = 0.005\left(1 - \left(\frac{28 - 25}{20}\right)\right) = 0.00425$$

$$q' = 18.5 \times 15 = 277.5\,\text{kN/m}^2$$

From Eq. (3.10):

$$\Delta = 0.00425\left(\frac{277.5}{100}\right) = 0.0117$$

Now, we can calculate the reduced rigidity index from Eq. (3.8):

$$I_{\mathrm{rr}} = \frac{62.75}{1 + (62.75 \times 0.0117)} = 36.18$$

Since $I_{\mathrm{rr}} < I_{\mathrm{cr}}$, we use Eq. (3.6) to calculate the F_{qc}:

$$F_{qc} = \exp\left\{(-3.8\tan(28)) + \left[\frac{(3.07\sin(28))(\log_{10}(2 \times 36.18))}{1 + \sin 28}\right]\right\} = 0.821$$

From Eq. (3.3):

$$F_{qs} = 1 + \tan(28) = 1.531$$

From Eq. (3.4):

$$F_{q\mathrm{d}} = 1 + 2\tan(28)(1 - \sin(28))^2 \times \tan^{-1}\underbrace{\left[\frac{15}{1.2}\right]}_{\text{radians}} = 1.446$$

From Eq. (3.2):

$$Q_{\mathrm{p}} = (1.13) \times [(18.5 \times 15)(14.72 - 1)(1.531)(1.446)(0.821)] = 7{,}819.55\,\mathrm{kN}$$

Step 3:
Determine Q_{s} in sandy soil.

First, we need to find L' which can be acquired from Eq. (3.18):

$$L' = 16.5 \times 1.2 = 19.8\,\mathrm{m}$$

At $z = 0\,\mathrm{m}$ we are between 0 and L'; therefore, we use Eq. (3.16):

$$\sigma_o' = 0$$
$$f = 0$$

At $z = 15\,\mathrm{m}$ we are between 0 and L'; therefore, we use Eq. (3.17):

$$\sigma_o' = 15 \times 18.5 = 277.5\,\mathrm{kN/m^2}$$
$$f = (1 - \sin(28)) \times 277.5\tan(21) = 56.51\,\mathrm{kN/m^2}$$

We use Eq. (3.15):

$$Q_s = \frac{f_{z=0} + f_{z=15}}{2} \times (3.76) \times 15$$

$$Q_s = \frac{0 + 56.51}{2} \times (3.76) \times 15 = 1,593.58 \text{ kN}$$

Step 4:
The ultimate load based on bearing capacity analyses, Q_u, is determined using Eq. (3.1) as follows:

$$Q_u = 7,819.55 + 1,593.58 = 9,413.1 \text{ kN}$$

Then, the allowable load, Q_{all}, can be found from Eq. (3.26):

$$Q_{all} = \frac{7,819.55 + 1,593.58}{3.0} = 3,137.71 \text{ kN}$$

Step 5:
In order to find the working loads from the elastic settlement analyses, we must first find a ratio between Q_s and Q_p. This relationship can be established through Eq. (3.27):

$$\eta = \frac{7,826.29}{1,593.58} = 4.90$$

From Eq. (3.28):

$$0.025 = Q_{wp}\left(\frac{\left(1 + \frac{0.6}{4.90}\right)15}{(1.13) \times (23,000,000)} + \frac{1.2(1 - 0.35^2)(0.85)}{(1.13)(25,000)} + \frac{1.2(1 - 0.35^2)3.237}{(4.90)(\pi \times 1.2)(15)(25,000)}\right)$$

Thus,

$Q_{wp} = 783.77$ kN

Therefore,

The total working load from Eq. (3.29):

$$Q_w = 783.77\left(1 + \frac{1}{4.90}\right) = 943.73 \text{ kN}$$

Step 6:
The above steps can be repeated for other shaft lengths.

3.4.1.2 foundationPro Solution

After launching the Shaft-1 application of foundationPro, the four sections (General, Drilled Shaft Information, Soil Properties, and OUTPUT) will appear in the main screen as shown in Fig. 3.4.

General Section

In this section, general information can be entered such as user name and project name and this information is optional. Additional, units to be used throughout the analysis must be specified and the factor of safety for bearing capacity must be entered in the provided textbox. For the calculations of the skin friction in sand, the user must input the required parameters as shown in Fig. 3.5. The critical depth ratio is usually 15–20. The effective earth pressure coefficient is usually 1 unless otherwise stated.

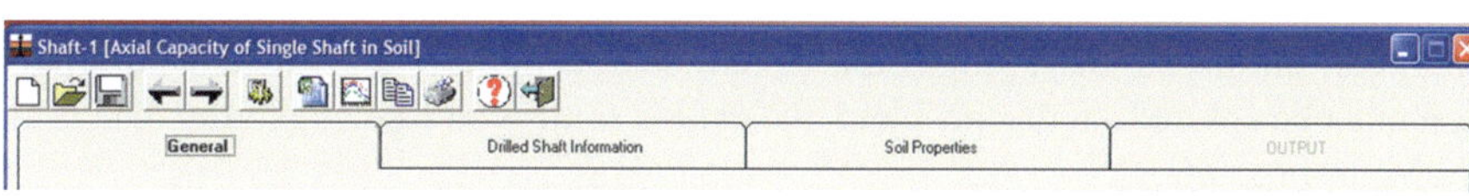

Fig. 3.4 Main sections of Shaft-1 application

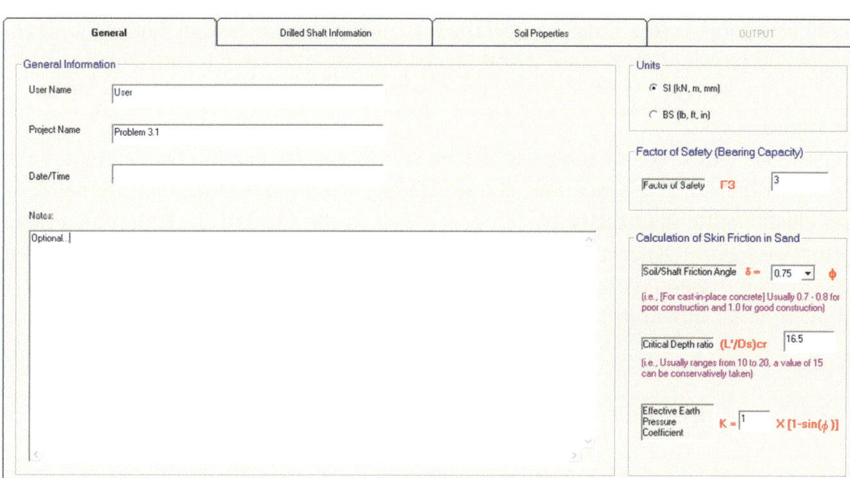

Fig. 3.5 The General information section of Shaft-1 application

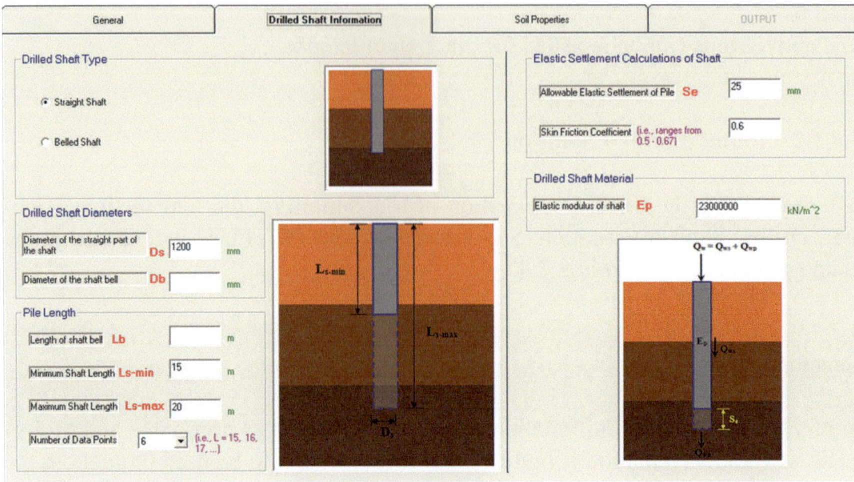

Fig. 3.6 The Drilled Shaft Information section for Shaft-1 application

Drilled Shaft Information Section

In this section, drilled shaft type must be specified (straight or belled), and its diameter, minimum shaft length, maximum shaft length, and settlement information must be entered. In this problem, the drilled shaft is straight and other data is as shown in Fig. 3.6.

Soil Properties Section

Number of soil layers must be specified as in the combo box in Fig. 3.7. Thicknesses and physical properties of soil layers must be entered in the provided table as well. In this problem, we have one soil layer and the other properties are as shown in the figure.

Now we can save our progress and hit run to get the results. One can view the results in the OUTPUT section. Axial capacity results based on bearing capacity and elastic settlement analyses can be viewed in the OUTPUT section as a table format (see Fig. 3.8) or as a chart format (see Fig. 3.9). The total working load for this problem based on elastic settlement analyses can be extracted from the Shaft-1 application as shown in Fig. 3.10.

3.4.2 Straight Shaft in Homogeneous Clayey Soil

Develop axial capacity design charts (bearing capacity and elastic settlement) for a 2-m-diameter straight drilled shaft in homogeneous clayey soil. Use a safety factor of 3.0 for bearing capacity computations. Other information is provided in Table 3.3.

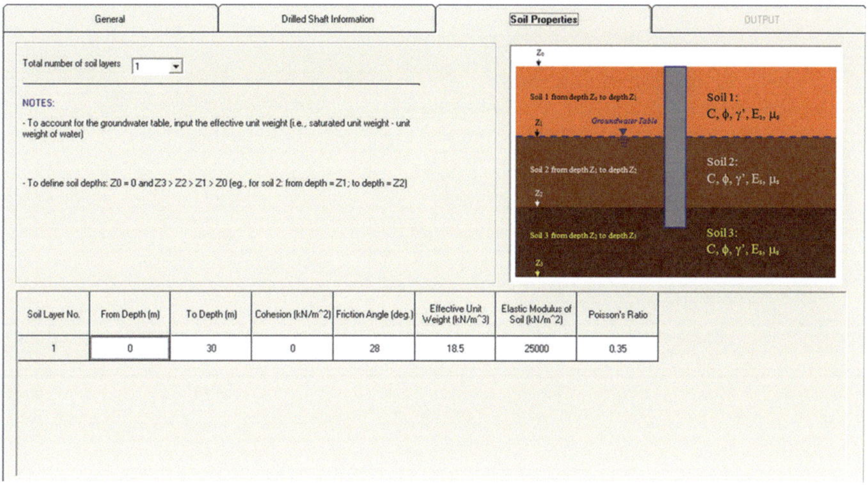

Soil Layer No.	From Depth (m)	To Depth (m)	Cohesion (kN/m^2)	Friction Angle (deg.)	Effective Unit Weight (kN/m^3)	Elastic Modulus of Soil (kN/m^2)	Poisson's Ratio
1	0	30	0	28	18.5	25000	0.35

Fig. 3.7 Soil Properties section for Shaft-1 application

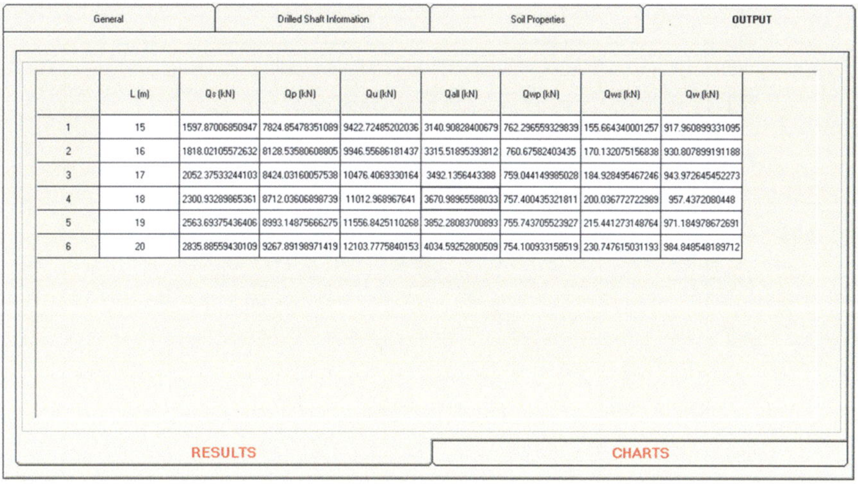

	L (m)	Qs (kN)	Qp (kN)	Qu (kN)	Qall (kN)	Qwp (kN)	Qws (kN)	Qw (kN)
1	15	1597.87006850947	7824.85478351089	9422.72485202036	3140.90828400679	762.296559329839	155.664340001257	917.960899331095
2	16	1818.02105572632	8128.53580608805	9946.55686181437	3315.51895393812	760.67582403435	170.132075156838	930.807899191188
3	17	2052.37533244103	8424.03160057538	10476.4069330164	3492.1356443388	759.044149985028	184.928495467246	943.972645452273
4	18	2300.93289865361	8712.03606898739	11012.968967641	3670.98965588033	757.400435321811	200.036772722989	957.4372080448
5	19	2563.69375436406	8993.14875666275	11556.8425110268	3852.28083700893	755.743705523927	215.441273148764	971.184978672691
6	20	2835.88559430109	9267.89198971419	12103.7775840153	4034.59252800509	754.100933158519	230.747615031193	984.848548189712

Figure 3.8 Axial capacity results—OUTPUT section (Table format)

3.4.2.1 Hand Solution

Axial capacity design charts can be developed by following the steps below:

Step 1:
Assume shaft length, $L = 20$ m

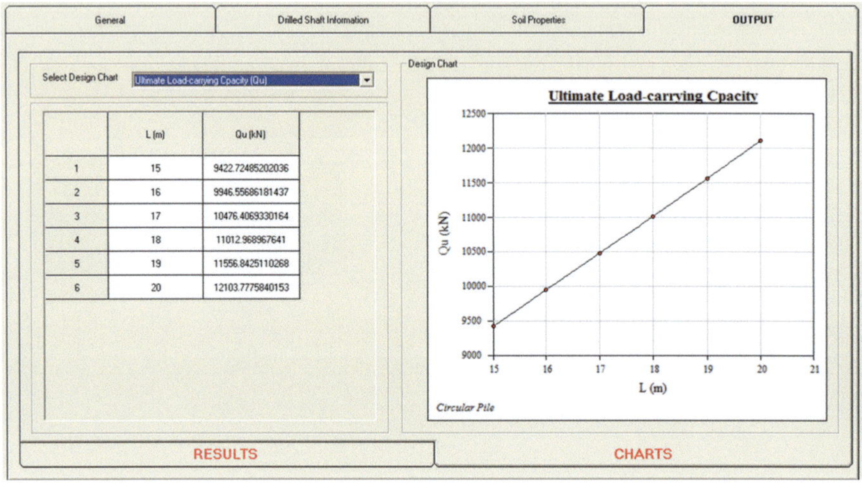

Fig. 3.9 Ultimate axial capacity design chart—OUTPUT section (Chart format)

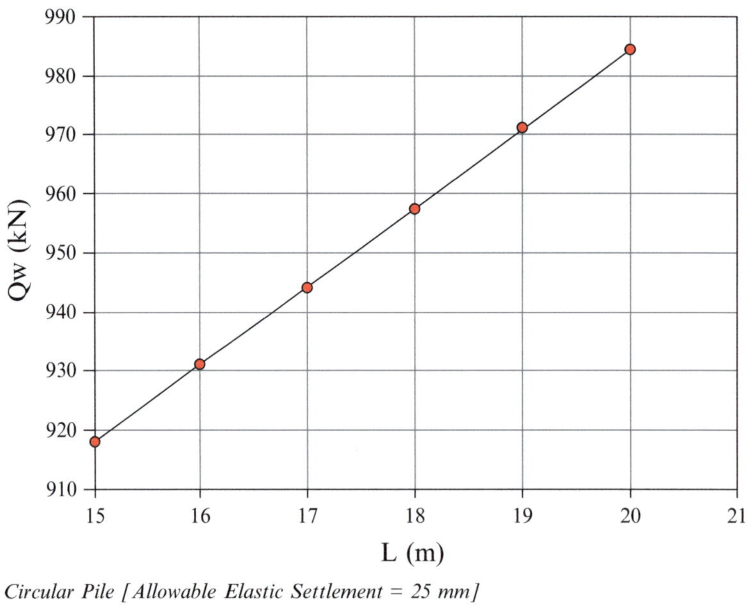

Fig. 3.10 Total working load design chart

Table 3.3 Soil properties and other design parameters

Property	Value	Unit
c	65	kN/m^2
ϕ	0	°
γ	19.8	kN/m^3
E_s	8,500	kN/m^2
μ_s	0.3	–
E_p	21,500,000	kN/m^2
S_e	0.025	m
ε	0.6	–

Step 2:
To determine end bearing capacity, first, we find N_c^* using Eq. (3.13) as follows:

$$N_c^* = 1.33\left[\left(\ln\left(\frac{8,500}{3 \times 65}\right)\right) + 1\right] = 6.350$$

Thus,
We use Eq. (3.12) to calculate Q_p:

$$Q_p = (3.14) \times (65) \times (6.350) = 1,296.035 \text{ kN}$$

Step 3:
In order to find the skin friction resistance we first use Eq. (3.20) to find α^*:

$$\alpha^* = 0.21 + 0.25\left(\frac{100}{65}\right) = 0.594$$

Next, we can find the skin friction resistance using Eq. (3.19):

$$Q_s = (0.594) \times (65) \times 6.28 \times 20 = 4,854.4 \text{ kN}$$

Step 4:
The ultimate load based on bearing capacity analyses, Q_u, is determined using Eq. (3.1) as follows:

$$Q_u = 1,296.0 + 4,854.4 = 6,150.4 \text{ kN}$$

Then, the allowable load, Q_{all}, can be found from Eq. (3.26):

$$Q_{all} = \frac{6,150.4}{3.0} = 2,050.1 \text{ kN}$$

Step 5:
First, η can be calculated using Eq. (3.27) as follows:

$$\eta = \frac{1,296.035}{4,854.43} = 0.26$$

Then, from Eq. (3.28):

$$0.025 = Q_{\mathrm{wp}} \left(\frac{\left(1 + \frac{0.6}{0.26}\right) 20}{(3.14) \times (21,500,000)} + \frac{2.0(1 - 0.30^2)(0.85)}{(3.14)(8,500)} + \frac{2.0(1 - 0.30^2) 3.10}{(0.26)(6.28)(20)(8,500)} \right)$$

Solving the above equation for Q_{wp}:
Thus,

$$Q_{\mathrm{wp}} = 316.55 \text{ kN}$$

From Eq. (3.29):

$$Q_{\mathrm{w}} = 316.55\left(1 + \frac{1}{0.26}\right) = 1,534.06 \text{ kN}$$

Step 6:
The above steps can be repeated for other shaft lengths. The solution below is for another assumed shaft length.

Step 1 (second shaft length):
Assume $L = 25$ m

Step 2 (second shaft length):
From Eq. (3.12):
To determine end bearing capacity, first, we find N_c^* using Eq. (3.13) as follows:

$$N_c^* = 1.33\left[\left(\ln\left(\frac{8,500}{3 \times 65}\right)\right) + 1\right] = 6.350$$

Thus, Q_{p} can be calculated using Eq. (3.12) as follows:

$$Q_{\mathrm{p}} = (3.14) \times (65) \times (6.350) = 1,296.035 \text{ kN}$$

Step 3 (second shaft length):
In order to find the skin friction resistance, we first use Eq. (3.20) to find α^*:

$$\alpha^* = 0.21 + 0.25\left(\frac{100}{65}\right) \leq 1$$

Next, we can find the skin friction resistance using Eq. (3.19):

$$Q_s = (0.594) \times (65) \times 6.28 \times 25 = 6,061.77 \text{ kN}$$

Step 4 (second shaft length):
First, we find η using Eq. (3.27):

$$\eta = \frac{1,296.035}{6,061.77} = 0.213$$

Then, from Eq. (3.28):

$$0.025 = Q_{wp}\left(\frac{\left(1+\frac{0.585}{8.01}\right)25}{(3.14)\times(21,500,000)} + \frac{2.0(1-0.30^2)(0.85)}{(3.14)(8,500)} + \frac{2.0(1-0.30^2)}{(0.213)(6.28)(25)(8,500)}\left(2+0.35\left(\sqrt{\frac{25}{2}}\right)\right)\right)$$

So, solving the above equation for Q_{wp}:

$$Q_{wp} = 312.7 \text{ kN}$$

The total working load is then determined using Eq. (3.29) as follows:

$$Q_w = 312.7 \times \left(1 + \frac{1}{0.213}\right) = 1,780.5 \text{ kN}$$

3.4.2.2 foundationPro Solution

General Section

In this General section, input data is entered as shown in Fig. 3.11. The SI unit was selected. The effective earth pressure coefficient is entered as 1 since it was not provided in the problem statement.

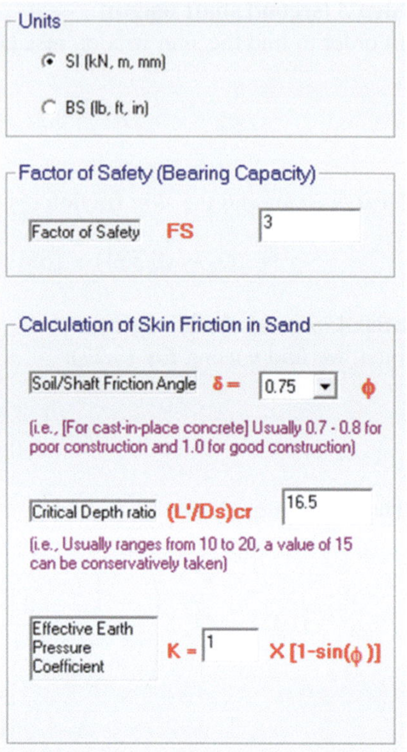

Fig. 3.11 Input data for General section

Drilled Shaft Information Section

In this section, we select the straight shaft type, we enter the diameter of the shaft as 2,000 mm, we enter the shaft length variation that we wish to consider (15–25 m with 11 points), we enter the allowable elastic settlement of the shaft, we enter the skin friction coefficient (0.6), and finally we enter the elastic modulus of the shaft as shown in Fig. 3.12.

Soil Properties Section

We enter the number of soil layer we will be dealing with (we have one in this problem). We enter the given depth parameters for the soil layer. If we are dealing with a single layer of soil and the depth parameters are not given, we select a reasonable depth (we choose 30 m) to consider as long as the maximum shaft length (25 m) is less than that depth. Finally, we enter the other soil properties as shown in Fig. 3.13.

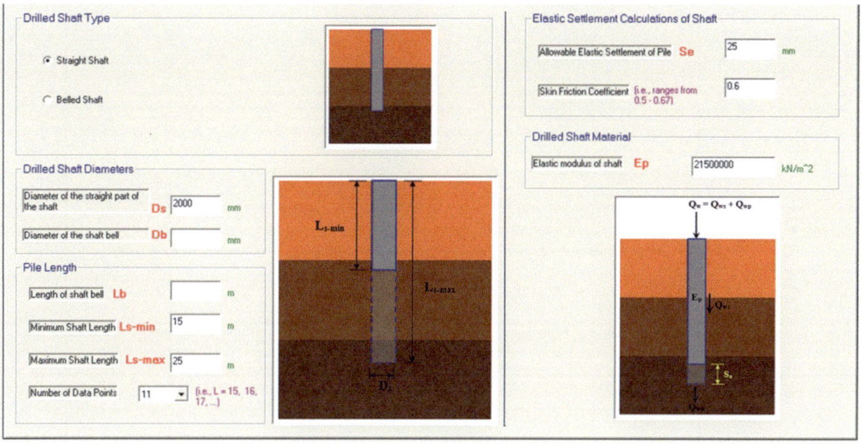

Fig. 3.12 Drilled shaft information

Soil Layer No.	From Depth (m)	To Depth (m)	Cohesion (kN/m^2)	Friction Angle (deg.)	Effective Unit Weight (kN/m^3)	Elastic Modulus of Soil (kN/m^2)	Poisson's Ratio
1	0	30	65	0	19.8	8500	0.3

Fig. 3.13 Soil properties in Shaft-1 application

Table 3.4 Axial capacity results for different shaft lengths

L (m)	Q_s (kN)	Q_p (kN)	Q_u (kN)	Q_{all} (kN)	Q_{wp} (kN)	Q_{ws} (kN)	Q_w (kN)
15	3,642.677	1,300.047	4,942.724	1,647.575	323.1615	905.4847	1,228.646
16	3,885.522	1,300.047	5,185.569	1,728.523	322.0527	962.5364	1,284.589
17	4,128.367	1,300.047	5,428.414	1,809.471	320.9632	1,019.235	1,340.198
18	4,371.212	1,300.047	5,671.259	1,890.42	319.8905	1,075.583	1,395.474
19	4,614.057	1,300.047	5,914.104	1,971.368	318.8324	1,131.583	1,450.415
20	4,856.902	1,300.047	6,156.95	2,052.317	317.7871	1,187.234	1,505.022
21	5,099.747	1,300.047	6,399.795	2,133.265	316.753	1,242.54	1,559.293
22	5,342.592	1,300.047	6,642.64	2,214.213	315.7287	1,297.499	1,613.228
23	5,585.438	1,300.047	6,885.485	2,295.162	314.713	1,352.112	1,666.825
24	5,828.283	1,300.047	7,128.33	2,376.11	313.7049	1,406.38	1,720.085
25	6,071.128	1,300.047	7,371.175	2,457.058	312.7032	1,460.302	1,773.005

Now we can save our progress and hit run to get the results. Axial capacity results based on bearing capacity and elastic settlement analyses at various shaft lengths are summarized in Table 3.4. Also, ultimate end bearing capacity design chart is shown in Fig. 3.14. A snapshot from the OUTPUT section is also shown in Fig. 3.15.

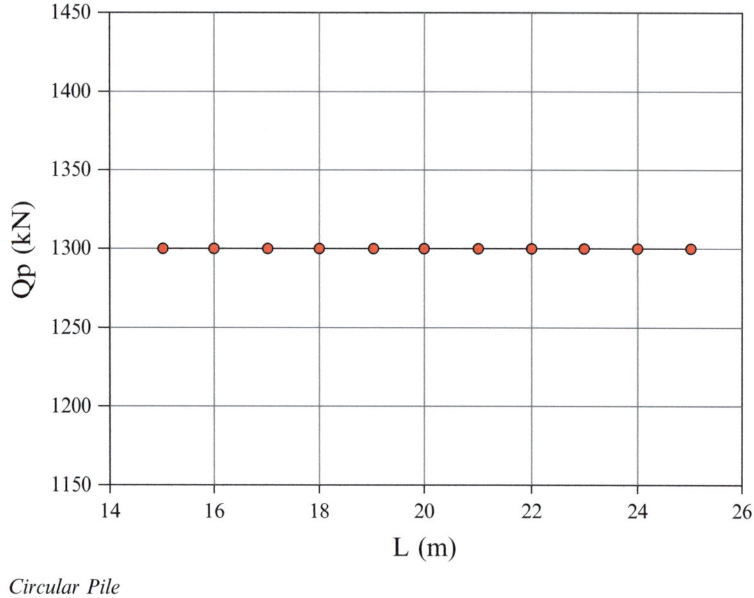

Fig. 3.14 Ultimate end bearing capacity design chart

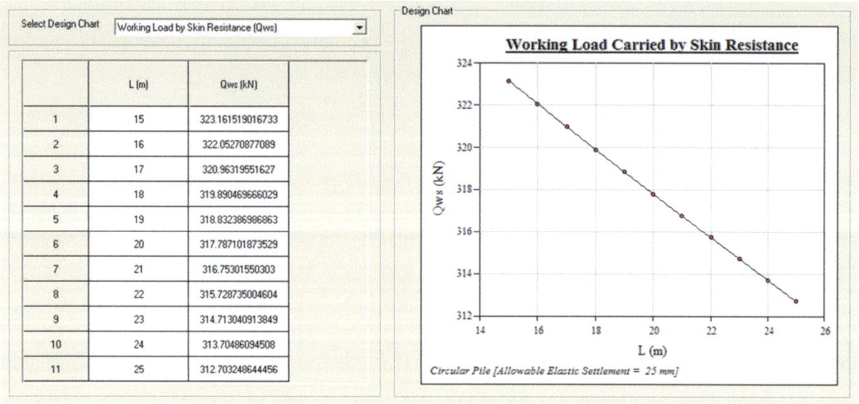

	L (m)	Qws (kN)
1	15	323.161519016733
2	16	322.05270877089
3	17	320.96319551627
4	18	319.890469666029
5	19	318.832386986863
6	20	317.787101873529
7	21	316.75301550303
8	22	315.728735004604
9	23	314.713040913849
10	24	313.70486094508
11	25	312.703248644456

Fig. 3.15 Results of working load carried by skin friction component

3.4.3 *Straight Shaft in Nonhomogeneous Clayey Soil*

Develop axial capacity design charts (bearing capacity and elastic settlement) for a straight drilled shaft (diameter = 2.5 ft) in nonhomogeneous clay soil with properties as in Table 3.5. Use a safety factor of 4.0 for bearing capacity calculations. Other useful information can be found in Table 3.6.

Table 3.5 Physical properties and thicknesses of clay layers

Soil layer	Thickness (ft)	Cohesion (lb/ft^2)	Friction angle (°)	Effective unit weight (lb/ft^3)	Elastic modulus of soil (lb/ft^2)	Poisson's ratio
1	40	835	0	47.6	135,000	0.3
2	60	1,350	0	53	166,000	0.3

Table 3.6 Other required design parameters

Property	Value	Unit
E_p	470,000,000	lb/ft^2
S_e	1	in.
ε	0.6	–

3.4.3.1 Hand Solution

We will follow the step-by-step procedure in Sect. 3.3 to develop the required design charts. Details of each step are presented below:

Step 1:
Assume shaft length, $L_1 = 20$ ft

Step 2:
N_c^* can be determined using Eq. (3.13) as follows:

$$N_c^* = 1.33\left[\left(\ln\left(\frac{135,000}{3 \times 835}\right)\right) + 1\right] = 6.63$$

Then, the end bearing capacity (Q_p) can be obtained from Eq. (3.12):

$$Q_p = (4.90) \times (835) \times (6.30) = 27,137.43 \text{ lb}$$

Step 3:
α* can be calculated from Eq. (3.20):

$$\alpha^* = 0.21 + 0.25\left(\frac{2,000}{835}\right) = 0.8$$

Therefore, Q_s can be found using Eq. (3.19):

$$Q_s = 0.80 \times (835) \times (\pi \times 2.5) \times 20 = 106,083.72 \text{ lb}$$

Step 4:
The ultimate load-carrying capacity is now obtained using Eq. (3.1) as follows:

$$Q_u = 27,137.4 + 106,083.7 = 133,221 \text{ lb}$$

So,

The allowable load-carrying capacity Q_{all} can be obtained using Eq. (3.26):

$$Q_{\text{all}} = \frac{133,221}{4} = 33,305.25 \text{ lb}$$

Step 5:

In order to find the working loads from the elastic settlement analyses, we must first find η using Eq. (3.27) as follows:

$$\eta = \frac{27,137.43}{106,083.72} = 0.25$$

Then, one can write Eq. (3.28):

$$0.0833 = Q_{\text{wp}}\left(\frac{\left(1+\left(\frac{0.6}{0.25}\right)\right)20}{(4.90)\times(470,000,000)} + \frac{2.5(1-0.30^2)(0.85)}{(4.90)(135,000)} + \frac{2.5(1-0.30^2)2.98}{(0.25)(\pi\times 2.5)(20)(135,000)}\right)$$

After solving the above equation, one can find $Q_{\text{wp}} = 19,473.93$ kN

Thus,

The total working load is determined from Eq. (3.29) as follows:

$$Q_{\text{w}} = 19,473.93\left(1+\frac{1}{0.24}\right) = 100,615.317 \text{ lb}$$

Step 6:

The above steps can be repeated for another shaft length. The solution below is for another assumed shaft length.

Step 1 (second shaft length):

Assume $L = 50$ ft

Step 2 (second shaft length):

From Eq. (3.13), one can find:

$$N_c^* = 1.33\left[\left(\ln\left(\frac{135,000}{3\times 1,350}\right)\right)+1\right] = 5.99$$

Then, the ultimate end bearing capacity is found from Eq. (3.12):

$$Q_p = (4.90) \times (1{,}350) \times (5.99) = 39{,}719.17 \text{ lb}$$

Step 3 (second shaft length):
From Eq. (3.20):

$$\alpha^* = 0.21 + 0.25\left(\frac{2{,}000}{1{,}350}\right) = 0.58$$

Therefore,
Q_s is found from Eq. (3.19) as follows:

$$Q_s = 0.58 \times \pi \times 2.5(835 \times 40 + 1{,}350 \times 10) = 213{,}780.43 \text{ lb}$$

Step 4 (second shaft length):
The ultimate load-carrying capacity can be determined from Eq. (3.1):

$$Q_u = 39{,}623 + 213{,}780 = 253{,}404 \text{ lb}$$

Step 5 (second shaft length):
From Eq. (3.27):

$$\eta = \left(\frac{39{,}623.85}{213{,}780.43}\right) = 0.185$$

And from Eq. (3.28):

$$0.083 = Q_{wp}\left(\frac{\left(1 + \frac{0.6}{0.185}\right)50}{4.90 \times 470{,}000{,}000} + \frac{2.5(1 - 0.3^2)0.85}{4.90 \times 166{,}000} + \frac{2.5(1 - 0.3^2)3.48}{0.185 \times \pi \times 2.5 \times 50 \times 166{,}000}\right)$$

Solving the above equation yields

$$Q_{wp} = 26{,}733.71 \text{ lb}$$

Thus,
The total working load using Eq. (3.29):

$$Q_w = 26{,}733.71\left(1 + \frac{1}{0.185}\right) = 171{,}240.30 \text{ lb}$$

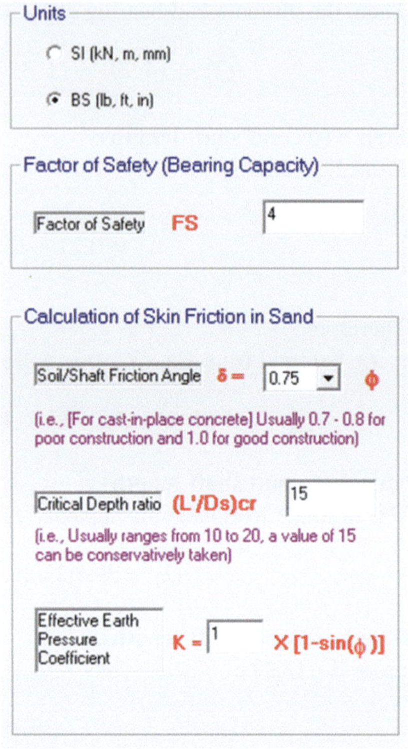

Fig. 3.16 Input data for the General section

3.4.3.2 foundationPro Solution

General Section

In this section, we select the BS unit to be used in the problem and we enter the Safety Factor of 4 in the provided textbox as shown in Fig. 3.16. For this problem, skin friction in sand section does not apply; therefore, the values we enter will not be considered because there is no sand surrounding the drilled shaft based on the given soil profile.

Drilled Shaft Information Section

A straight shaft is selected for this problem. 30 in. is entered for the shaft diameter. The range of shaft lengths considered in this problem is 20 ft to 50 ft with 6 points (20, 26, 32, . . ., 50 ft). Elastic settlement properties and shaft modulus are shown in Fig. 3.17.

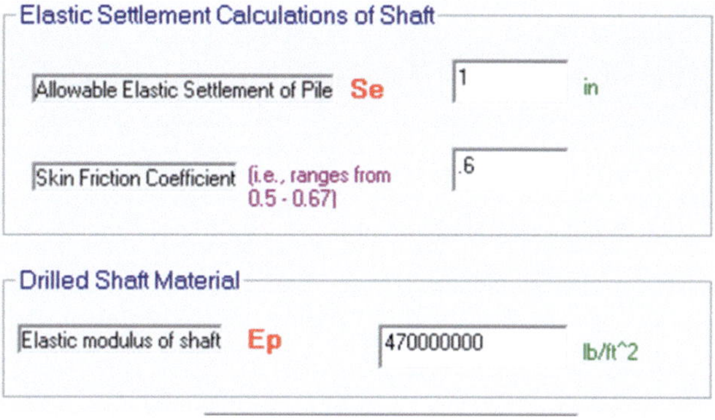

Fig. 3.17 Elastic settlement and shaft properties

Soil Layer No.	From Depth (ft)	To Depth (ft)	Cohesion (lb/ft^2)	Friction Angle (deg.)	Effective Unit Weight (lb/ft^3)	Elastic Modulus of Soil (lb/ft^2)	Poisson's Ratio
1	0	40	835	0	47.6	135000	.3
2	40	100	1350	0	53	166000	.3

Fig. 3.18 Soil layer properties and thicknesses

General | Drilled Shaft Information | Soil Properties | **OUTPUT**

	L (ft)	Qs (lb)	Qp (lb)	Qu (lb)	Qall (lb)	Qwp (lb)	Qws (lb)	Qw (lb)
1	20	106083.729943945	27254.1890252848	133337.918969229	33334.4797423073	19862.3418623039	77311.8329890551	97174.174851359
2	26	137908.848927129	27254.1890252848	165163.037952414	41290.7594881035	19515.7247445257	98751.4665360001	118267.191280526
3	32	169733.967910312	27254.1890252848	196988.156935597	49247.0392338993	19193.0952294892	119530.990511564	138724.085741054
4	38	201559.086893496	27254.1890252848	228813.275918781	57203.3189796952	18885.2513160487	139666.383302968	158551.634619016
5	44	236747.862007426	41645.1860395063	278393.048046932	69598.2620117331	25726.5471734701	146252.319160596	171978.866334066
6	50	272702.405999116	41645.1860395063	315348.592037623	78837.1480094057	26267.732302606	166066.368461702	191334.000764207

Fig. 3.19 A snapshot of the axial capacity results from the OUTPUT section

Soil Properties Section

Two layers are specified. Then, the given thicknesses and physical properties of the soil layers are entered as shown in Fig. 3.18.

Now, we can save our progress and hit run to get the results. A snapshot from the OUTPUT section for the axial capacity results based on bearing capacity and elastic settlement at different shaft lengths is shown in Fig. 3.19. The allowable load-carrying capacity design chart is also shown in Fig. 3.20.

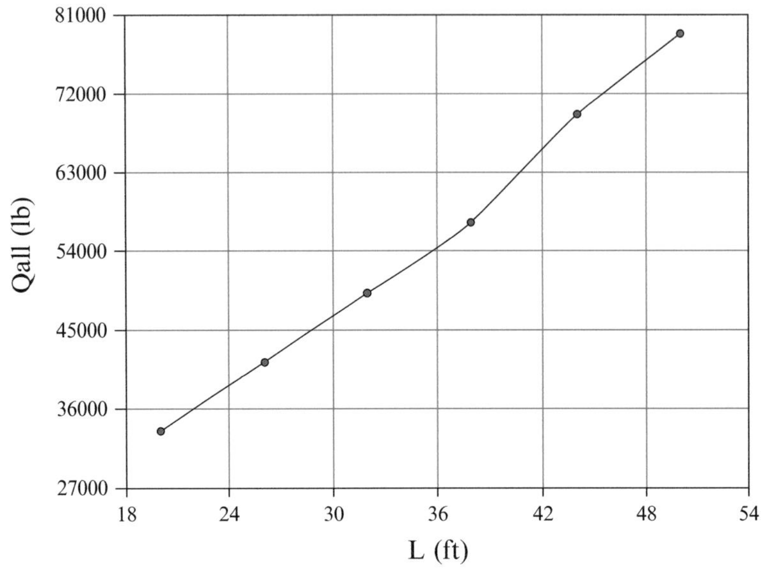

Circular Pile

Fig. 3.20 The allowable load-carrying capacity design chart

Table 3.7 Sand layer properties and thicknesses

Soil layer	Thickness (m)	Friction angle (°)	Effective unit weight (kN/m^3)	Elastic modulus of soil (kN/m^2)	Poisson's ratio
1	10	30	19	9,000	0.35
2	9	32	10	10,000	0.35
3	11	35	9.9	11,000	0.35

3.4.4 Bell Shaft in Nonhomogeneous Sandy Soil

Develop axial capacity design charts based on bearing capacity and elastic settlement for a belled drilled shaft (D_s = shaft diameter = 1.2 m and D_b = bell diameter = 1.5 m) in nonhomogeneous cohesionless sandy soil with properties as given in Table 3.7, and length of the shaft bell is 0.85 m. Use a factor of safety of 3.0 for bearing capacity calculations. Table 3.8 provides additional useful information to solve this problem.

Table 3.8 Design parameters

Property	Value	Unit
E_p	21,000,000	kN/m^2
δ	0.75ϕ	–
S_e	0.025	m
ε	0.6	–
$(L'/D_s)_{cr}$	18	–

3.4.4.1 Hand Solution

One must follow the steps below to determine the required axial capacity design charts:

Step 1:
Assume total shaft length, $L = 7$ m

Step 2:
Calculate the critical (I_{cr}) rigidity index from Eq. (3.7):

$$I_{cr} = 0.5\exp\left[2.85\cot\left(45 - \frac{30}{2}\right)\right] = 69.63$$

Before we can calculate the reduced rigidity index, we must first calculate the rigidity index from Eq. (3.9):

$$I_r = \text{soil rigidity index} = \frac{9,000}{2(1 + 0.35) \times (19 \times 7) \times \tan(30)} = 43.40$$

Then, we must calculate n from Eq. (3.11):

$$n = 0.005\left(1 - \left(\frac{30 - 25}{20}\right)\right) = 0.00375$$

The effective stress at the shaft base is found as follows:

$$q' = 19 \times 7 = 133\,\text{kN/m}^2$$

Also, we must calculate Δ from Eq. (3.10) as follows:

$$\Delta = 0.00375 \times \left(\frac{133}{100}\right) = 0.004985$$

Now, we may calculate the reduced rigidity index from Eq. (3.8):

$$I_{\mathrm{rr}} = \frac{43.40}{1 + (43.40 \times 0.004985)} = 35.68$$

Since $I_{\mathrm{rr}} < I_{\mathrm{cr}}$ we use Eq. (3.6) to determine compressibility factor:

$$F_{qc} = \exp\left\{(-3.8 \tan(30)) + \left[\frac{(3.07 \sin(30))(\log_{10}(2 \times 35.68))}{1 + \sin 30}\right]\right\} = 0.742$$

From Eq. (3.3):

$$F_{qs} = 1 + \tan(30) = 1.577$$

From Eq. (3.4):

$$F_{qd} = 1 + 2\tan(30)(1 - \sin(30))^2 \times \tan^{-1}\underbrace{\left[\frac{7}{1.5}\right]}_{\text{radians}} = 1.39$$

Thus, the end bearing capacity is calculated from Eq. (3.2):

$$Q_{\mathrm{p}} = (1.13) \times [(19 \times 7)(18.40 - 1)(1.577)(1.39)(0.742)] = 7{,}207\,\mathrm{kN}$$

Step 3:
Determine Q_{s} frictional resistance (skin friction).
To find the critical length, we use Eq. (3.18):

$$L' = 18 \times 0.8 = 14.4\,\mathrm{m}$$

At $z = 0\,\mathrm{m}$ we are between 0 and L':

$$\sigma_o' = 0$$

From Eq. (3.16):

$$f = 0$$

At $z = 7\,\mathrm{m}$ we are between 0 and L':

$$\sigma_o' = 7 \times 19 = 133\,\mathrm{kN/m^2}$$

From Eq. (3.17):

$$f = (1 - \sin(30)) \times 133 \tan(22.5) = 27.54\ \mathrm{kN/m^2}$$

From Eq. (3.15):

$$Q_s = \frac{0 + 27.54}{2} \times (3.141 \times 1.2) \times 7 = 363.45\,\text{kN}$$

Step 4:
From Eq. (3.1):

$$Q_u = 7,207 + 363.45 = 7,570\,\text{kN}$$

From Eq. (3.26):

$$Q_{all} = \frac{7,570}{3} = 2,523\,\text{kN}$$

Step 5:
From Eq. (3.27):

$$\eta = \left(\frac{7,570}{363}\right) = 20.8$$

From Eq. (3.28):

$$0.025 = Q_{wp}\left(\frac{\left(1 + \frac{0.6}{16.92}\right)7}{1.13 \times 21 \times 10^6} + \frac{1.5(1 - 0.35^2)0.85}{1.76 \times 9,000} + \frac{1.5(1 - 0.35^2) \times 2.84}{20.8 \times 3.76 \times 7 \times 9,000}\right)$$

$$Q_{wp} = 349.08\,\text{kN}$$

From Eq. (3.29):

$$Q_w = 349.08\left(1 + \frac{1}{16.92}\right) = 369.72\,\text{kN}$$

Step 6:
Repeat the above steps for as many shaft lengths as needed to have enough data points for the development of the required capacity design charts.

Step 1 (second shaft length):
Assume $L = 13$ m

Step 2 (second shaft length):
Calculate the critical (I_{cr}) rigidity index from Eq. (3.7):

$$I_{\mathrm{cr}} = 0.5\exp\left[2.85\cot\left(45 - \frac{32}{2}\right)\right] = 85.48$$

Before we can calculate the reduced rigidity index, we must first calculate the rigidity index from Eq. (3.9):

$$I_{\mathrm{r}} = \frac{10,000}{2(1+0.35)\times(10\times 13)\times\tan(32)} = 45.59$$

Also, we must calculate n from Eq. (3.11):

$$n = 0.005\left(1 - \left(\frac{32-25}{20}\right)\right) = 0.00325$$

We must also calculate Δ from Eq. (3.10) as follows:

$$\Delta = 0.00325\times\left(\frac{220}{100}\right) = 0.00715$$

Now, we may calculate the reduced rigidity index from Eq. (3.8):

$$I_{\mathrm{rr}} = \frac{45.59}{1+(45.59\times 0.00715)} = 34.38$$

Since $I_{\mathrm{rr}} < I_{\mathrm{cr}}$ we use Eq. (3.6) to calculate the compressibility factor:

$$F_{qc} = \exp\left\{(-3.8\tan(32)) + \left[\frac{(3.07\sin(32))\left(\log_{10}(2\times 34.38)\right)}{1+\sin 32}\right]\right\} = 0.656$$

From Eq. (3.3):

$$F_{qs} = 1 + \tan(32) = 1.62$$

From Eq. (3.4):

$$F_{qd} = 1 + 2\tan(32)(1-\sin(32))^2 \times \tan^{-1}\underbrace{\left[\frac{13}{1.5}\right]}_{\text{radians}} = 1.40$$

Therefore,

The end bearing capacity is determined from Eq. (3.2) as follows:

$$Q_{\mathrm{p}} = (1.13)\times[(220)(23.18-1)(1.62)(1.40)(0.656)] = 10,888\,\mathrm{kN}$$

Step 3 (second shaft length):
The critical depth can be estimated using Eq. (3.18) as follows:

$$L^{'} = 18 \times 0.8 = 14.4\,\text{m}$$

At $z = 0\,\text{m}$ we are between 0 and L':

$$\sigma_o^{'} = 0$$

From Eq. (3.16):

$$f = 0$$

At $z = 10\,\text{m}$ we are between 0 and L':

$$\sigma_o^{'} = 10 \times 19 = 190\,\text{kN/m}^2$$

From Eq. (3.17):

$$f = (1 - \sin(30)) \times 190 \tan(22.5) = 39.35\,\text{KN/m}^2$$

From Eq. (3.15):

$$Q_\text{s} = \frac{0 + 39.35}{2} \times (3.141 \times 1.2) \times 10 = 741.73\,\text{kN}$$

At $z = 13\,\text{m}$ we are between 0 and L':

$$\sigma_o^{'} = 190 + 3 \times 10 = 220\,\text{kN/m}^2$$

From Eq. (3.16):

$$f = (1 - \sin(32)) \times 217.8 \tan(24) = 46.04\,\text{kN/m}^2$$

From Eq. (3.15):

$$Q_\text{s} = \frac{25.45 + 29.35}{2} \times (3.141 \times 1.2) \times 3 = 482.89\,\text{kN}$$

Thus,

$$Q_\text{s} = 741.73 + 482.89 = 1{,}224\,\text{kN}$$

Step 4 (second shaft length):
From Eq. (3.1):

$$Q_u = 10,888 + 1,224 = 12,116\,\text{kN}$$

From Eq. (3.26):

$$Q_{all} = \frac{12,116}{3} = 4,038\,\text{kN}$$

Step 5 (second shaft length):
From Eq. (3.27):

$$\eta = \left(\frac{10,888}{1,224}\right) = 8.9$$

From Eq. (3.28):

$$0.025 = Q_{wp}\left(\frac{\left(1+\frac{0.6}{10.389}\right)13}{1.13\times 21,000,000} + \frac{1.5\left(1-0.35^2\right)\times 0.85}{1.76\times 10,000} + \frac{1.13\left(1-0.35^2\right)\times 3.151}{8.9\times\pi\times 1.2\times 13\times 10,000}\right)$$

$$Q_{wp} = 386.5\,\text{kN}$$

From Eq. (3.29):

$$Q_w = 386.5\times\left(1+\frac{1}{8.9}\right) = 430\,\text{kN}$$

Step 6 (second shaft length):
Steps 1 through 5 can be repeated for as many shaft lengths as needed to develop the required design charts.

3.4.4.2 foundationPro Solution

General Section

SI unit is selected for this problem with a safety factor of 3 as provided. Other required design parameters for the skin friction in sand are entered as shown in Fig. 3.21.

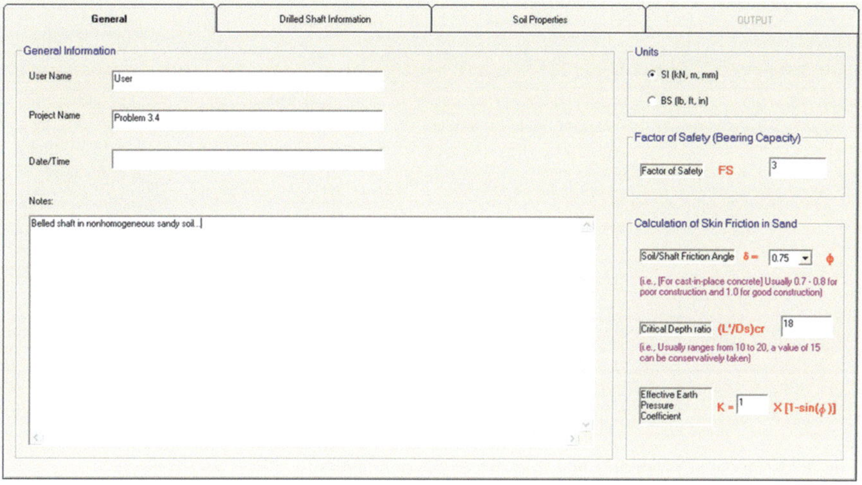

Fig. 3.21 General information section

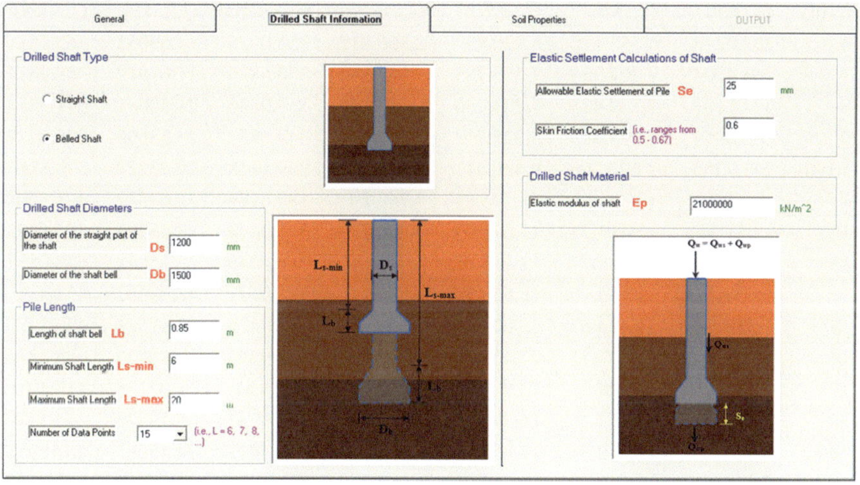

Fig. 3.22 Drilled shaft information

Drilled Shaft Information Section

In this section, the belled shaft type is selected. The shaft diameter and the bell diameter are also entered as 1,200 and 1,500 mm, respectively. The length of the shaft bell is entered in the provided textbox as 0.85 m as given in the problem. For the development of the design charts, various shaft lengths are considered from 6 m to 20 m with a total number of shaft lengths of 15 including the minimum and the maximum shaft lengths (i.e., 6, 7, 8, . . ., 20 m). Other required shaft information is shown in Fig. 3.22.

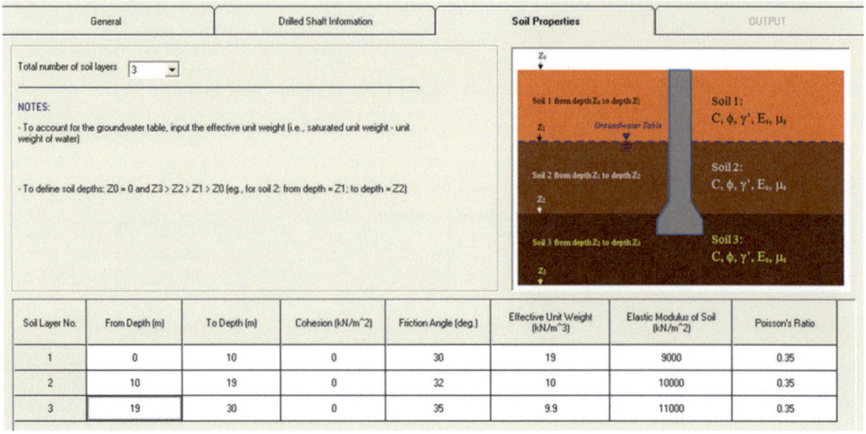

Fig. 3.23 Physical properties and thicknesses of sand layers

Table 3.9 Axial capacity results for different shaft lengths

L (m)	Q_s (kN)	Q_p (kN)	Q_u (kN)	Q_{all} (kN)	Q_{wp} (kN)	Q_{ws} (kN)	Q_w (kN)
6	267.0248	6,653.426	6,920.45	2,306.817	351.0925	14.09055	365.1831
7	363.4504	7,207.334	7,570.785	2,523.595	350.5825	17.67912	368.2617
8	474.7107	7,729.425	8,204.136	2,734.712	350.0846	21.50081	371.5854
9	600.8057	8,224.907	8,825.713	2,941.904	349.5957	25.53696	375.1327
10	741.7355	8,697.706	9,439.441	3,146.48	349.1138	29.77223	378.886
11	895.6489	10,331.22	11,226.87	3,742.29	387.6166	33.60382	421.2204
12	1,057.443	10,613.24	11,670.69	3,890.229	387.0444	38.56291	425.6073
13	1,227.178	10,888.77	12,115.95	4,038.649	386.4949	43.55848	430.0533
14	1,404.403	11,153.95	12,558.35	4,186.116	385.9621	48.59682	434.559
15	1,589.59	11,414.87	13,004.46	4,334.82	385.4426	53.67521	439.1179
16	1,783.894	11,677.17	13,461.06	4,487.02	384.9323	58.80525	443.7376
17	1,984.246	11,924.2	13,908.44	4,636.148	384.4305	63.97113	448.4016
18	2,193.97	12,173.26	14,367.23	4,789.076	383.9331	69.19574	453.1289
19	2,410.134	12,412.31	14,822.44	4,940.813	383.4412	74.4539	457.8951
20	2,636.408	15,553.53	18,189.94	6,063.312	421.9616	71.52478	493.4864

Soil Properties Section

One must select a total of three layers. Then, physical properties and thicknesses of sand layers must be entered as shown in Fig. 3.23.

Now we can save our progress and hit run to get the results. Axial capacity results for this belled shaft in nonhomogeneous sand at various shaft lengths are summarized in Table 3.9. Ultimate skin friction (bearing capacity analyses) design chart is also shown in Fig. 3.24. A snapshot from Shaft-1 application for the total working load results is shown in Fig. 3.25.

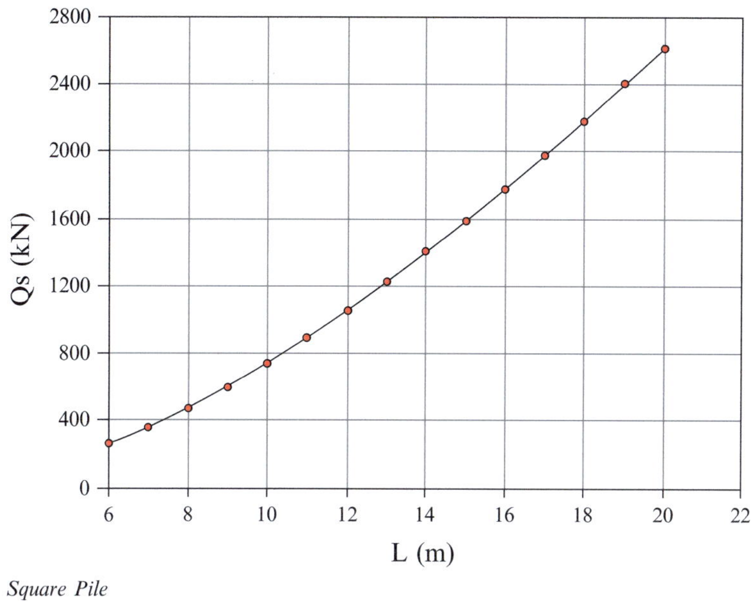

Fig. 3.24 Ultimate skin friction design chart

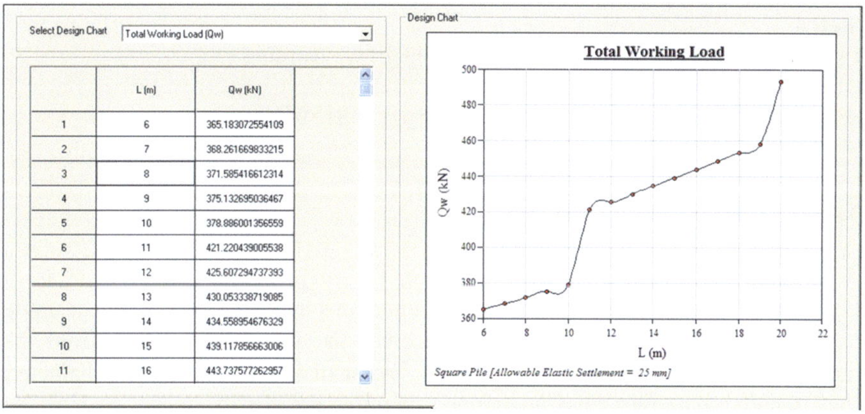

	L (m)	Qw (kN)
1	6	365.183072554109
2	7	368.261669833215
3	8	371.585416612314
4	9	375.132695036467
5	10	378.886001356559
6	11	421.220439005538
7	12	425.607294737393
8	13	430.053338719085
9	14	434.558954676329
10	15	439.117856663006
11	16	443.737577262957

Fig. 3.25 A snapshot from the OUTPUT section of the Shaft-1 application

3.4.5 *Straight Shaft in Rock*

Determine the axial capacity of a straight drilled shaft with a diameter of 1 m and shaft length of 4 m in rock (smooth socket). Consider a factor of safety of 3.0 and unconfined compression strength of 4,000 kN/m^2.

3.4.5.1 Hand Solution

Step 1:
First, we determine the force that the soil under the shaft tip can carry from Eq. (3.14):

The unconfined compression is 4,000 kN/m^2 and is equal to 4 MN/m^2.

For SI units, ω is 4.83.

Therefore,

$$Q_{\mathrm{p}} = \left[4.83 \times (4)^{0.51}\left(\frac{\pi}{4} \times 1^2\right)\right] = 7.693\,\mathrm{MN}$$

Step 2:
We determine the force that can be carried by the soil surrounding the shaft; this force is due to the friction between soil and the shaft surface. From Eq. (3.21):

$$Q_{\mathrm{s}} = 1.2566 \times 1 \times 4 \times \sqrt{4} = 10.053\,\mathrm{MN}$$

Step 3:
We use Eq. (3.1) to find the sum of the loads that the shaft foundation can withstand:

$$Q_{\mathrm{u}} = 7.693 + 10.053 = 17.746\,\mathrm{MN}$$

Step 4:
Using the factor of safety given in the problem we can find Q_{all}:

$$Q_{\mathrm{all}} = 17.746/3 = 5.915\,\mathrm{MN}$$

3.4.5.2 foundationPro Solution

After launching Shaft-2 of foundationPro, the main screen will appear as in Fig. 3.26. For the axial capacity of shaft foundations in rock, we only have a single page in foundationPro. In this page we enter the given information from the problem statement as shown in the figure: factor of safety, units, shaft diameter, shaft length, shaft type, and unconfined compression for the rock (we must be careful in entering this value in kN/m^2).

Then, we hit run! The results are shown in Fig. 3.27. One can see that the Shaft-2 application will calculate soil resistance at the shaft tip (Q_{p}), as well as the skin friction resistance (Q_{s}). In addition, the program will calculate the ultimate (Q_{u}) and allowable (Q_{all}) force that the foundation can resist.

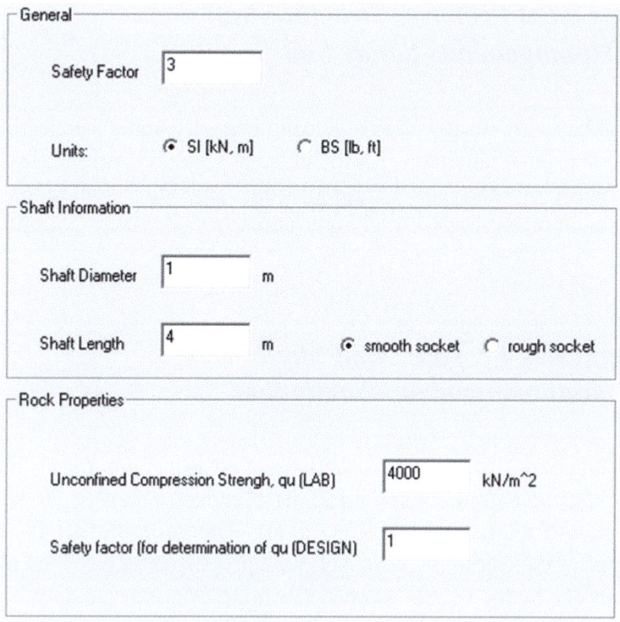

Fig. 3.26 Input section for the straight shaft in rock

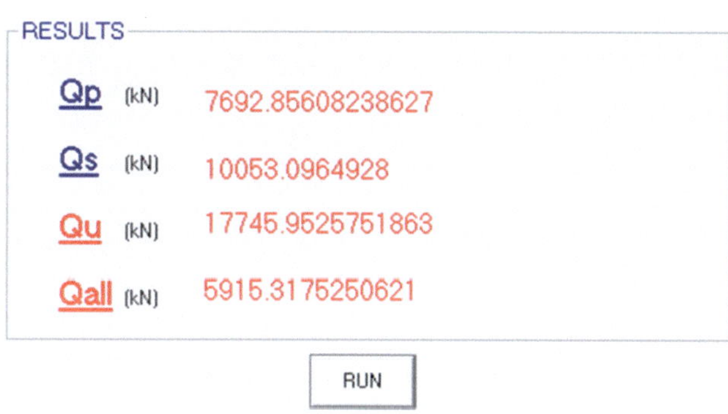

Fig. 3.27 Axial capacity results

3.5 Suggested Projects

In this section, you will find some suggested design projects to allow the reader practice the concepts and the ideas discussed in the previous sections. These suggested projects will cover a variety of drilled shaft types in rock and in homogeneous and nonhomogeneous clayey and sandy soils.

3.5.1 Suggested Projects: Straight Shaft in Homogeneous Sandy Soil

Develop axial capacity design charts (bearing capacity and elastic settlement) for a straight drilled shaft (diameter = 1.5 m) in homogeneous cohesionless sandy soil. Consider a factor of safety of 3.5 for bearing capacity calculations. Table 3.10 provides physical properties of sand layer and other necessary parameters.

3.5.2 Suggested Projects: Bell Shaft in Nonhomogeneous Sandy Soil

Develop bearing capacity and elastic settlement design charts for the axial capacity of a drilled shaft (shaft diameter = 1.2 m) that has a bell at its tip (length of bell = 0.75 m and diameter of bell = 1.5 m). The drilled shaft is embedded in nonhomogeneous cohesionless sandy soil with properties as provided in Table 3.11. Consider a safety factor of 3.0 when performing bearing capacity computations. Other useful information is also provided in Table 3.12.

Table 3.10 Soil properties and other design parameters

Property	Value	Unit
c	0	kN/m^2
ϕ	30	°
γ	20	kN/m^3
E_s	25,000	kN/m^2
μ_s	0.35	–
E_p	23,000,000	kN/m^2
δ	0.75ϕ	–
S_e	0.025	m
$(L'/D_s)_{cr}$	16	–
ε	0.6	–

Table 3.11 Thicknesses and physical properties of sand layers

Soil layer	Thickness (m)	Friction angle (°)	Effective unit weight (kN/m^3)	Elastic modulus of soil (kN/m^2)	Poisson's ratio
1	5	30	17	9,000	0.35
2	10	35	15	10,000	0.35
3	15	40	13	11,000	0.35

Table 3.12 Additional required design parameters

Property	Value	Unit
E_p	21,000,000	kN/m^2
δ	0.75ϕ	–
S_e	0.025	m
ε	0.6	–
$(L'/D_s)_{cr}$	18	–

3.5.3 Suggested Projects: Straight Shaft in Rock

Determine the axial design capacity of a straight drilled shaft with a diameter of 2 m and a length of 8 m in rock with rough socket. Consider a safety factor of 5.0 for bearing capacity analysis and unconfined compression strength of 7,000 kN/m^2.

References

Chen Y-J, Kulhawy FH. Case history evaluation of the behavior of drilled shafts under axial and lateral loading. Final report, project 1493-04, EPRI TR-104691, Geotechnical group, Cornell University, Ithaca, NY; December, 1994.

Das B. Principles of foundation engineering. Stamford, CT, Cengage Learning; 2010.

Kulhawy FH, Jackson CS. Some observations on un-drained side resistance of drilled shafts. Proceedings, Foundation Engineering: Current Principles and Practices, ASCE Jour. of Soil Mech. 1989;2:1011–25.

Meyerhof G. The bearing capacity of foundation under eccentric and inclined loads. Journal of Soil Mechanics and Foundations Division, ASCE. 1963;97:95–117.

O'Neill MW, Reese LC. Drilled shafts: construction procedure and design methods. FHWA report no. IF-99-025. 1999.

Zhang L, Einstein H. End bearing capacity of drilled shafts in rock. Journal of Geotechnical and Geoenvironmental Engineering, American Society of Civil Engineers. 1998;124(7):574–84.

Further Readings

Bowles JE. Foundation analysis and design. New York: McGraw-Hill; 1996.

Coduto D. Foundation design principle and practices. New Jersey: Prentice-Hall Inc.; 2001.

O'Neill MW, Reese LC. Drilled shafts: construction procedure and design methods. FHWA report no. IF-99-025 http://trid.trb.org/view.aspx?id=740721 http://isddc.dot.gov/OLPFiles/FHWA/009752.pdf 1999.

Vesic S. Design of pile foundations. National cooperative highway research program synthesis of practice no. 42. Washington, DC: Transportation Research Board; 1977.

Chapter 4
Design of Mechanically Stabilized Earth Retaining Walls

Abstract This chapter deals with the design of mechanically stabilized earth retaining wall using strip reinforcement. Internal and external stability requirements of an MSE wall with strip reinforcement are discussed in details in this chapter. For internal stability of an MSE wall, safety against pullout and breakage of reinforcing strips is checked. For external stability, safety against overturning, sliding, and bearing capacity is also checked. The overall safety of the MSE wall (internal and external) was performed considering various design parameters such as constant/varying strip length and spacing between the strips (along the wall height and the wall length) under different types of applied surcharge loading conditions (line load, strip load, and embankment load). Additionally, a step-by-step procedure was introduced in this chapter to perform internal and external stability requirements for safe and economical designs. A number of design problems are also presented in this chapter and its solutions are explained in details. These problems are first hand-solved, and then, resolved using the MSE Wall-1 application of the foundationPro program. Finally, two design projects are suggested at the end of this chapter to allow the reader practice the concepts learned herein.

Keywords MSE wall • Reinforcing strips • Pullout • Breakage • Bearing capacity • Sliding • Overturning • foundationPro

4.1 Introduction

This chapter deals with the design of mechanically stabilized earth retaining wall structures using reinforcing strips. As these MSE walls with strip reinforcement are constructed to retain soil behind it, thin and long reinforcing strips (ties) are one of its main design components. The design of these walls mainly is of twofold: (1) internal stability which can be generally satisfied when the reinforcing strips (a) resist pullout forces as a result of the developed horizontal stresses on the wall from the soil and any additional applied surcharges, and (b) withstand the vertical forces and don't break as a result of the vertical stresses on them from the soil and any additional applied loads, and (2) external stability which can be satisfied by ensuring that the wall with the surrounding soil and the attached reinforcing strips to the wall act as one unit and this unit (a) doesn't slide due to the horizontal

M. Yamin, *Problem Solving in Foundation Engineering using foundationPro*,
DOI 10.1007/978-3-319-17650-5_4

stresses, (b) doesn't overturn due to developed overturning moments, and (c) doesn't make the foundation soil (soil underneath this unit) to fail and not to bear the unit weight. Procedures and equations to satisfy the above aspects required for a safe and economical design are discussed in details in this chapter.

Chief design parameters that are responsible for satisfying internal and external stabilities of MSE walls are thoroughly explained herein, which include strip lengths (L), strip thickness (t), strip width (w), number of strip levels and their depths (n and Z_i) along the wall height, and horizontal spacing between the reinforcing strips (S_H) across the wall at different strip levels (depths). The effect of various types of additional applied surcharge loading on top of soil is also included in the design process.

A step-by-step procedure is introduced in this chapter to allow safe and economical design of an MSE wall in a systematic way through considering equal or unequal strip lengths, equal or unequal vertical distances between strip levels along the wall height, and constant or varying horizontal spacing at different strip levels along the wall length.

Several design problems are presented in this chapter. First, these design problems are hand-solved and their solution was explained in details, and then the foundationPro program was used to resolve these problems to replicate and verify the hand solution. Also, the program was used to investigate a wider and detailed solution and design alternatives for the hand-solved problems. Since the foundationPro includes a set of several applications, the MSE Wall-1 application of the foundationPro is the responsible application to perform MSE Wall analysis and design. Two design projects are suggested at the end of this chapter to allow the reader to practice and apply the learned concepts.

4.2 Theory

In this section, three main topics are introduced: how to estimate the different types of stresses (horizontal and vertical) due to earth pressures and additional surcharge loading on top of soil the MSE wall must be designed for, and how to perform internal and external stability calculations of mechanically stabilized earth walls.

First, let us consider designing an MSE wall with a height of H as depicted in Fig. 4.1 to retain a granular backfill soil behind it that has a unit weight of γ_1 and friction angle of ϕ_1'. Also, let us give the unit weight, friction angle, and cohesion of the foundation soil (soil underneath the wall) as γ_2, ϕ_2', and c_2', respectively. To safely design this wall and prevent the retained soil from moving and pushing the wall, one needs to arrange the reinforcing strips at a horizontal spacing of $S_H : (S_{H(1)}, S_{H(2)}, S_{H(3)})$ across the wall at different levels (depths), where $S_{H(1)}$ is the horizontal spacing of the strips at level 1, $S_{H(2)}$ is the horizontal spacing of the strips at level 2, and so on (see Fig. 4.1). Additionally, one must select the depths (levels) where the strips must be placed vertically along the wall height. These depths or levels are Z_1, Z_2,

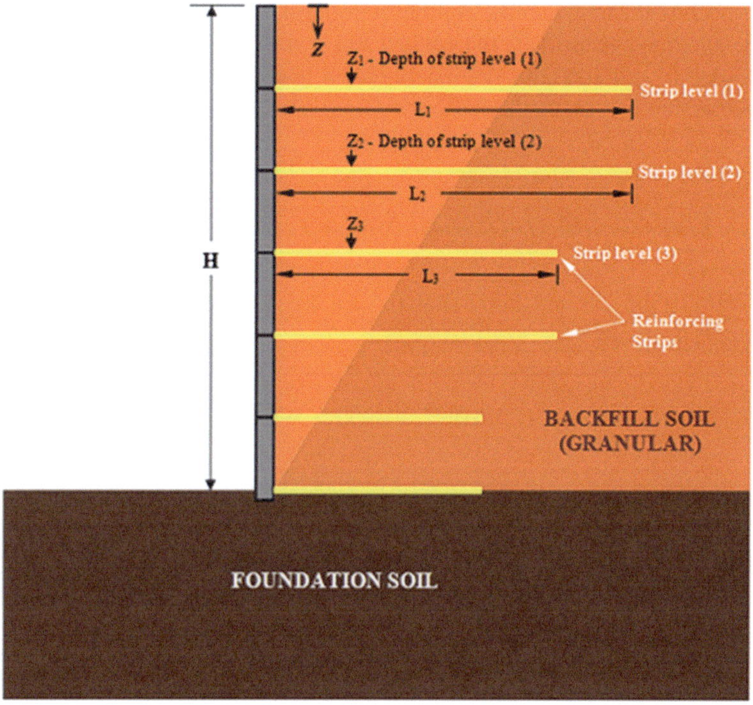

Fig. 4.1 Analysis of MSE wall

..., Z_i, ..., Z_n, where n is the total number of strip levels and Z_i is the depth from the ground surface to the reinforcing strip at level i. Therefore, the effective vertical distances for each reinforcing strip at each strip level, S_v : ($S_{v(1)}$, $S_{v(2)}$, ..., $S_{v(i)}$, ..., $S_{v(n)}$), can be calculated using Eqs. (4.1) through (4.3), where $S_{v(1)}$ is the effective vertical distance the reinforcing strips at strip level 1 (top) will be responsible for and can be found using Eq. (4.1), $S_{v(i)}$ is the effective vertical distance the reinforcing strips at any level i (between levels 1 and n) will be responsible for and can be found using Eq. (4.3), $S_{v(n)}$ is the effective vertical distance the reinforcing strips at strip level n (bottom) will be responsible for and can be found using Eq. (4.3), and so on:

$$S_{v(1)} = \frac{Z_1 + Z_2}{2} \tag{4.1}$$

$$S_{v(i)} = \frac{Z_{i+1} - Z_{i-1}}{2} \tag{4.2}$$

$$S_{v(n)} = \frac{2H - Z_n - Z_{n-1}}{2} \tag{4.3}$$

After selecting the total number of reinforcing strip levels and their depths along the wall height, one can estimate the vertical and horizontal stresses at each strip level (depth) due to soil and any additional applied loading as will be discussed in the next subsection.

4.2.1 *Vertical and Horizontal Stresses*

Total vertical ($\sigma'_{v(T)}$) and horizontal ($\sigma'_{h(T)}$) stresses can be obtained using the following equations:

$$\sigma'_{v(T)} = \sigma'_{v(s)} + \sigma'_{v(load)} \tag{4.4}$$

$$\sigma'_{h(T)} = \sigma'_{h(s)} + \sigma'_{h(load)} \tag{4.5}$$

where

$\sigma'_{v(s)}$ and $\sigma'_{h(s)}$ are the vertical and horizontal stresses due to the soil, respectively

$\sigma'_{v(load)}$ and $\sigma'_{h(load)}$ are the vertical and horizontal stresses due to the additional applied surcharge loading on top of soil, respectively

These different types of stresses can be found as in the following subsections.

4.2.1.1 Stresses Due to Soil

Vertical and horizontal stresses due to the soil can be calculated using the following equations:

$$\sigma'_{v(s)} = \gamma_1 z \tag{4.6}$$

$$\sigma'_{h(s)} = K_a \gamma_1 z \tag{4.7}$$

where

K_a is the Rankine's active lateral earth pressure coefficient and can be found using Eq. (4.8):

$$K_a = \tan^2\left(45 - \frac{\phi'_1}{2}\right) \tag{4.8}$$

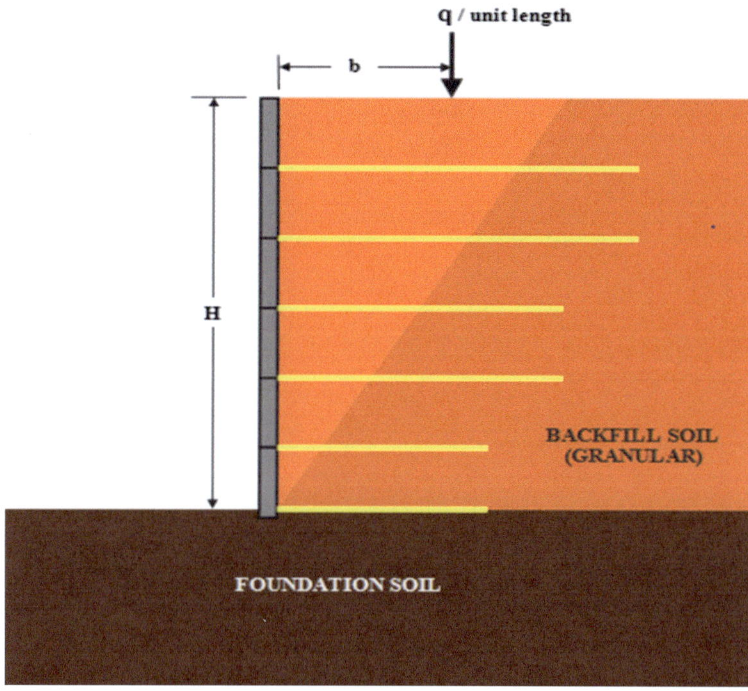

Fig. 4.2 Analysis of MSE wall with a line load

4.2.1.2 Stresses Due to Surcharge Loading

Several surcharge loading cases on top of soil are considered herein, such as line load, strip load, and embankment load. Vertical and horizontal stresses due to these different cases of loading are estimated as below.

Line load (see Fig. 4.2)

Based on the theory of elasticity, the vertical and horizontal stresses due to line load as derived by (Das 1997) can be found as follows:

$$\sigma'_{\mathrm{v(load)}} = \frac{2qz^3}{\pi\left(b^2 + z^2\right)^2} \tag{4.9}$$

$$\sigma'_{\mathrm{h(load)}} = \frac{2qb^2z}{\pi\left(b^2 + z^2\right)^2} \tag{4.10}$$

However, Das (2010) suggested that horizontal stress can be taken as in the following equations to account for the plastic behavior of soil:

$$\sigma'_{h(\text{load})} = \frac{4m_1^3 m_2}{\pi H\left(m_1^2 + m_2^2\right)} \quad (\text{For } m_1 > 0.4) \tag{4.11}$$

$$\sigma_{h(\text{load})} = \frac{0.203 q m_2}{H\left(0.16 + m_2^2\right)^2} \quad (\text{For } m_1 \leq 0.4) \tag{4.12}$$

where

$$m_1 = \frac{b}{H} \tag{4.13}$$

$$m_2 = \frac{z}{H} \tag{4.14}$$

– Strip load (see Fig. 4.3)

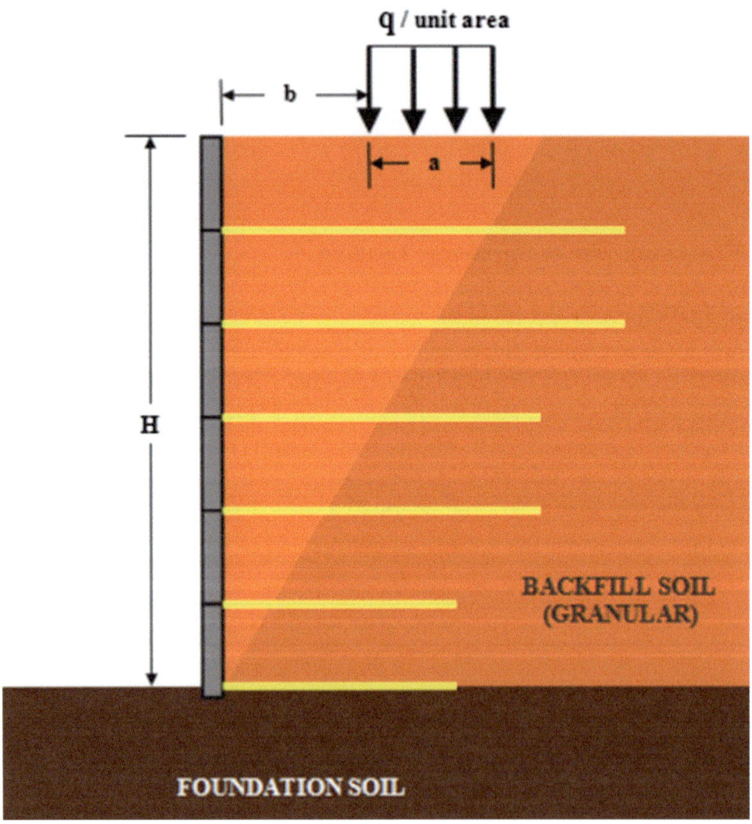

Fig. 4.3 Analysis of MSE wall with a strip load

Vertical and horizontal stresses due to strip loading can be determined according to (Laba and Kennedy 1986) as in the following equations:

The vertical stress due to strip load is expressed by

$$\sigma'_{\mathrm{v(load)}} = \frac{qa}{a+z} \quad (\text{For } z \leq 2b) \tag{4.15}$$

$$\sigma'_{\mathrm{v(load)}} = \frac{qa}{a+\frac{z}{2}+b} \quad (\text{For } z > 2b) \tag{4.16}$$

Also, the horizontal stress due to strip load is expressed by

$$\sigma'_{\mathrm{h(load)}} = M\left[\frac{2q}{\pi}(\beta - \sin\beta\cos 2\alpha)\right] \tag{4.17}$$

where

$$M = 1.4 - \frac{0.4b}{0.14H} \geq 1 \tag{4.18}$$

$$\alpha = \tan^{-1}\left(\frac{b+\frac{a}{2}}{z}\right) \tag{4.19}$$

$$\beta = \tan^{-1}\left(\frac{b+a}{z}\right) - \tan^{-1}\left(\frac{b}{z}\right) \tag{4.20}$$

– Embankment load (see Fig. 4.4)

The vertical stress due to embankment load can be determined using the following equation:

$$\sigma'_{\mathrm{v(load)}} = \sigma'_{\mathrm{v(load1)}} + \sigma'_{\mathrm{v(load2)}} + \sigma'_{\mathrm{v(load3)}} \tag{4.21}$$

where

$$\sigma'_{\mathrm{v(load1)}} = q^*b^*\frac{\left(\tan^{-1}\left(\frac{Z}{-b}\right) - \tan^{-1}\left(\frac{Z}{-b-a_1}\right)\right)}{(a_1{}^*\pi)} - q^*Z^*\frac{\left(\frac{(-b-a_1)}{(-b-a_1)^2+Z^2}\right)}{\pi} \tag{4.22}$$

$$\sigma'_{\mathrm{v(load2)}} = q^*x^*\frac{\left(\tan^{-1}\left(\frac{Z}{x}\right) - \tan^{-1}\left(\frac{Z}{x-a_3}\right)\right)}{(-a_3{}^*\pi)} - q^*Z^*\frac{\left(\frac{(X-a_3)}{(X-a_3)^2+Z^2}\right)}{\pi} \tag{4.23}$$

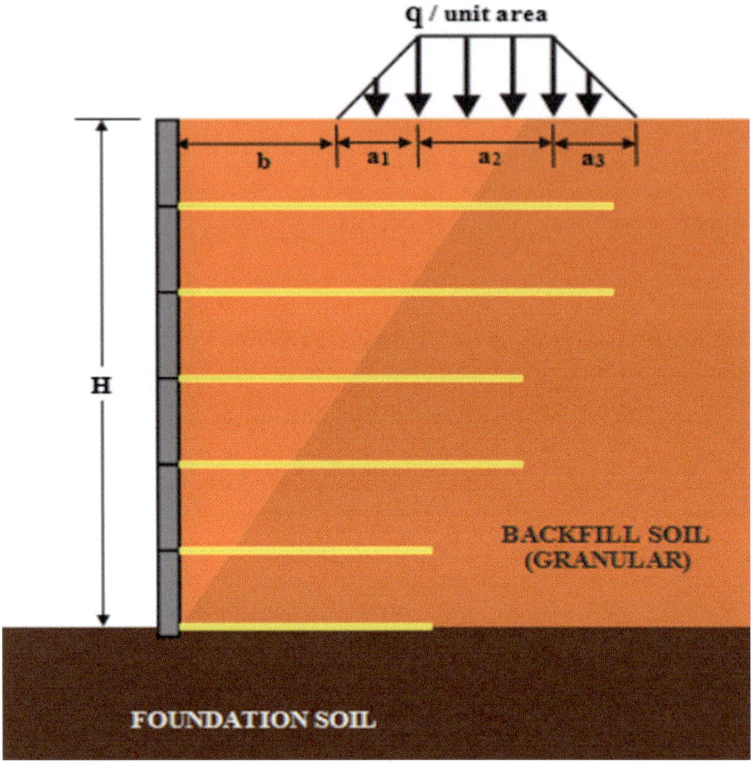

Fig. 4.4 Analysis of MSE wall with an embankment load

The variable x above is described in the following equations:

$$x = a_1 + a_{2-} + a_3 + b \tag{4.24}$$

$$\sigma'_{\mathrm{v(load3)}} = \frac{q^* a_2}{a_2 + Z} \quad \text{For } Z \le 2(a_1 + b) \tag{4.25}$$

$$\sigma'_{\mathrm{v(load3)}} = \frac{q^* a_2}{a_1 + a_2 + b + \frac{Z}{2}} \quad \text{For } Z > 2(a_1 + b) \tag{4.26}$$

Additionally, the horizontal stress due to embankment load is expressed in the following equation:

$$\sigma'_{\mathrm{h(load)}} = \sigma'_{\mathrm{h(load1)}} + \sigma'_{\mathrm{h(load2)}} + \sigma'_{\mathrm{h(load3)}} \tag{4.27}$$

where

$$\sigma'_{\mathrm{h(load1)}} = \boldsymbol{Z}^*\boldsymbol{q}^*\frac{\mathrm{Log}\left(\frac{(-\boldsymbol{b}-\boldsymbol{a}_1)^2+\boldsymbol{Z}^2}{(-\boldsymbol{b})^2+(\boldsymbol{Z})^2}\right)}{(\boldsymbol{\pi}^*\boldsymbol{a}_1)} + (\boldsymbol{b})^*\boldsymbol{q}^*\frac{\left(\tan^{-1}\left(\frac{\boldsymbol{Z}}{-\boldsymbol{b}+\boldsymbol{a}_1}\right) - \tan^{-1}\left(\frac{\boldsymbol{Z}}{-\boldsymbol{b}}\right)\right)}{(\boldsymbol{\pi}^*\boldsymbol{a}_1)} + \boldsymbol{q}^*\boldsymbol{Z}^*\frac{\left(\frac{-\boldsymbol{b}-\boldsymbol{a}_1}{(-\boldsymbol{b}-\boldsymbol{a}_1)^2+\boldsymbol{Z}^2}\right)}{\boldsymbol{\pi}} \tag{4.28}$$

$$+\sigma'_{\mathrm{h(load2)}} = \boldsymbol{Z}^*\boldsymbol{q}^*\frac{\mathrm{Log}\left(\frac{(\boldsymbol{x}-\boldsymbol{a}_3)^2+\boldsymbol{Z}^2}{(\boldsymbol{x})^2+(\boldsymbol{Z})^2}\right)}{(\boldsymbol{\pi}^*\boldsymbol{a}_3)} - (\boldsymbol{x})^*\boldsymbol{q}^*\frac{\left(\tan^{-1}\left(\frac{\boldsymbol{Z}}{\boldsymbol{x}+\boldsymbol{a}_3}\right) - \tan^{-1}\left(\frac{\boldsymbol{Z}}{\boldsymbol{x}}\right)\right)}{(\boldsymbol{\pi}^*\boldsymbol{a}_3)} + \boldsymbol{q}^*\boldsymbol{Z}^*\frac{\left(\frac{\boldsymbol{x}-\boldsymbol{a}_3}{(\boldsymbol{x}-\boldsymbol{a}_3)^2+\boldsymbol{Z}^2}\right)}{\boldsymbol{\pi}} \tag{4.29}$$

The variable x above is described in Eq. (4.24):

$$\sigma'_{\mathrm{h(load3)}} = m\left[\frac{2q}{\pi}(\beta - \sin\beta\cos 2\alpha)\right] \tag{4.30}$$

where α, β, and m in Eq. (4.30) can be determined from the following equations:

$$\alpha = \tan^{-1}\left(\frac{a_1+b+\frac{a_2}{2}}{Z}\right) \tag{4.31}$$

$$\beta = \tan^{-1}\left(\frac{a_1+b+a_2}{Z}\right) - \tan^{-1}\left(\frac{a_1+b}{Z}\right) \tag{4.32}$$

$$m = 1.4 - \frac{0.4^*(a_1+b)}{0.14^*H} \geq 1 \tag{4.33}$$

4.2.2 *Internal Stability*

Internal stability of an MSE wall is acquired when the breakage and pullout of the reinforcing strips at each strip level (depth) are prevented. First, the maximum horizontal spacing ($S_{\text{H-max}}$) at each strip level is obtained by satisfying the breakage condition. The actual design horizontal spacing ($S_{\text{H-design}}$) can be taken as a single value for all strip levels or can be varying with strip levels (e.g., one horizontal spacing value for strip levels 1, 2, and 3, and another horizontal spacing value for the remaining strip levels). In any case, the design horizontal spacing ($S_{\text{H-design}}$) at a strip level must not exceed the maximum horizontal spacing ($S_{\text{H-max}}$) at that strip level. Then, after selecting the horizontal spacing at each strip levels, the minimum required length (L_{min}) of the reinforcing strips at each level is obtained by satisfying the pullout condition. Once again, the actual design length (L_{design}) can be taken as a single value for all strip levels or can be varying with strip levels (e.g., one strip length for strip levels 1, 2, and 3, and another strip length for the remaining strip levels). In any case, the design length (L_{design}) at a strip level must not be less than the minimum required length (L_{min}) of the reinforcing strips at that strip level.

The following subsections summarize the equations needed to determine horizontal spacing between reinforcing strips and length of reinforcing strips at each strip level to satisfy internal stability of the MSE wall.

4.2.2.1 Breakage of Reinforcing Strips

For an assumed strip thickness (t) and width (w), the maximum horizontal spacing between reinforcing strips ($S_{\text{H-max}}$) at each strip level to meet the minimum factor of safety against breakage for any strip ($\text{FS}_{\text{B-min}}$) is defined by

$$S_{\text{H-max}} = \frac{wtf_{\text{y}}}{\sigma'_{\text{h(T)}} S_{\text{V}} \text{FS}_{\text{B-min}}} \tag{4.34}$$

where

f_{y} is the yielding strength of the reinforcing strip material

After calculating the maximum horizontal spacing ($S_{\text{H-max}}$) at each strip level, the design horizontal spacing ($S_{\text{H-design}}$) at each strip level can be selected to account for practical and economical considerations.

One can also determine the actual applied breakage/pullout force (F_{BP}) at each strip level and the maximum breaking force ($F_{\text{B-max}}$) the reinforcing strips can withstand using the following equations:

$$F_{\text{BP}} = \sigma'_{\text{h(T)}} S_{\text{V}} S_{\text{H-design}} \tag{4.35}$$

$$F_{\text{B-max}} = wtf_{\text{y}} \tag{4.36}$$

The actual factor of safety against breakage (FS_B) at each strip level based on the selected horizontal spacing ($S_{\text{H-design}}$) between strips is defined by

$$FS_B = \frac{F_{\text{B-max}}}{F_{\text{BP}}} \tag{4.37}$$

4.2.2.2 Pullout of Reinforcing Strips

The minimum strip length ($L_{\min}$) at each level (depth) to meet the factor of safety against strip pullout (FS_P) is expressed by

$$L_{\min} = \frac{(H-z)}{\tan\left(45 + \frac{\phi_1'}{2}\right)} + \frac{FS_P F_{\text{BP}}}{2w\sigma'_{\text{v(T)}} \tan \phi'_\mu} \tag{4.38}$$

where

ϕ'_μ is the friction angle between the soil and the reinforcing strip

F_{BP} is the actual applied breakage/pullout force at each strip level and obtained from Eq. (4.35)

After calculating the minimum strip length ($L_{\min}$) at each strip level, the design strip length (L_{design}) at each level can be selected to account for practical and economical considerations.

One can also determine the maximum pullout force ($F_{\text{P-max}}$) any reinforcing strip can withstand at each strip level as a result of the friction between the soil and the strips using the following equation:

$$F_{\text{P-max}} = \left(2w\sigma'_{\text{v(T)}} \tan\left(\phi'_\mu\right)\right) \times \left(L_{\text{design}} - \frac{H-z}{\tan\left(45 + \frac{\phi'}{2}\right)}\right) \tag{4.39}$$

Therefore, the actual factor of safety against pullout (FS_P) at each strip level based on the final selections of the horizontal spacing ($S_{\text{H-design}}$) between strips and the design strip length (L_{design}) is obtained from

$$FS_P = \frac{F_{\text{P-max}}}{F_{\text{BP}}} \tag{4.40}$$

4.2.3 *External Stability*

After selecting the final design horizontal spacing between strips and strip length at each strip level to meet the internal stability requirements ($FS_{\text{P-min}}$ and $FS_{\text{B-min}}$), the wall with the attached reinforcing strips and the retained soil behind the wall are considered to act as a one unit/system. The stability of this one unit must be checked to avoid sliding, overturning, and bearing capacity. The following subsections discuss how to check the external stability of the MSE wall system against the mentioned three conditions (sliding, overturning, and bearing capacity).

4.2.3.1 Overturning

The overturning safety factor (FS_O) for the MSE wall must be greater than a minimum specified factor of safety for overturning ($FS_{O\text{-min}}$) to avoid wall overturning. One can determine FS_O through dividing the resisting moments (M_R) by the overturning moment (M_O) as in the following equation:

$$FS_O = \frac{M_R}{M_O} \geq FS_{O\text{-min}} \tag{4.41}$$

Thus, the overturning and resisting moments can be determined using the following equations:

$$M_O = \sum_{i=1}^{n} \left[\sigma'_{h(T)} \times S_{v(i)} \times (H - Z_i) \right] \tag{4.42}$$

$$M_R = M_{R(soil)} + M_{R(load)} \tag{4.43}$$

where

$$M_{R(soil)} = \frac{\gamma'_1}{2} \sum_{i=1}^{n} \left[S_{v(i)} \left(L_{design(i)} \right)^2 \right] \tag{4.44}$$

The resisting moment due to additional applied surcharge loading can be found depending on the loading type as follows: ($M_{R(load)} = 0$ when no surcharge loading is applied on top of soil):

$$M_{R(load)} = qb \text{ (For line load)} \tag{4.45}$$

$$M_{R(load)} = qa_2 \left[a_1 + \frac{a_2}{2} + b \right] \text{ (For strip load)} \tag{4.46}$$

$$M_{R(load)} = q \left[a_2 \left(a_1 + \frac{a_2}{2} + b \right) + \frac{a_1}{2} \left(\frac{2a_1}{3} + b \right) + \frac{a_3}{2} \left(a_1 + a_2 + \frac{a_3}{2} + b \right) \right]$$
$$\text{(For embankment load)} \tag{4.47}$$

4.2.3.2 Sliding

The sliding safety factor (FS_S) for the MSE wall must be greater than a minimum specified factor of safety for sliding ($FS_{O\text{-min}}$) to avoid wall sliding. One can determine FS_S through dividing the resisting/driving forces (F_R) by the driving force (F_D) as in the following equation:

$$FS_S = \frac{F_R}{F_D} \geq FS_{S\text{-min}} \tag{4.48}$$

So, the driving and resisting forces can be determined using the following equations:

$$F_D = \sum_{i=1}^{n} \left[\sigma'_{h(T)} \times S_{v(i)} \right] \tag{4.49}$$

$$F_R = F_{R(soil)} + F_{R(load)} \tag{4.50}$$

where

$$F_{R(soil)} = \left(\gamma'_1 \tan\left(\frac{2\phi'_1}{3} \right) \right) \left(\sum_{i=1}^{n} \left[S_{v(i)} L_{design(i)} \right] \right) \tag{4.51}$$

The resisting force due to additional applied surcharge loading can be found depending on the loading type as follows ($F_{R(load)} = 0$ when no surcharge loading is applied on top of soil):

$$F_{R(load)} = q \tan\left(\frac{2\phi'_1}{3} \right) \text{ (For line load)} \tag{4.52}$$

$$F_{R(load)} = qa \tan\left(\frac{2\phi'_1}{3} \right) \text{ (For strip load)} \tag{4.53}$$

$$F_{R(load)} = q\left(\frac{a_1}{2} + a_2 + \frac{a_3}{2} \right) \tan\left(\frac{2\phi'_1}{3} \right) \text{ (For embankment load)} \tag{4.54}$$

4.2.3.3 Bearing Capacity

The bearing capacity safety factor (FS_{BC}) for the MSE wall must be greater than a minimum specified factor of safety for bearing capacity ($FS_{BC\text{-min}}$) to avoid failure of foundation soil (soil beneath MSE wall). One can determine FS_{BC} through dividing the ultimate bearing capacity (q_u as in Chap. 1—Eq. (1.1)) by the total vertical stress at the base of the MSE wall (depth $z = H$) ($\sigma'_{\text{v(T) at } z=H}$) as in the following equation:

$$\text{FS}_{\text{BC}} = \frac{c_2 N_c + \frac{1}{2}\gamma_2 N_\gamma L_{\text{design}(n)}}{\sigma'_{\text{v(T) at } z=H}} \geq \text{FS}_{\text{BC-min}} \qquad (4.55)$$

In the above equation,

N_c and N_γ are bearing capacity factors corresponding to the friction angle ϕ'_2 of the foundation soil (soil beneath the MSE wall) and can be calculated from Eqs. (1.3) and (1.4), respectively.

$\sigma'_{\text{v(T) at } z=H}$ is the total vertical stress at the base of the MSE wall (at depth $z = H$) due to soil and additional surcharge loading and can be determined from Eq. (4.4).

4.3 Step-by-Step Procedure

In order to design an MSE wall with strip reinforcement, one can follow the step-by-step procedure described below which will enable us to determine required design parameters which satisfy internal and external stability of the MSE wall.

Step 1:
Obtain properties of granular backfill soil (γ_1 and ϕ'_1), properties of foundation soils (γ_2, ϕ'_2, and c'_2), height of wall (H), and information about the surcharge load if exists (q, a, b, etc.).

Step 2:
Establish minimum values for the safety factors to meet while designing of the MSE wall with reinforcing strips:

- For internal stability: $\text{FS}_{\text{B-min}}$ to avoid breakage, and $\text{FS}_{\text{P-min}}$ to avoid pullout
- For external stability: $\text{FS}_{\text{O-min}}$ to avoid overturning, $\text{FS}_{\text{S-min}}$ to avoid sliding, and $\text{FS}_{\text{BC-min}}$ to avoid bearing capacity failure

A safety factor of 3.0 could be used as a suggested value for the above minimum safety factors.

Step 3:
Assume the dimensions of the reinforcing strips (w and t). Then, obtain the yielding strength of the strips (f_y) and the friction angle between the strips and the backfill soil (ϕ'_μ).

Step 4:
Assume the number of reinforcing strip levels (n) along the wall height and the depths for which each strip level must be placed at from the ground surface (Z_i).

Step 5:
Calculate the vertical effective distances at each strip level ($S_{v(i)}$) using Eqs. (4.1) through (4.3).

Step 6:
Calculate the total vertical ($\sigma'_{v(T)}$) and horizontal ($\sigma'_{h(T)}$) stresses at each reinforcing strip level due to soil and additional surcharge loads using Eqs. (4.4) and (4.5), respectively.

Step 7:
Calculate the maximum horizontal spacing ($S_{H\text{-}max}$) between the strips at each strip level to satisfy $FS_{B\text{-}min}$ using Eq. (4.34). Then, select appropriate and practical design values for the horizontal spacing ($S_{H\text{-}design}$) between the strips at each strip level to be considered.

Step 8:
Calculate the maximum breakage force the reinforcing strip can withstand using Eq. (4.36). Also, calculate the actual applied breakage/pullout force and the actual factor of safety against breakage at each strip level using Eqs. (4.35) and (4.37), respectively.

Step 9:
Calculate the minimum strip length (L_{min}) at each level to satisfy $FS_{P\text{-}min}$ using Eq. (4.38). Then, select appropriate and practical design values for the strip length (L_{design}) at each strip level to be considered.

Step 10:
Calculate the maximum pullout force any reinforcing strip can withstand at each strip level due to friction using Eq. (4.39). Also, calculate the actual factor of safety against pullout at each strip level using Eq. (4.40).

Step 11:
Check external stability to avoid overturning, sliding, and bearing capacity failure using Eqs. (4.41), (4.48), and (4.55), respectively. Overturning and resisting moments could be calculated using Eqs. (4.42) and (4.43), respectively. Additionally, driving and resisting forces could be calculated using Eqs. (4.49) and (4.50), respectively.

4.4 Design Problems

Several design problems are presented in this section to help the reader reiterate the theory and integrate the MSE Wall-1 application of the foundationPro program in the design process of the mechanically stabilized earth retaining wall using reinforcing strips. These examples were selected to give exposure to a wide variety of challenges and design circumstances that can be faced in designing an MSE wall with strip reinforcement while helping us iron out the finer details of the theories introduced earlier.

4.4.1 MSE Wall with Constant Length and Spacing (Horizontal and Vertical) of Strips

Design an MSE wall that is 10 m high with galvanized steel reinforcement strips to retain a granular backfill behind it. Physical properties of the backfill soil, the foundation soil, and the strip reinforcement are listed in Table 4.1.

4.4.1.1 Hand Solution

To design the 10 m high MSE wall, one must follow the steps below:

Step 1:
The following are given in the problem statement:

$$H = 10\text{m};\ \gamma_1 = 16.5\ \text{kN/m}^3;\ \phi_1^{'} = 36^{\text{o}};\ \gamma_2 = 17.3\ \text{kN/m}^3;$$
$$\phi_2^{'} = 28^{\text{o}};\ \text{and}\ c_2^{'} = 50\ \text{kN/m}^2$$

No surcharge load is applied on top of soil.

Table 4.1 Physical properties and design parameters

Item	Property	Value	Unit
Backfill soil	ϕ_1	36	°
	γ_1	16.5	kN/m^3
Foundation soil	ϕ_2	28	°
	γ_2	17.3	kN/m^3
	c_2	50	kN/m^2
Reinforcing strips	w	75	mm
	t	6	mm
	f_y	240,000	kN/m^2
	ϕ_μ	20	°

Step 2:
The following minimum values for the safety factors are considered to satisfy internal and external stability requirements for the 10 m high MSE wall:

For internal stability:

$$FS_{B\text{-min}} = 3.0 \text{ and } FS_{P\text{-min}} = 3.0$$

For external stability:

$$FS_{O\text{-min}} = 3.5;\ FS_{S\text{-min}} = 4.0;\ \text{and } FS_{BC\text{-min}} = 5.0$$

Step 3:
A suggestion for the dimensions of the reinforcing strips is given in the problem statement as follows:

$$w = 75 \text{ mm and } t = 6 \text{ mm}.$$

Also,
$f_y = 240,000\ \text{kN/m}^2$ and $\phi'_\mu = 20°$.

Step 4:
Let's assume 13 strip levels along the wall height (vertically) with depths as in Table 4.2 below:

Step 5:
Using Eqs. (4.1) through (4.3), one can calculate $S_{v(i)}$ as follows:
At the first strip level,

$$S_{v(1)} = \frac{0.75 + 1.5}{2} = 1.125 \text{ m}$$

Table 4.2 Suggested strip levels

Strip level (i)	Depth, Z_i (m)
1	0.75
2	1.5
3	2.25
4	3.0
5	3.75
6	4.50
7	5.25
8	6.0
9	6.75
10	7.5
11	8.25
12	9.0
13	9.75

Also, at the third strip level,

$$S_{v(3)} = \frac{Z_4 - Z_2}{2} = \frac{3.0 - 1.5}{2} = 0.75 \text{ m}$$

At the last strip level, $i = 13$:

$$S_{v(13)} = \frac{2 \times 10 - Z_{13} - Z_{12}}{2} = \frac{2 \times 10 - 9.75 - 9.0}{2} = 0.625 \text{ m}$$

The effective vertical distances, $S_{v(i)}$, are summarized in Table 4.3 below:

Step 6:
The total vertical stresses are calculated using Eq. (4.4) as follows:

$$\sigma'_{v(T)} = \sigma'_{v(soil)} + \sigma'_{v(load)}$$

Since $\sigma'_{v(load)} = 0$, $\sigma'_{v(T)}$ can be calculated using the following equation:

$$\sigma'_{v(T)} = \sigma'_{v(soil)} = 16.5 \times Z_i$$

Calculations of the total vertical effective stresses at each strip level are summarized in Table 4.4.

However, the total horizontal stresses are calculated using Eq. (4.5) as follows:

$$\sigma'_{h(T)} = \sigma'_{h(soil)} + \sigma'_{h(load)}$$

Table 4.3 Effective vertical distances ($S_{v(i)}$)

Strip level (i)	$S_{v(i)}$ (m)
1	1.125
2	0.75
3	0.75
4	0.75
5	0.75
6	0.75
7	0.75
8	0.75
9	0.75
10	0.75
11	0.75
12	0.75
13	0.625

Table 4.4 Total vertical effective stresses ($\sigma'_{v(T)}$) at each strip level

Strip level (i)	Z_i (m)	$\sigma'_{v(T)}$ (kN/m²)
1	0.75	12.375
2	1.5	24.75
3	2.25	37.125
4	3.0	49.5
5	3.75	61.815
6	4.50	74.25
7	5.25	86.625
8	6.0	99
9	6.75	111.365
10	7.5	123.75
11	8.25	136.125
12	9.0	148.5
13	9.75	160.875

Since $\sigma'_{h(load)} = 0$, $\sigma'_{h(T)}$ can be calculated using the following equation:

$$\sigma'_{h(T)} = \sigma'_{h(soil)} = \tan^2\left(45 - \frac{36}{2}\right) \times 16.5 \times Z_i = 0.26 \times 16.5 \times Z_i = 4.29 \times Z_i$$

Calculations of the total horizontal effective stresses at each strip level are summarized in Table 4.5.

Step 7:
The maximum horizontal spacing ($S_{H\text{-}max}$) between the strips at each strip level to satisfy $FS_{B\text{-}min} = 3.0$ can be calculated using Eq. (4.34) as follows:

At $Z_i = 0.75$ m (first strip level, $i = 1$):

$$S_{H\text{-}max} = \frac{0.75 \times 0.006 \times 240{,}000}{3.21 \times 1.125 \times 3} = 9.96 \text{ m}$$

At $Z_i = 9.75$ m (last strip level, $i = 13$):

$$S_{H\text{-}max} = \frac{0.75 \times 0.006 \times 240{,}000}{41.82 \times 0.625 \times 3} = 0.62 \text{ m}$$

Calculations of the maximum horizontal spacing between the reinforcing strips at each strip level are summarized in Table 4.6.

Let's use a design horizontal spacing between the strips at all strip levels with a value of 1.0 m. Therefore,

$S_{H\text{-}design} = 1.0$ m at all strip levels.

Table 4.5 Total horizontal effective stresses ($\sigma'_{h(T)}$) at each strip level

Strip level (i)	Z_i (m)	$\sigma'_{h(T)}$ (kN/m²)
1	0.75	3.21
2	1.5	6.435
3	2.25	9.65
4	3.0	12.87
5	3.75	16.08
6	4.50	19.305
7	5.25	22.5225
8	6.0	25.74
9	6.75	28.95
10	7.5	32.175
11	8.25	35.39
12	9.0	38.61
13	9.75	41.82

Table 4.6 Maximum horizontal spacing (S_{H-max}) between the strips at each strip level

Strip level (i)	Z_i (m)	S_{H-max} (m)
1	0.75	9.96
2	1.5	7.5
3	2.25	4.87
4	3.0	3.72
5	3.75	2.98
6	4.50	2.486
7	5.25	2.13
8	6.0	1.86
9	6.75	1.65
10	7.5	1.49
11	8.25	1.35
12	9.0	1.24
13	9.75	0.62

Step 8:

The maximum breakage force the reinforcing strip can withstand is calculated using Eq. (4.36) as follows:

$$F_{\text{B-max}} = 0.75 \times 0.006 \times 240{,}000 = 108 \text{ kN}$$

Also, the actual applied breakage/pullout force and the actual factor of safety against breakage at each strip level can be determined using Eqs. (4.35) and (4.37), respectively:

At $Z_i = 2.25$ m (third strip level, $i = 3$):

Using Eq. (4.35),

$$F_{BP} = 9.96 \times 0.75 \times 1.0 = 7.23 \text{ kN};$$

Using Eq. (4.37),

$$FS_B = \frac{108}{7.23} = 14.9$$

Calculations of the actual breakage/pullout forces and safety factors against breakage at each strip level are summarized in Table 4.7.

Step 9:
The minimum strip length (L_{min}) at each level to satisfy $FS_{P\text{-}min} = 3.0$ can be calculated using Eq. (4.38) as follows:

For example, at $Z_i = 2.25$ m (third strip level, $i = 3$):

$$L_{min} = \frac{(10 - 2.25)}{\tan\left(45 + \frac{36^o}{2}\right)} + \frac{3.0 \times 7.23}{2 \times 0.075 \times 37.125 \times \tan 20^o} = 14.65 \text{ m}$$

Calculations of the minimum strip length at each strip level are summarized in Table 4.8.

Let's use a constant design strip length of 21 m at all strip levels along the wall height. Therefore,

$L_{design} = 21$ m at all strip levels.

Step 10:
The maximum pullout force any reinforcing strip can withstand due to friction can be calculated using Eq. (4.39) as follows:
For example, at $Z_i = 2.25$ m (third strip level, $i = 3$):

Table 4.7 Actual breakage/pullout forces and safety factors against breakage at each strip level

Strip level (i)	Z_i (m)	F_{BP} (kN)	FS_B
1	0.75	3.61	29.88
2	1.5	4.82	22.41
3	2.25	7.23	14.94
4	3.0	9.64	11.21
5	3.75	12.05	8.96
6	4.50	14.46	7.47
7	5.25	16.87	6.40
8	6.0	19.28	5.60
9	6.75	21.69	4.98
10	7.5	24.10	4.48
11	8.25	26.51	4.07
12	9.0	28.91	3.74
13	9.75	26.10	4.14

Table 4.8 Minimum strip length (L_{min}) at each strip level

Strip level (i)	Z_i (m)	L_{min} (m)
1	0.75	20.76
2	1.5	15.03
3	2.25	14.65
4	3.0	14.27
5	3.75	13.88
6	4.50	13.50
7	5.25	13.12
8	6.0	12.74
9	6.75	12.36
10	7.5	11.97
11	8.25	11.59
12	9.0	11.21
13	9.75	9.04

Table 4.9 Actual maximum pullout/friction forces and safety factors against pullout at each strip level

Strip level (i)	Z_i (m)	$F_{P\text{-}max}$ (kN)	FS_P
1	0.75	11.00	3.04
2	1.5	22.52	4.67
3	2.25	34.56	4.78
4	3.0	47.11	4.89
5	3.75	60.18	5.00
6	4.50	73.77	5.10
7	5.25	87.87	5.21
8	6.0	102.49	5.32
9	6.75	117.62	5.42
10	7.5	133.27	5.53
11	8.25	149.44	5.64
12	9.0	166.13	5.75
13	9.75	183.33	7.02

Using Eq. (4.39),

$$F_{\text{P-max}} = (2 \times 0.075 \times 37.125 \times \tan 20^{\circ}) \times \left(21 - \frac{10 - 2.25}{\tan\left(45 + \frac{36^{\circ}}{2}\right)}\right) = 34.56 \text{ kN}$$

Using Eq. (4.40),

$$\text{FS}_\text{P} = \frac{34.56}{7.23} = 4.48$$

Calculations of the actual maximum pullout/friction forces and safety factors against pullout at each strip level are summarized in Table 4.9.

Step 11:
Check external stability to avoid overturning, sliding, and bearing capacity failure using Eqs. (4.41), (4.48), and (4.55), respectively. Overturning and resisting moments can be calculated using Eqs. (4.42) and (4.43), respectively. Additionally, driving and resisting forces can be calculated using Eqs. (4.49) and (4.50), respectively.

– Check overturning (Eq. (4.41))

To determine the safety factor against overturning and compare it with the minimum safety factor, first, we need to calculate the overturning and resisting moments as follows:
Equation (4.42) can be used to determine the overturning moment (M_O):

$$M_O = \sum_{i=1}^{13} \left[\sigma'_{h(T)} \times S_{v(i)} \times (10 - Z_i)\right] = \underbrace{3.21 \times 1.125 \times (10 - 0.75)}_{i=1} + \cdots$$

$$+ \underbrace{12.57 \times 0.75 \times (10 - 3)}_{i=4} + \cdots$$

Thus,

$$M_O = 713.9\,\text{m kN}$$

Equation (4.43) can be used to determine the resisting moment (M_R).
Since $M_{R(\text{load})} = 0$

$$M_R = M_{R(\text{soil})} = \frac{16.5}{2}\sum_{i=1}^{13} \left[S_{v(i)}(21)^2\right] = \underbrace{\frac{16.5 \times 21^2}{2} \times 1.125}_{i=1} + \cdots$$

$$+ \underbrace{\frac{16.5 \times 21^2}{2} \times 0.75}_{i=4} + \cdots$$

Therefore,

$$M_R = 36{,}382\,\text{m kN}$$

So, using Eq. (4.41),

$$\text{FS}_O = \frac{36{,}382}{713.9} = 51 \geq \text{FS}_{\text{O-min}} = 3.5\ \text{O.K.}$$

– Check sliding (Eq. (4.48))

To determine the safety factor against sliding and compare it with the minimum safety factor for sliding, first, we need to calculate the driving and resisting forces as follows:

Equation (4.49) can be used to determine the driving force (F_{D}):

$$F_{\text{D}} = \sum_{i=1}^{13} \left[\sigma'_{\text{h(T)}} \times S_{\text{v}(i)} \right] = \underbrace{3.21 \times 1.125}_{i=1} + \cdots + \underbrace{12.57 \times 0.75}_{i=4} + \cdots$$

Thus,

$$F_{\text{D}} = 214.2 \text{ kN/m}$$

Equation (4.50) can be used to determine the resisting moment (F_{R}) noting that $F_{\text{R(load)}} = 0$:

$$F_{\text{R}} = F_{\text{R(soil)}} = \left(16.5 \times \tan\left(\frac{2 \times 36}{3} \right) \right) \left(\sum_{i=1}^{13} \left[S_{\text{v}(i)} \times 21 \right] \right) = 154.27 \times \sum_{i=1}^{13} S_{\text{v}(i)}$$

Therefore,

$$F_{\text{R}} = 1{,}542.7 \text{ kN/m}$$

So, using Eq. (4.48),

$$\text{FS}_{\text{S}} = \frac{1{,}542.7}{214.2} = 7.2 \geq \text{FS}_{\text{S-min}} = 4.0 \text{ O.K.}$$

– Check bearing capacity (Eq. (4.55))

Bearing capacity factors can be obtained as follows:

$N_c = 25.80$ using Eq. (1.3) with $\phi' = 28°$.
$N_\gamma = 16.72$ using Eq. (1.4) with $\phi' = 28°$.

Also, the total vertical stress at depth, $Z = 10$ m, can be obtained from Eq. (4.4) (keeping in mind that no surcharge load is applied) as follows:

$$\sigma'_{\text{v(T)}} = 16.5 \times 10 = 165 \text{ kN/m}^2$$

Therefore, Eq. (4.55) can be used to determine the safety factor against bearing capacity and compare it with the minimum safety factor for bearing capacity as follows:

$$FS_{BC} = \frac{50 \times 25.8 + \frac{1}{2} \times 17.3 \times 16.72 \times 21}{165} = 25.3 \geq FS_{BC\text{-min}} = 5.0 \text{ O.K.}$$

4.4.1.2 foundationPro Solution

After launching the MSE Wall-1 application, the input/output screen will show up as shown in Fig. 4.5 with the main five sections (General, Reinforcing Strips, Soil Properties, Additional Surcharge, and DESIGN/OUTPUT).

General Information Section

In the General Information section, all minimum values of safety factors to satisfy internal and external stability requirements are entered herein. Also, the units (BS or SI) to be used and the wall height are required in this section. For our design problem, wall height and safety factors are entered as shown in Fig. 4.6.

Fig. 4.5 Main sections of MSE Wall-1 application of the foundationPro

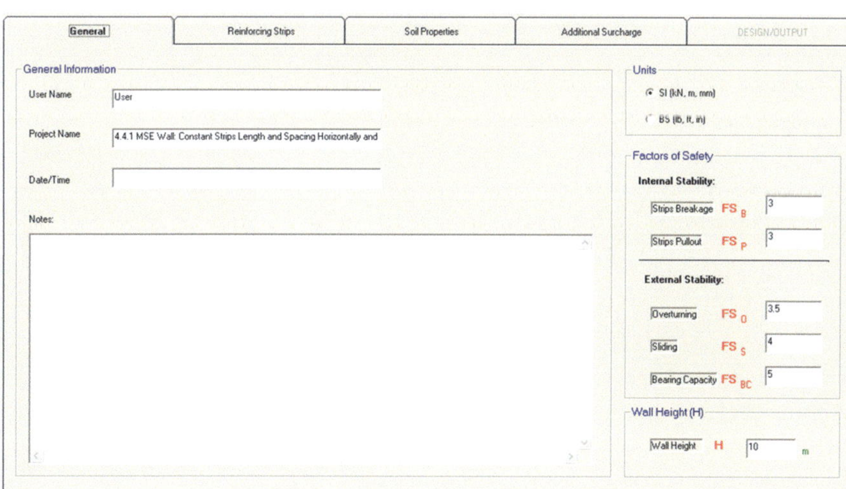

Fig. 4.6 General Information section

Reinforcing Strips Section

Dimensions and strength of the reinforcing strips are entered in the provided textboxes as shown in Fig. 4.7. Also, the number of strip levels which is 13 is selected herein. The depths from the ground surface to each strip level are also entered in the provided table with yellow cells as shown in the figure.

Soil Properties Section

Physical properties of the granular backfill to be retained behind the wall and the foundation soil to carry the MSE wall are entered in this section. For our design problem, the physical properties are entered as shown in Fig. 4.8.

Additional Surcharge Section

In this section, one can provide any additional surcharge loading information if it exists. For this design problem, no surcharge load is applied on top of soil. Therefore, none is selected as shown in Fig. 4.9.

DESIGN/OUTPUT Section (Internal Stability)

In this section, one can perform the design and also view the output results in a table and chart format. After providing all necessary input information in the previous

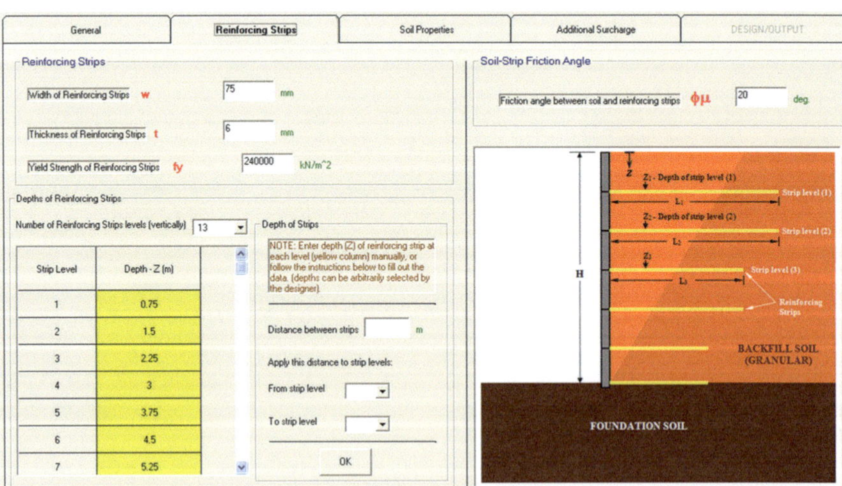

Fig. 4.7 Reinforcing Strips section

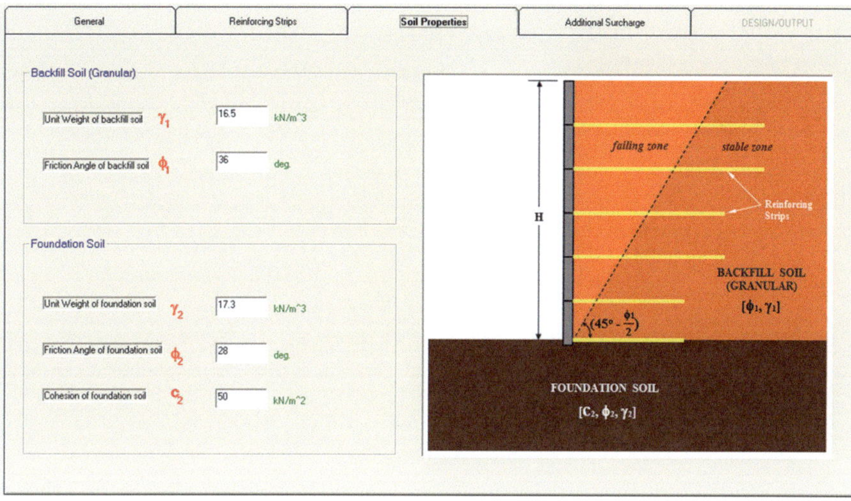

Fig. 4.8 Soil Properties section

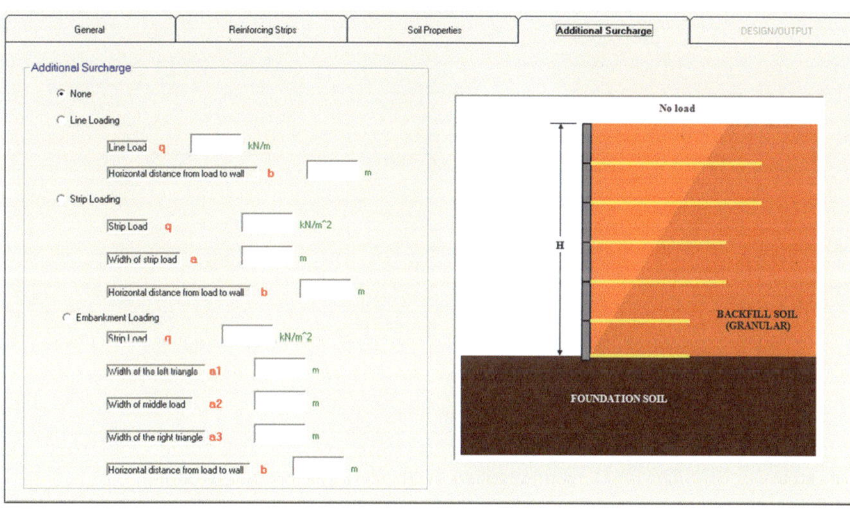

Fig. 4.9 Additional Surcharge Loading section

section and the information is saved, the application will check the data for errors. In the case no errors are found, one can perform the analysis and the first design screen (Internal Stability) will show up as in Fig. 4.10 to allow the user to enter design values for horizontal spacing between the reinforcing strips at each strip level based on the calculated horizontal spacing values. These design values are entered in the column labeled ($S_{\text{H-design}}$) with yellow cells as shown in Fig. 4.10. For

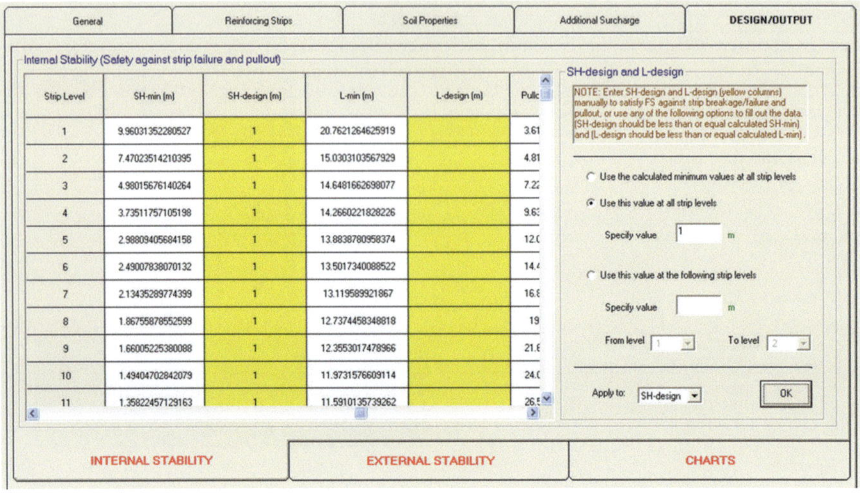

Fig. 4.10 DESIGN/OUTPUT section (design horizontal spacing—internal stability)

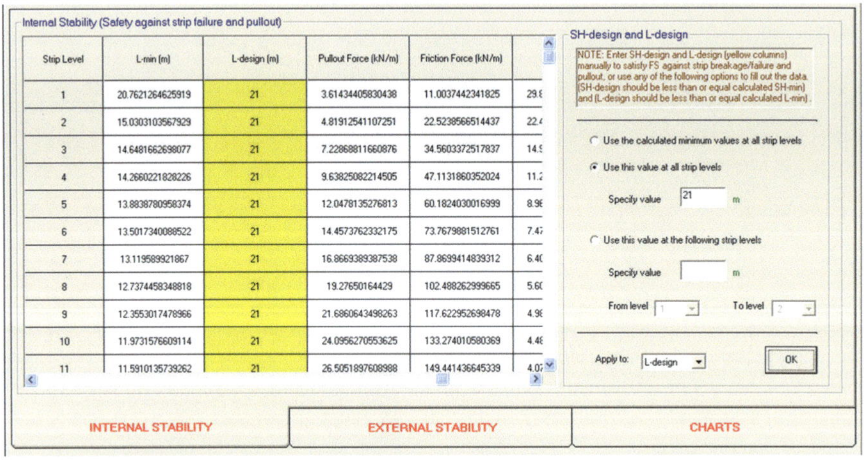

Fig. 4.11 DESIGN/OUTPUT section (design of strip length—internal stability)

our problem, the design horizontal spacing was selected to 1 m at all strip levels. After the design values for the horizontal spacing are selected, the user is required to enter the design strip length based on the calculated minimum strip length at each level. These design values must be entered in the column labeled (L_{design}) with yellow cells as shown in Fig. 4.11. For this design problem, the design length was selected as 21 m at all strip levels. In this first design screen, the various forces and safety factors will be automatically calculated based on the selected design values (horizontal spacing and length) as shown in Fig. 4.12.

Internal Stability (Safety against strip failure and pullout)

Strip Level	Pullout Force (kN/m)	Friction Force (kN/m)	FS-B	FS-P
3	7.22868811660876	34.5603372517837	14.9404702842079	4.7809971455785
4	9.63825082214505	47.1131860352024	11.2053527131559	4.88814691634235
5	12.0478135276813	60.1824030016999	8.96428217052472	4.99529668710622
6	14.4573762332175	73.7679881512761	7.47023514210396	5.10244645787011
7	16.8669389387538	87.8699414839312	6.40305869323195	5.20959622863397
8	19.27650164429	102.488262999665	5.60267635657797	5.31674599939785
9	21.6860643498263	117.622952698478	4.98015676140263	5.42389577016173
10	24.0956270553625	133.274010580369	4.48214108526238	5.5310455409256
11	26.5051897608988	149.441436645339	4.07467371387488	5.63819531168946
12	28.9147524664351	166.125230893388	3.73511757105197	5.74534508245332
13	26.1035959766427	183.325393324515	4.13736100178066	7.02299382386063

INTERNAL STABILITY | EXTERNAL STABILITY

Fig. 4.12 DESIGN/OUTPUT section (various forces and safety factors at each strip level—internal stability)

DESIGN/OUTPUT Section (External Stability)

The second screen in the DESIGN/OUTPUT section is for the external stability check. The application will not allow the user to view this screen unless the internal stability is satisfied based on the selected design parameters (horizontal spacing between strips and strip length) at each strip level. In this screen, one can view the external stability checks (overturning, sliding, and bearing capacity). If any of these requirements is violated, the application will give a message to inform the user with that regard. Details of all checks and calculations are shown in this screen as depicted in Fig. 4.13. The three safety factors in our design problem were satisfied and shown in the figure with the highlighted cells.

DESIGN/OUTPUT Section (CHARTS)

In this third screen (CHARTS) of the DESIGN/OUTPUT section, one can view any required output results in table and chart formats. For example, the total horizontal stress with depth is plotted and data was summarized as shown in Fig. 4.14.

Other charts and tables can be obtained to be used in another application like Word or Excel. For example, distributions of various forces (maximum pullout or breakage/pullout) with depth are shown in Figs. 4.15 and 4.16. Also, distributions of safety factors against breakage and pullout are shown in Figs. 4.17 and 4.18, respectively.

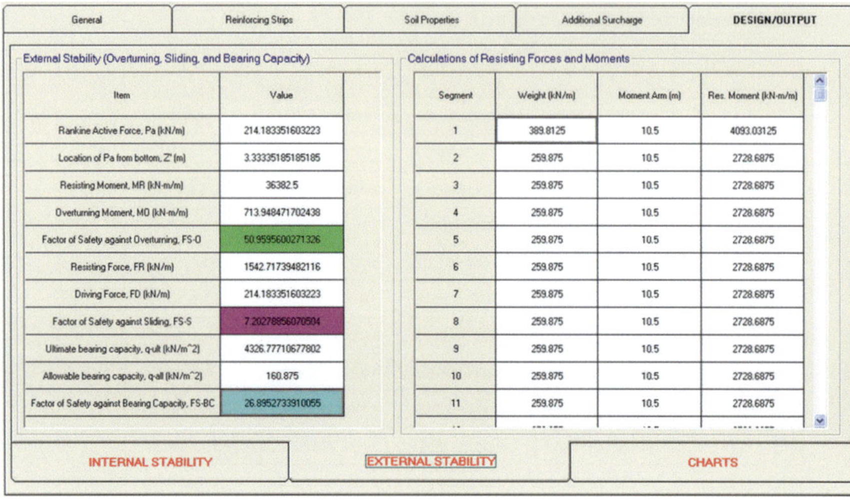

Fig. 4.13 DESIGN/OUTPUT section (external stability)

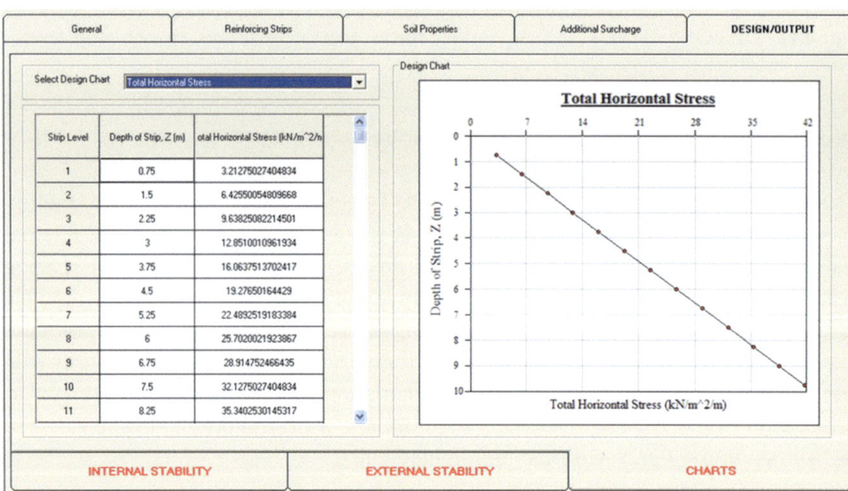

Fig. 4.14 DESIGN/OUTPUT section (charts)

4.4.2 MSE Wall with Varying Length and Spacing (Horizontal and Vertical) of Strips

Design an MSE wall that is 10 m high with galvanized steel reinforcement strips to retain a granular backfill behind it. The physical properties of the backfill soil, the foundation soil, and the strip reinforcement are listed in Table 4.10.

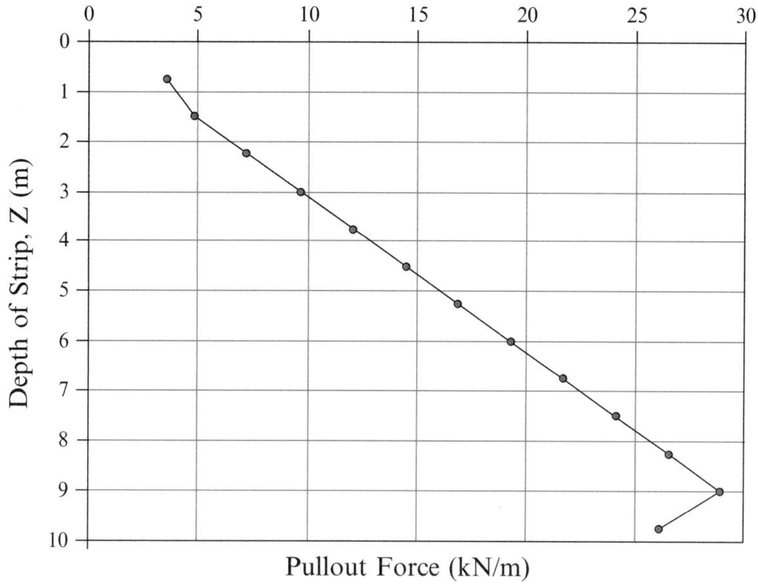

Fig. 4.15 Distribution of breakage/pullout forces

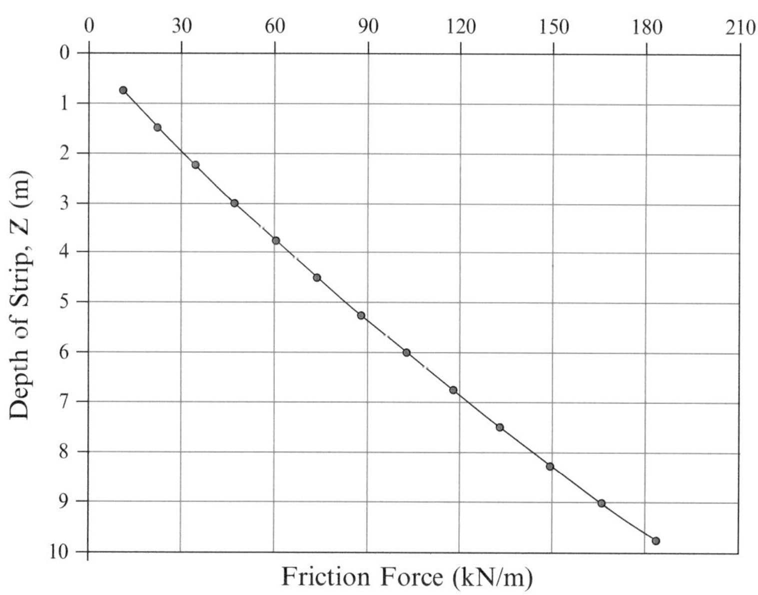

Fig. 4.16 Distribution of the maximum pullout/friction forces

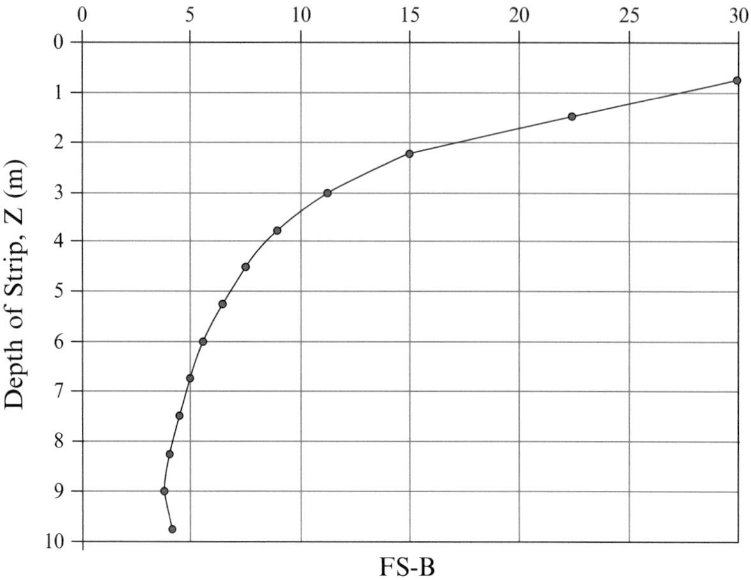

Fig. 4.17 Distribution of safety factor against breakage for the selected design parameters

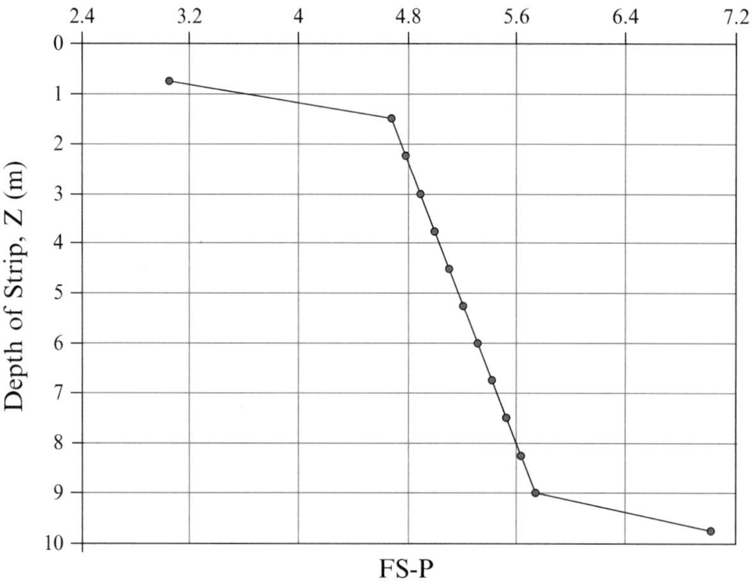

Fig. 4.18 Distribution of safety factor against pullout for the selected design parameters

Table 4.10 Physical properties and design parameters

Item	Property	Value	Unit
Backfill soil	ϕ_1	36	°
	γ_1	16.5	kN/m^3
Foundation soil	ϕ_2	28	°
	γ_2	17.3	kN/m^3
	c_2	50	kN/m^2
Reinforcing strips	w	75	mm
	t	6	mm
	f_y	240,000	kN/m^2
	ϕ_μ	20	°

4.4.2.1 Hand Solution

To design the 10 m high MSE wall, one must follow the steps below:

Step 1:
The following are given in the problem statement:

$$H = 10\text{ m};\ \gamma_1 = 16.5\text{ kN/m}^3;\ \phi_1' = 36^{\circ};\ \gamma_2 = 17.3\text{ kN/m}^3;$$
$$\phi_2' = 28^{\circ};\text{ and } c_2' = 50\text{ kN/m}^2.$$

No surcharge load is applied on top of soil.

Step 2:
The following minimum values for the safety factors are considered to satisfy internal and external stability requirements for the 10 m high MSE wall:

For internal stability:

$$\text{FS}_{\text{B-min}} = 3.0 \text{ and } \text{FS}_{\text{P-min}} = 3.0$$

For external stability:

$$\text{FS}_{\text{O-min}} = 3.5;\ \text{FS}_{\text{S-min}} = 4.0;\text{ and } \text{FS}_{\text{BC-min}} = 5.0.$$

Step 3:
A suggestion for the dimensions of the reinforcing strips is given in the problem statement as follows:

$$w = 75\text{ mm and } t = 6\text{ mm}.$$

Also,

$$f_y = 240,000\text{ kN/m}^2 \text{ and } \phi_\mu' = 20^{\circ}.$$

Step 4:
Let's assume 12 strip levels along the wall height (vertically) with depths as in Table 4.11 below:

Table 4.11 Suggested strip levels

Strip level (i)	Depth, Z_i (m)
1	0.75
2	1.5
3	2.25
4	3.0
5	3.75
6	4.50
7	5.25
8	6.0
9	7.0
10	8.0
11	9.0
12	10.0

Step 5:
Using Eqs. (4.1) through (4.3), one can calculate $S_{v(i)}$ as follows:
At the first strip level,

$$S_{v(1)} = \frac{0.75 + 1.5}{2} = 1.125 \text{ m}$$

Also, at the third strip level,

$$S_{v(3)} = \frac{Z_4 - Z_2}{2} = \frac{3.0 - 1.5}{2} = 0.75 \text{ m}$$

At the tenth strip level, $i = 10$:

$$S_{v(10)} = \frac{Z_{11} - Z_9}{2} = \frac{9 - 7}{2} = 1 \text{ m}$$

At the last strip level, $i = 12$:

$$S_{v(12)} = \frac{2 \times 10 - Z_{12} - Z_{11}}{2} = \frac{2 \times 10 - 10 - 9.0}{2} = 0.5 \text{ m}$$

The effective vertical distances, $S_{v(i)}$, are summarized in Table 4.12 below:

Step 6:
The total vertical stresses are calculated using Eq. (4.4) as follows:

$$\sigma'_{v(T)} = \sigma'_{v(soil)} + \sigma'_{v(load)}$$

Since $\sigma'_{v(load)} = 0$, $\sigma'_{v(T)}$ can be calculated using the following equation:

$$\sigma'_{v(T)} = \sigma'_{v(soil)} = 16.5 \times Z_i$$

Table 4.12 Effective vertical distances ($S_{v(i)}$)

Strip level (i)	$S_{v(i)}$ (m)
1	1.125
2	0.75
3	0.75
4	0.75
5	0.75
6	0.75
7	0.75
8	0.875
9	1.0
10	1.0
11	1.0
12	0.5

Table 4.13 Total vertical effective stresses ($\sigma'_{v(T)}$) at each strip level

Strip level (i)	Z_i (m)	$\sigma'_{v(T)}$ (kN/m²)
1	0.75	12.375
2	1.5	24.75
3	2.25	37.125
4	3.0	49.5
5	3.75	61.875
6	4.50	74.25
7	5.25	86.625
8	6.0	99
9	7.0	115.5
10	8.0	132
11	9.0	148.5
12	10.0	165

Calculations of the total vertical effective stresses at each strip level are summarized in Table 4.13.

However, the total horizontal stresses are calculated using Eq. (4.5) as follows:

$$\sigma'_{h(T)} = \sigma'_{h(soil)} + \sigma'_{h(load)}$$

Since $\sigma'_{h(load)} = 0$, $\sigma'_{h(T)}$ can be calculated using the following equation:

$$\sigma'_{h(T)} = \sigma'_{h(soil)} = \tan^2\left(45 - \frac{36}{2}\right) \times 16.5 \times Z_i = 0.26 \times 16.5 \times Z_i = 4.29 \times Z_i$$

Calculations of the total horizontal effective stresses at each strip level are summarized in Table 4.14.

Table 4.14 Total horizontal effective stresses ($\sigma'_{h(T)}$) at each strip level

Strip level (i)	Z_i (m)	$\sigma'_{h(T)}$ (kN/m²)
1	0.75	3.21
2	1.5	6.43
3	2.25	9.64
4	3.0	12.85
5	3.75	16.06
6	4.50	19.28
7	5.25	22.49
8	6.0	25.70
9	7.0	29.99
10	8.0	34.27
11	9.0	38.55
12	10.0	42.84

Step 7:
The maximum horizontal spacing ($S_{H\text{-max}}$) between the strips at each strip level to satisfy $FS_{B\text{-min}} = 3.0$ can be calculated using Eq. (4.34) as follows:
At $Z_i = 0.75$ m (first strip level, $i = 1$):

$$S_{\text{H-max}} = \frac{0.75 \times 0.006 \times 240{,}000}{3.21 \times 1.125 \times 3} = 9.96 \text{ m}$$

At $Z_i = 3.0$ m (fourth strip level, $i = 4$):

$$S_{\text{H-max}} = \frac{0.75 \times 0.006 \times 240{,}000}{41.82 \times 0.75 \times 3} = 3.73 \text{ m}$$

At $Z_i = 9.0$ m (eleventh strip level, $i = 11$):

$$S_{\text{H-max}} = \frac{0.75 \times 0.006 \times 240{,}000}{41.82 \times 1.0 \times 3} = 0.933 \text{ m}$$

Calculations of the maximum horizontal spacing between the reinforcing strips at each strip level are summarized in Table 4.15.

Based on the calculated maximum horizontal spacing, one can select the following design values for the horizontal spacing between strips at different strip levels:
$S_{\text{H-design}} = 1.0$ m at all strips levels from 1 through 8.
$S_{\text{H-design}} = 0.75$ m at all strips levels from 9 through 12.

Step 8:
The maximum breakage force the reinforcing strip can withstand is calculated using Eq. (4.36) as follows:

$$F_{\text{B-max}} = 0.75 \times 0.006 \times 240{,}000 = 108 \text{ kN}$$

Table 4.15 Maximum horizontal spacing ($S_{H\text{-}max}$) between the strips at each strip level

Strip level (i)	Z_i (m)	$S_{H\text{-}max}$ (m)
1	0.75	9.96
2	1.5	7.47
3	2.25	4.98
4	3.0	3.74
5	3.75	2.99
6	4.50	2.49
7	5.25	2.13
8	6.0	1.60
9	7.0	1.20
10	8.0	1.05
11	9.0	0.93
12	10.0	1.68

Table 4.16 Actual breakage/pullout forces and safety factors against breakage at each strip level

Strip level (i)	Z_i (m)	F_{BP} (kN)	FS_B
1	0.75	3.61	29.9
2	1.5	4.82	22.4
3	2.25	7.23	14.9
4	3.0	9.64	11.2
5	3.75	12.05	9.0
6	4.50	14.46	7.5
7	5.25	16.87	6.4
8	6.0	22.49	4.8
9	7.0	22.49	4.8
10	8.0	25.70	4.2
11	9.0	28.91	3.7
12	10.0	16.06	6.7

Also, the actual applied breakage/pullout force and the actual factor of safety against breakage at each strip level can be found using Eqs. (4.35) and (4.37), respectively:

At $Z_i = 3$ m (fourth strip level, $i - 4$):
Using Eq. (4.35),

$$F_{BP} = 12.85 \times 0.75 \times 1.0 = 9.64 \text{ kN};$$

Using Eq. (4.37),

$$FS_B = \frac{108}{9.64} = 11.2$$

Calculations of the actual breakage/pullout forces and safety factors against breakage at each strip level are summarized in Table 4.16.

Table 4.17 Minimum strip length ($L_{\min}$) at each strip level

Strip level (i)	Z_i (m)	$L_{\min}$ (m)
1	0.75	20.76
2	1.5	15.03
3	2.25	14.65
4	3.0	14.27
5	3.75	13.88
6	4.50	13.50
7	5.25	13.12
8	6.0	14.52
9	7.0	12.23
10	8.0	11.72
11	9.0	11.21
12	10.0	5.35

Step 9:
The minimum strip length ($L_{\min}$) at each level to satisfy $FS_{\text{P-min}} = 3.0$ can be calculated using Eq. (4.38) as follows:

For example, at $Z_i = 3$ m (fourth strip level, $i = 4$):

$$L_{\min} = \frac{(10-3)}{\tan\left(45 + \frac{36^{\circ}}{2}\right)} + \frac{3.0 \times 9.64}{2 \times 0.075 \times 49.5 \times \tan 20^{\circ}} = 14.27 \text{ m}$$

Calculations of the minimum strip length at each strip level are summarized in Table 4.17.

Based on the calculated minimum strip lengths, one can use the following varying design lengths of strips:

$L_{\text{design}} = 22$ m for the upper four strip levels (from strip levels 1 to 4).
$L_{\text{design}} = 15$ m for strip levels from 5 to 12.

Step 10:
The maximum pullout force any reinforcing strip can withstand due to friction can be calculated using Eq. (4.39) as follows:

For example, at $Z_i = 3$ m (fourth strip level, $i = 4$):
Using Eq. (4.39),

$$F_{\text{P-max}} = (2 \times 0.075 \times 49.5 \times \tan 20^{\circ}) \times \left(22 - \frac{10-3}{\tan\left(45 + \frac{36^{\circ}}{2}\right)}\right) = 49.8 \text{ kN}$$

Using Eq. (4.40),

Table 4.18 Actual maximum pullout/friction forces and safety factors against pullout at each strip level

Strip level (i)	Z_i (m)	$F_{\mathrm{P\text{-}max}}$ (kN)	$\mathrm{FS_P}$
1	0.75	11.68	3.23
2	1.5	23.88	4.95
3	2.25	36.59	5.06
4	3.0	49.82	5.17
5	3.75	39.91	3.31
6	4.50	49.45	3.42
7	5.25	59.49	3.53
8	6.0	70.06	3.12
9	7.0	84.95	3.78
10	8.0	100.76	3.92
11	9.0	117.48	4.06
12	10.0	135.12	8.41

$$\mathrm{FS_P} = \frac{49.8}{9.64} = 5.17$$

Calculations of the actual maximum pullout/friction forces and safety factors against pullout at each strip level are summarized in Table 4.18.

Step 11:

Check external stability to avoid overturning, sliding, and bearing capacity failure using Eqs. (4.41), (4.48), and (4.55), respectively. Overturning and resisting moments can be calculated using Eqs. (4.42) and (4.43), respectively. Additionally, driving and resisting forces can be calculated using Eqs. (4.49) and (4.50), respectively.

– Check overturning (Eq. (4.41))

To determine the safety factor against overturning and compare it with the minimum safety factor, first, we need to calculate the overturning and resisting moments as follows:

Equation (4.42) can be used to determine the overturning moment (M_O):

$$M_\mathrm{O} = \sum_{i=1}^{12} \left[\sigma'_{\mathrm{h(T)}} \times S_{\mathrm{v}(i)} \times (10 - Z_i)\right] = \underbrace{3.21 \times 1.125 \times (10 - 0.75)}_{i=1} + \cdots$$

$$+ \underbrace{12.85 \times 0.75 \times (10 - 3)}_{i=4} + \cdots$$

Thus,

$$M_\mathrm{O} = 713.9\ \mathrm{m\,kN}$$

Equation (4.43) can be used to determine the resisting moment (M_R).

Since $M_{R(load)} = 0$

$$M_R = M_{R(soil)} = \frac{16.5}{2}\left(\sum_{i=1}^{4}\left[S_{v(i)}(22)^2\right] + \sum_{i=5}^{12}\left[S_{v(i)}(15)^2\right]\right)$$
$$= \underbrace{\frac{16.5 \times 22^2}{2} \times 1.125}_{i=1} + \cdots + \underbrace{\frac{16.5 \times 15^2}{2} \times 0.75}_{i=5} + \cdots$$

Therefore,

$$M_R = 25,774\ \mathrm{m\,kN}$$

So, using Eq. (4.41),

$$\mathrm{FS_O} = \frac{25,774}{713.9} = 36.1 \geq \mathrm{FS_{O\text{-}min}} = 3.5\ \text{O.K.}$$

– Check sliding (Eq. 4.48)

To determine the safety factor against sliding and compare it with the minimum safety factor for sliding, first, we need to calculate the driving and resisting forces as follows:

Equation (4.49) can be used to determine the driving force (F_D):

$$F_D = \sum_{i=1}^{12}\left[\sigma'_{h(T)} \times S_{v(i)}\right] = \underbrace{3.21 \times 1.125}_{i=1} + \cdots + \underbrace{12.85 \times 0.75}_{i=4} + \cdots$$
$$+ \underbrace{25.7 \times 0.875}_{i=8} + \cdots$$

Thus,

$$F_D = 214.2\ \mathrm{kN/m}$$

Equation (4.50) can be used to determine the resisting moment (F_R) noting that $F_{R(load)} = 0$:

$$F_R = F_{R(soil)} = \left(16.5 \times \tan\left(\frac{2 \times 36}{3}\right)\right)\left(\sum_{i=1}^{4}\left[S_{v(i)} \times 22\right] + \sum_{i=4}^{12}\left[S_{v(i)} \times 15\right]\right)$$
$$= 161.6 \times \sum_{i=1}^{4} S_{v(i)} + 110.2 \times \sum_{i=5}^{12} S_{v(i)}$$

Therefore,

$$F_R = 1{,}275.5 \text{ kN/m}$$

So, using Eq. (4.48),

$$\text{FS}_\text{S} = \frac{1{,}275.5}{214.2} = 6 \geq \text{FS}_{\text{S-min}} = 4 \text{ O.K.}$$

– Check bearing capacity (Eq. (4.55))

Bearing capacity factors can be obtained as follows:

$N_c = 25.80$ using Eq. (1.3) with $\phi' = 28°$.
$N_\gamma = 16.72$ using Eq. (1.4) with $\phi' = 28°$.

Also, the total vertical stress at depth, $Z = 10$ m, can be obtained from Eq. (4.4) (keeping in mind that no surcharge load is applied on top on soil) as follows:

$$\sigma'_{\text{v(T)}} = 16.5 \times 10 = 165 \text{ kN/m}^2$$

Therefore, Eq. (4.55) can be used to determine the safety factor against bearing capacity and compare it with the minimum safety factor for bearing capacity as follows:

$$\text{FS}_{\text{BC}} = \frac{50 \times 25.8 + \frac{1}{2} \times 17.3 \times 16.72 \times 15}{165} = 21 \geq \text{FS}_{\text{BC-min}} = 5 \text{ O.K.}$$

4.4.2.2 foundationPro Solution

General Information Section

In the General Information section, the SI unit is selected and all minimum values of safety factors to satisfy internal and external stability requirements are entered as shown in Fig. 4.19.

Reinforcing Strips Section

Dimensions and strength of the reinforcing strips are entered in the provided textboxes as shown in Fig. 4.20. Also, the number of strip levels which is 12 is selected herein. The depths from the ground surface to each strip level are also entered in the provided table with yellow cells as shown in the figure.

Soil Properties Section

In this design problem, the physical properties are entered as shown in Fig. 4.21.

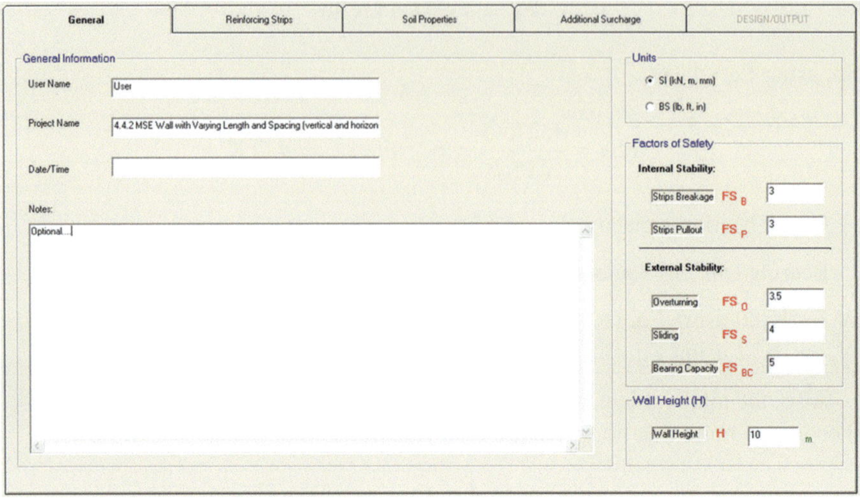

Fig. 4.19 General Information section

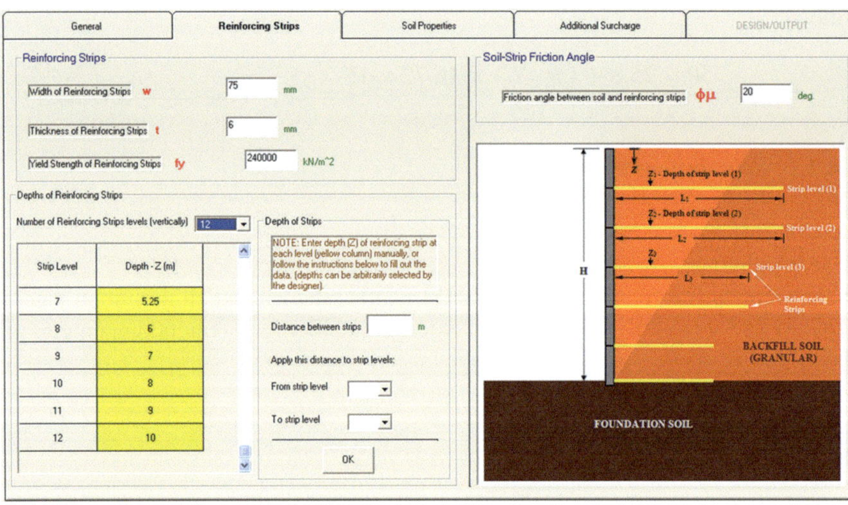

Fig. 4.20 Reinforcing Strips section

Additional Surcharge Section

For this design problem, no surcharge load is applied on top of soil. Therefore, none is selected as shown in Fig. 4.22.

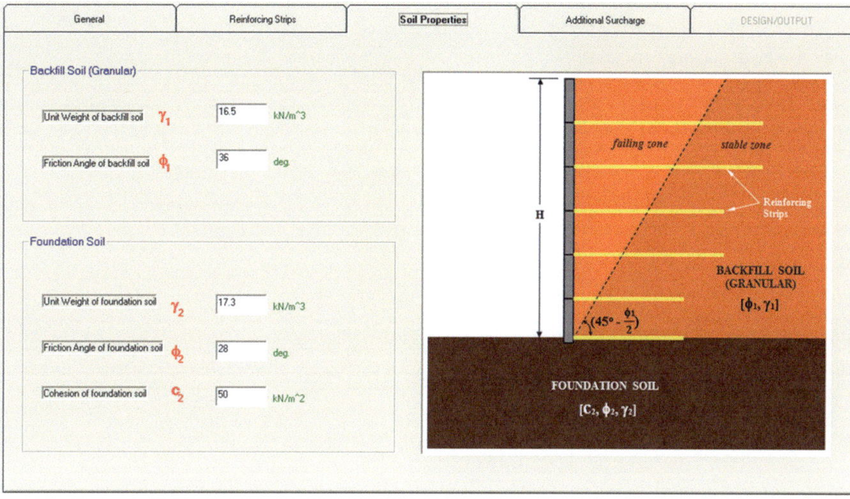

Fig. 4.21 Soil Properties section

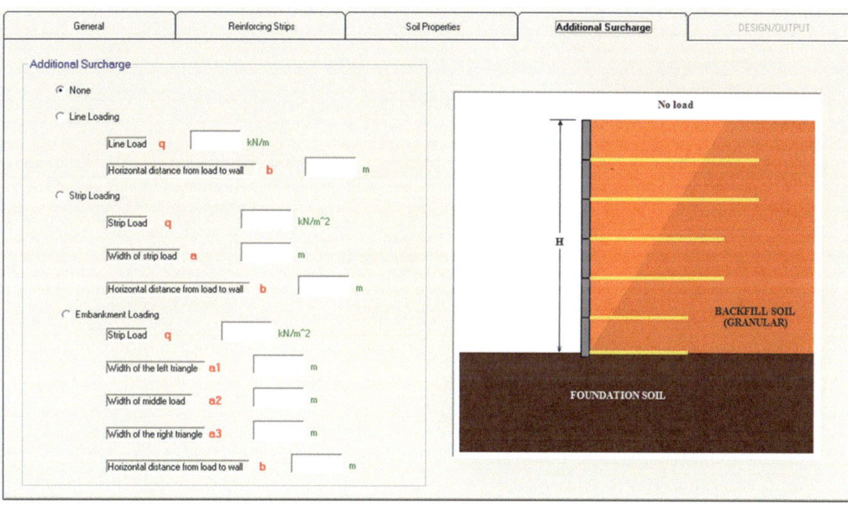

Fig. 4.22 Additional Surcharge Loading section

DESIGN/OUTPUT Section (Internal Stability)

For our problem, two design horizontal spacing values were selected to be 1 m from strip level 1 to 8 and 0.75 m from strip level 9 to 12 as shown in Fig. 4.23. Additionally, the design length was varied along the wall height and selected as 22 m for the upper four strip levels (1–4) and 15 m for strip levels from 5 to 12 as shown in Fig. 4.24.

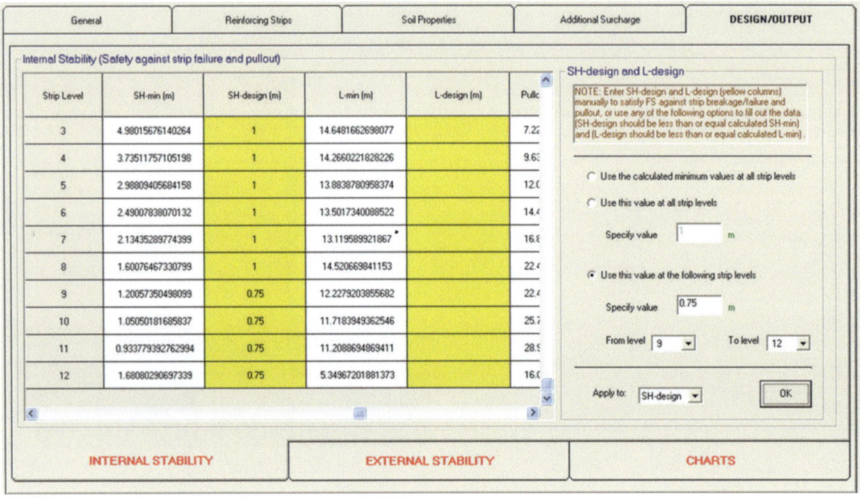

Fig. 4.23 DESIGN/OUTPUT section (design of horizontal spacing—internal stability)

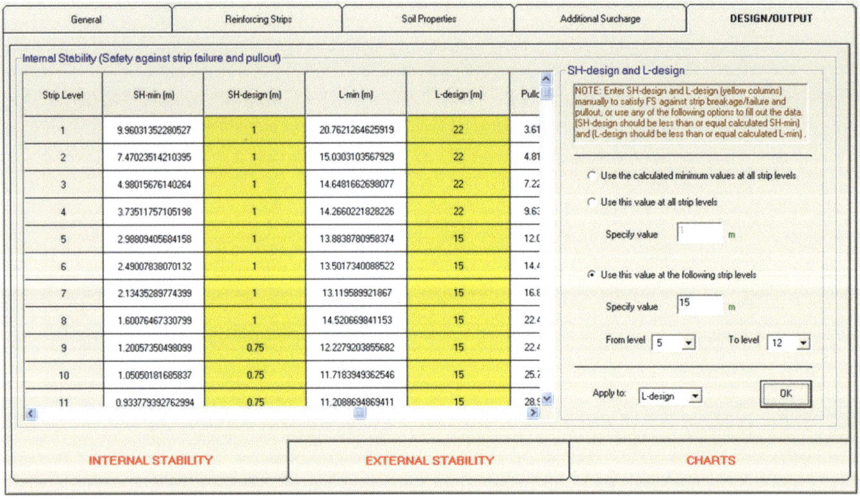

Fig. 4.24 DESIGN/OUTPUT section (design of strip length—internal stability)

DESIGN/OUTPUT Section (External Stability)

Details of all external stability checks and calculations are shown in this screen as depicted in Fig. 4.25. The three safety factors in our design problem were satisfied and shown in the figure with the highlighted cells.

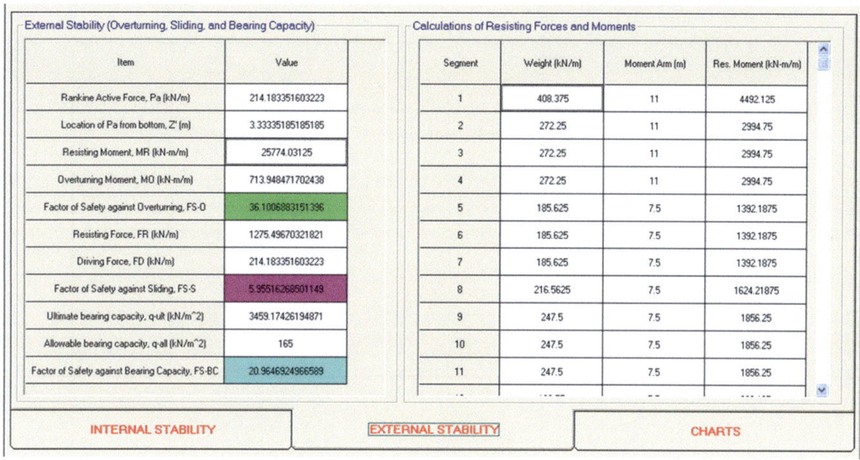

Item	Value
Rankine Active Force, Pa (kN/m)	214.183351603223
Location of Pa from bottom, Z' (m)	3.33335185185185
Resisting Moment, MR (kN-m/m)	25774.03125
Overturning Moment, MO (kN-m/m)	713.948471702438
Factor of Safety against Overturning, FS-O	36.1006883151396
Resisting Force, FR (kN/m)	1275.49670321821
Driving Force, FD (kN/m)	214.183351603223
Factor of Safety against Sliding, FS-S	5.95516268501149
Ultimate bearing capacity, q-ult (kN/m^2)	3459.17426194871
Allowable bearing capacity, q-all (kN/m^2)	165
Factor of Safety against Bearing Capacity, FS-BC	20.9646924966589

Segment	Weight (kN/m)	Moment Arm (m)	Res. Moment (kN-m/m)
1	408.375	11	4492.125
2	272.25	11	2994.75
3	272.25	11	2994.75
4	272.25	11	2994.75
5	185.625	7.5	1392.1875
6	185.625	7.5	1392.1875
7	185.625	7.5	1392.1875
8	216.5625	7.5	1624.21875
9	247.5	7.5	1856.25
10	247.5	7.5	1856.25
11	247.5	7.5	1856.25

Fig. 4.25 DESIGN/OUTPUT section (external stability)

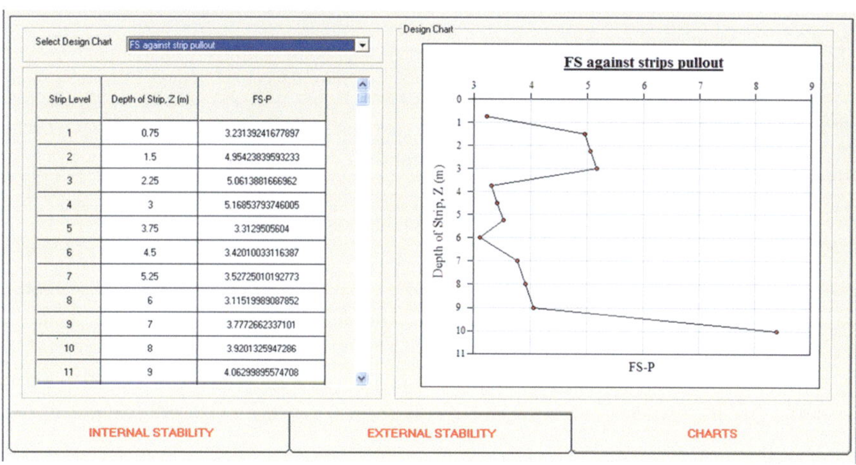

Strip Level	Depth of Strip, Z (m)	FS-P
1	0.75	3.23139241677897
2	1.5	4.95423839593233
3	2.25	5.0613881666962
4	3	5.16853793746005
5	3.75	3.3129505604
6	4.5	3.42010033116387
7	5.25	3.52725010192773
8	6	3.11519989087852
9	7	3.7772662337101
10	8	3.9201325947286
11	9	4.06299895574708

Fig. 4.26 DESIGN/OUTPUT section (charts)

DESIGN/OUTPUT Section (CHARTS)

For example, the safety factor against pullout versus depth is plotted and data was summarized as shown in Fig. 4.26. Other charts and tables may also be obtained such as the distribution of the maximum pullout/friction force as shown in Fig. 4.27.

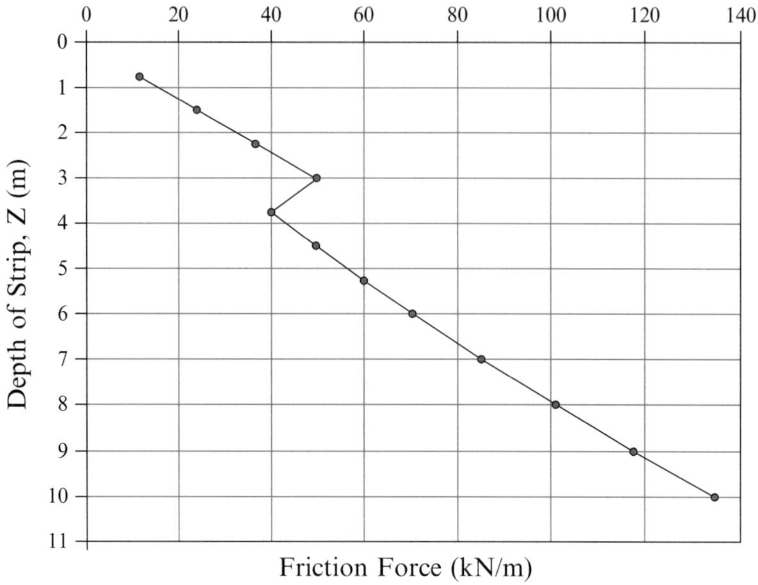

Fig. 4.27 Distribution of breakage/pullout forces

Table 4.19 Physical properties and design parameters

Item	Property	Value	Unit
Backfill soil	ϕ_1	32	°
	γ_1	108.5	lb/ft^3
Foundation soil	ϕ_2	30	°
	γ_2	112	lb/ft^3
	c_2	500	lb/ft^2
Reinforcing strips	w	3	in.
	t	0.25	in.
	f_y	5,000,000	lb/ft^2
	ϕ_μ	18	°

Table 4.20 Details of strip loading

Property	Value	Unit
q (strip)	1,400	lb/ft^2
a	20	ft
b	8	ft

4.4.3 MSE Wall with Strip Surcharge Loading and Constant Length and Spacing of Strips

Design an MSE wall that is 25 ft high with galvanized steel reinforcement strips to retain a granular backfill behind it. Physical properties of the backfill soil, the foundation soil, and the strip reinforcement are listed in Table 4.19. An additional surcharge strip loading is applied on top of soil with details as provided in Table 4.20.

4.4.3.1 Hand Solution

To design the 25 ft high MSE wall, one must follow the steps below:

Step 1:
The following are given in the problem statement:

$H = 25$ ft; $\gamma_1 = 108.5\ \text{lb/ft}^3$; $\phi_1' = 32^\circ$; $\gamma_2 = 112\ \text{lb/ft}^3$; $\phi_2' = 30^\circ$; $c_2' = 500\ \text{lb/ft}^2$; $q = 1{,}400\ \text{lb/ft}^2$; $a = 20$ ft; and $b = 8$ ft.

Step 2:
The following minimum values for the safety factors are considered to satisfy internal and external stability requirements in the design of the 25 ft high MSE wall:

For internal stability:

$$FS_{\text{B-min}} = 3.0 \text{ and } FS_{\text{P-min}} = 3.0.$$

For external stability:

$$FS_{\text{O-min}} = 3.5;\ FS_{\text{S-min}} = 3.5;\ \text{and } FS_{\text{BC-min}} = 4.0.$$

Step 3:
A suggestion for the dimensions of the reinforcing strips is given in the problem statement as follows:

$$w = 3 \text{ in. and } t = 0.25 \text{ in.}$$

Also,

$$f_y = 5{,}000{,}000\ \text{lb/ft}^2 \text{ and } \phi_\mu' = 18^\circ.$$

Step 4:
Let's assume ten strip levels along the wall height (vertically) equally spaced with depths as in Table 4.21 below:

Step 5:
Using Eqs. (4.1) through (4.3), one can calculate $S_{v(i)}$ as follows:

Table 4.21 Suggested depths for strip levels

Strip level (i)	Depth, Z_i (ft)
1	2.5
2	5.0
3	7.5
4	10.0
5	12.5
6	15.0
7	17.5
8	20.0
9	22.5
10	25.0

At the first strip level ($i = 1$),

$$S_{v(1)} = \frac{2.5 + 5.0}{2} = 3.75 \text{ ft}$$

Also, at the third strip level ($i = 3$),

$$S_{v(3)} = \frac{Z_4 - Z_2}{2} = \frac{10 - 5}{2} = 2.5 \text{ ft}$$

At the last strip level ($i = 10$),

$$S_{v(10)} = \frac{2 \times 25 - Z_{10} - Z_9}{2} = \frac{2 \times 25 - 25.0 - 22.5}{2} = 1.25 \text{ ft}$$

The effective vertical distances, $S_{v(i)}$, are summarized in Table 4.22 below:

Step 6:
The total vertical stresses are calculated using Eq. (4.4) as follows:

$$\sigma'_{v(T)} = \sigma'_{v(soil)} + \sigma'_{v(load)}$$

The vertical stresses due to soil ($\sigma'_{v(soil)}$) can be found using Eq. (4.6):

$$\sigma'_{v(soil)} = 108.5 \times Z_i$$

However, the vertical stresses due to the strip loading ($\sigma'_{v(load)}$) can be found using Eqs. (4.15) and (4.16):

$$\sigma'_{v(load)} = \frac{1,400 \times 20}{20 + Z_i} \quad (\text{For } Z_i \leq 2 \times 8 = 16 \text{ ft})$$

Table 4.22 Effective vertical distances ($S_{v(i)}$)

Strip level (i)	Z_i (ft)	$S_{v(i)}$ (ft)
1	2.5	3.75
2	5.0	2.5
3	7.5	2.5
4	10.0	2.5
5	12.5	2.5
6	15.0	2.5
7	17.5	2.5
8	20.0	2.5
9	22.5	2.5
10	25.0	1.25

Table 4.23 Total vertical effective stresses ($\sigma'_{v(T)}$) at each strip level

Strip level (i)	Z_i (ft)	$\sigma'_{v(soil)}$ (lb/ft^2)	$\sigma'_{v(load)}$ (lb/ft^2)	$\sigma'_{v(T)}$ (lb/ft^2)
1	2.5	271.25	1,244.44	1,515.69
2	5.0	542.5	1,120.00	1,662.50
3	7.5	813.75	1,018.18	1,831.93
4	10.0	1,085	933.33	2,018.33
5	12.5	1,356.25	861.54	2,217.79
6	15.0	1,627.5	800.00	2,427.50
7	17.5	1,898.75	761.90	2,660.65
8	20.0	2,170	736.84	2,906.84
9	22.5	2,441.25	713.38	3,154.63
10	25.0	2,712.5	691.36	3,403.86

$$\sigma'_{v(load)} = \frac{1,400 \times 20}{20 + \frac{Z_i}{2} + 8} \quad (\text{For } Z_i > 2 \times 8 = 16 \text{ ft})$$

Calculations of the total vertical effective stresses at each strip level are summarized in Table 4.23.

Additionally, the total horizontal stresses are calculated using Eq. (4.5) as follows:

$$\sigma'_{h(T)} = \sigma'_{h(soil)} + \sigma'_{h(load)}$$

The horizontal stresses due to soil can be found using Eq. (4.7):

$$\sigma'_{h(soil)} = \tan^2\left(45 - \frac{32}{2}\right) \times 108.5 \times Z_i = 0.307 \times 108.5 \times Z_i = 33.3 \times Z_i$$

However, the total horizontal stresses due to the strip loading on top of soil can be obtained as follows:

Using Eq. (4.18),

$$M = 1.4 - \frac{0.4 \times 8}{0.14 \times 25} = 0.486$$

But, M must be greater than or equal to 1.0. Then, $M = 1$.

Also, using Eqs. (4.19) and (4.20),

$$\alpha = \tan^{-1}\left(\frac{8 + \frac{20}{2}}{Z_i}\right) = \tan^{-1}\left(\frac{18}{Z_i}\right)$$

Table 4.24 Total horizontal effective stresses ($\sigma'_{h(T)}$) at each strip level

Strip level (i)	Z_i (ft)	$\sigma'_{h(soil)}$ (lb/ft^2)	$\sigma'_{h(load)}$ (lb/ft^2)	$\sigma'_{h(T)}$ (lb/ft^2)
1	2.5	83.34	372.56	455.90
2	5.0	166.69	624.93	791.62
3	7.5	250.03	734.15	984.19
4	10.0	333.38	740.23	1,073.60
5	12.5	416.72	689.39	1,106.11
6	15.0	500.06	614.27	1,114.34
7	17.5	583.41	533.81	1,117.22
8	20.0	666.75	457.57	1,124.32
9	22.5	750.09	389.58	1,139.67
10	25.0	833.44	330.91	1,164.35

$$\beta = \tan^{-1}\left(\frac{8+20}{Z_i}\right) = \tan^{-1}\left(\frac{8}{Z_i}\right)$$

Now, $\sigma'_{h(load)}$ can be obtained using Eq. (4.17):

$$\sigma'_{h(load)} = 1.0 \times \left[\frac{2 \times 1,400}{\pi}\left(\tan^{-1}\left(\frac{8}{Z_i}\right) - \sin\left(\tan^{-1}\left(\frac{8}{Z_i}\right)\right)\cos\left(2 \times \tan^{-1}\left(\frac{18}{Z_i}\right)\right)\right)\right]$$

Calculations of the total horizontal effective stresses at each strip level are summarized in Table 4.24.

Step 7:
The maximum horizontal spacing ($S_{H\text{-}max}$) between the strips at each strip level to satisfy $FS_{B\text{-}min} = 3.0$ can be calculated using Eq. (4.34) as follows:

At $Z_i = 2.5$ ft (first strip level, $i = 1$):

$$S_{H\text{-max}} = \frac{\frac{3}{12} \times \frac{0.25}{12} \times 5,000,000}{455.9 \times 3.75 \times 3} = 5.1 \text{ ft}$$

At $Z_i = 25$ ft (last strip level, $i = 10$):

$$S_{H\text{-max}} = \frac{\frac{3}{12} \times \frac{0.25}{12} \times 5,000,000}{1164.35 \times 1.25 \times 3} = 6 \text{ ft}$$

Calculations of the maximum horizontal spacing between the reinforcing strips at each strip level are summarized in Table 4.25.

Let's use a design horizontal spacing between the strips at all strip levels a value of 2.5 ft. Therefore,

$S_{H\text{-design}} = 2.5$ ft at all strip levels.

Table 4.25 Maximum horizontal spacing ($S_{H\text{-}max}$) between the strips at each strip level

Strip level (i)	Z_i (ft)	$S_{H\text{-}max}$ (ft)
1	2.5	5.1
2	5.0	4.4
3	7.5	3.5
4	10.0	3.2
5	12.5	3.1
6	15.0	3.1
7	17.5	3.1
8	20.0	3.1
9	22.5	3.0
10	25.0	6.0

Step 8:
The maximum breakage force the reinforcing strip can withstand is calculated using Eq. (4.36) as follows:

$$F_{\text{B-max}} = \frac{3}{12} \times \frac{0.25}{12} \times 5,000,000 = 2,6041.7 \text{ lb}$$

Also, the actual applied breakage/pullout force and the actual factor of safety against breakage at each strip level can be determined using Eqs. (4.35) and (4.37), respectively:

At $Z_i = 7.5$ ft (third strip level, $i = 3$):
Using Eq. (4.35),

$$F_{\text{BP}} = 984.19 \times 2.5 \times 2.5 = 6,151.2 \text{ lb;}$$

Using Eq. (4.37),

$$\text{FS}_{\text{B}} = \frac{26,041}{6,151.2} = 4.2$$

Calculations of the actual breakage/pullout forces and safety factors against breakage at each strip level are summarized in Table 4.26.

Step 9:
The minimum strip length ($L_{\min}$) at each level to satisfy $\text{FS}_{\text{P-min}} = 3.0$ can be calculated using Eq. (4.38) as follows:

For example, at $Z_i = 7.5$ ft (third strip level, $i = 3$):

$$L_{\min} = \frac{(25 - 7.5)}{\tan\left(45 + \frac{32^{\circ}}{2}\right)} + \frac{3.0 \times 6,151.16}{2 \times \frac{3}{12} \times 1,831.9 \times \tan 18^{\circ}} = 71.7 \text{ ft}$$

Calculations of the minimum strip length at each strip level are summarized in Table 4.27.

Table 4.26 Actual breakage/pullout forces and safety factors against breakage at each strip level

Strip level (i)	Z_i (ft)	F_{BP} (lb)	FS_B
1	2.5	4,274.11	6.1
2	5.0	4,947.63	5.3
3	7.5	6,151.16	4.2
4	10.0	6,710.01	3.9
5	12.5	6,913.19	3.8
6	15.0	6,964.60	3.7
7	17.5	6,982.61	3.7
8	20.0	7,027.00	3.7
9	22.5	7,122.96	3.7
10	25.0	3,638.60	7.2

Table 4.27 Minimum strip length (L_{min}) at each strip level

Strip level (i)	Z_i (ft)	L_{min} (ft)
1	2.5	64.5
2	5.0	66.0
3	7.5	71.7
4	10.0	69.7
5	12.5	64.5
6	15.0	58.5
7	17.5	52.6
8	20.0	47.4
9	22.5	43.1
10	25.0	19.7

Let's use a constant design strip length of 72 ft at all strip levels along the wall height. Therefore,

$L_{design} = 72$ ft at all strip levels.

Step 10:

The maximum pullout force any reinforcing strip can withstand due to friction can be calculated using Eq. (4.39) as follows:

For example, at $Z_i = 7.5$ m (third strip level, $i = 3$):

Using Eq. (4.39),

$$F_{P\text{-max}} = \left(2 \times \frac{3}{12} \times 1,831.9 \times \tan 18^{\circ}\right) \times \left(72 - \frac{25 - 7.5}{\tan\left(45 + \frac{32^{\circ}}{2}\right)}\right) = 18,541.3 \text{ lb}$$

Using Eq. (4.40),

$$FS_P = \frac{18,541.3}{6,151.2} = 3.0$$

Calculations of the actual maximum pullout/friction forces and safety factors against pullout at each strip level are summarized in Table 4.28.

Table 4.28 Actual maximum pullout/friction forces and safety factors against pullout at each strip level

Strip level (i)	Z_i (ft)	$F_{\text{P-max}}$ (lb)	FS_P
1	2.5	14,658.16	3.4
2	5.0	16,452.18	3.3
3	7.5	18,541.32	3.0
4	10.0	20,882.31	3.1
5	12.5	23,445.23	3.4
6	15.0	26,208.70	3.8
7	17.5	29,324.97	4.2
8	20.0	32,692.80	4.7
9	22.5	36,189.79	5.1
10	25.0	39,815.30	10.9

Step 11:

Check external stability to avoid overturning, sliding, and bearing capacity failure using Eqs. (4.41), (4.48), and (4.55), respectively. Overturning and resisting moments can be calculated using Eqs. (4.42) and (4.43), respectively. Additionally, driving and resisting forces can be calculated using Eqs. (4.49) and (4.50), respectively.

– Check overturning (Eq. (4.41))

To determine the safety factor against overturning and compare it with the minimum safety factor, first, we need to calculate the overturning and resisting moments as follows:

Equation (4.42) can be used to determine the overturning moment (M_O):

$$M_O = \sum_{i=1}^{10} \left[\sigma'_{h(T)} \times S_{v(i)} \times (25 - Z_i)\right]$$
$$= \underbrace{455.9 \times 3.75 \times (25 - 2.5)}_{i=1} + \cdots + \underbrace{1,073.6 \times 2.5 \times (25 - 10)}_{i=4} + \cdots$$

Thus,

$$M_O = 256,221.9 \text{ ft lb}$$

Equation (4.43) can be used to determine the resisting moment (M_R):

$$M_R = \underbrace{\left(\frac{108.5}{2}\sum_{i=1}^{10}\left[S_{v(i)}(72)^2\right]\right)}_{M_{R(\text{soil})}} + \underbrace{\left(1,400 \times 20 \times \left(8 + \frac{20}{2}\right)\right)}_{M_{R(\text{load})}}$$

$$M_R = \underbrace{\frac{108.5 \times 72^2}{2} \times 3.75}_{i=1} + \cdots + \underbrace{\frac{108.5 \times 72^2}{2} \times 2.5}_{i=4} + \cdots + 504,000$$

Therefore,

$$M_{\mathrm{R}} = 7,534,800 \text{ ft lb}$$

So, using Eq. (4.41),

$$\mathrm{FS_O} = \frac{7,534,800}{256,221.9} = 29.4 \geq \mathrm{FS_{O\text{-}min}} = 3.5 \text{ O.K.}$$

– Check sliding (Eq. (4.48))

To determine the safety factor against sliding and compare it with the minimum safety factor for sliding, first, we need to calculate the driving and resisting forces as follows:

Equation (4.49) can be used to determine the driving force (F_{D}):

$$F_{\mathrm{D}} = \sum_{i=1}^{10} \left[\sigma'_{\mathrm{h(T)}} \times S_{\mathrm{v}(i)}\right] = \underbrace{455.9 \times 3.75}_{i=1} + \cdots + \underbrace{1,073.6 \times 2.5}_{i=4} + \cdots$$

Thus,

$$F_{\mathrm{D}} = 23,817 \text{ lb/ft}$$

Equation (4.50) can be used to determine the resisting moment (F_{R}):

$$F_{\mathrm{R}} = \underbrace{\left(108.5 \times \tan\left(\frac{2 \times 32}{3}\right)\right)\left(\sum_{i=1}^{10}\left[S_{\mathrm{v}(i)} \times 72\right]\right)}_{F_{\mathrm{R(soil)}} - \text{ Eq. (4.51)}} + \underbrace{\left(1,400 \times 20 \times \tan\left(\frac{2 \times 32}{3}\right)\right)}_{F_{\mathrm{R(load)}} - \text{ Eq. (4.53)}}$$

$$F_{\mathrm{R}} = 3,051 \times \sum_{i=1}^{10}\left(S_{\mathrm{v}(i)}\right) + 10,935.5$$

Therefore,

$$F_{\mathrm{R}} = 87,210.7 \text{ lb/ft}$$

So, using Eq. (4.48),

$$\mathrm{FS_S} = \frac{87,210.7}{23,817} = 3.66 \geq \mathrm{FS_{S\text{-}min}} = 3.5 \text{ O.K.}$$

– Check bearing capacity (Eq. (4.55))

Bearing capacity factors can be obtained as follows:

$N_c = 30.10$ using Eq. (1.3) with $\phi' = 30°$.
$N_\gamma = 21.22$ using Eq. (1.4) with $\phi' = 30°$.

Also, the total vertical stress at depth ($Z = 25$ ft) can be obtained from Eq. (4.4) as follows (see also Step 6):

$$\sigma'_{v(T)} = \underbrace{(108.5 \times 25)}_{\sigma'_{v(soil)}} + \underbrace{\left(\frac{1,400 \times 20}{20 + \frac{25}{2} + 8}\right)}_{\sigma'_{v(load)}}$$

Thus,

$$\sigma'_{v(T)} = 3,403.9\,\text{lb/ft}^2$$

Therefore, Eq. (4.55) can be used to determine the safety factor against bearing capacity and compare it with the minimum safety factor for bearing capacity as follows:

$$FS_{BC} = \frac{500 \times 30.10 + \frac{1}{2} \times 112 \times 21.22 \times 72}{3,403.9} = 31 \geq FS_{BC\text{-min}} = 4.0 \text{ O.K.}$$

4.4.3.2 foundationPro Solution

General Information Section

For our design problem, the BS units are selected, and wall height and safety factors are entered as shown in Fig. 4.28.

Rcinforcing Strips Sсction

Dimensions and strength of the reinforcing strips are entered in the provided textboxes as shown in Fig. 4.29. Also, the number of strip levels which is 10 is selected herein. The depths from the ground surface to each strip level are also entered in the provided table with yellow cells as shown in the figure.

Fig. 4.28 General Information section

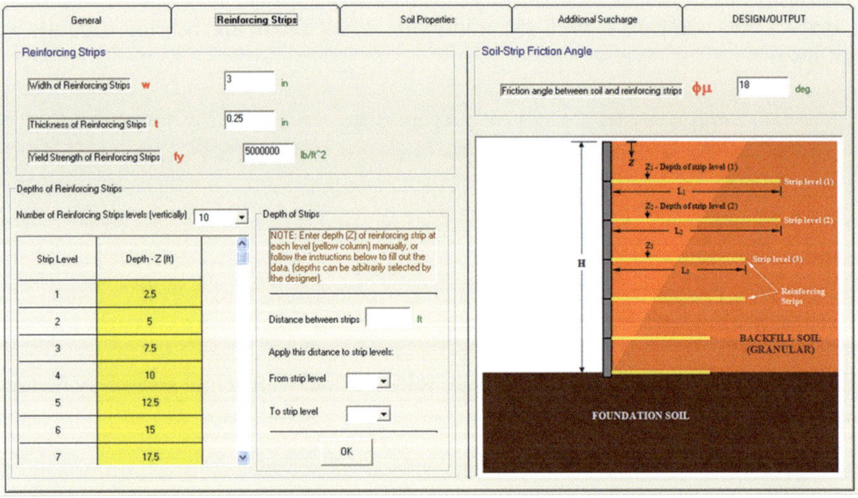

Fig. 4.29 Reinforcing Strips section

Soil Properties Section

Physical properties of the granular backfill to be retained behind the wall and the foundation soil to carry the MSE wall are entered in this section. For our design problem, the physical properties are entered as shown in Fig. 4.30.

Fig. 4.30 Soil Properties section

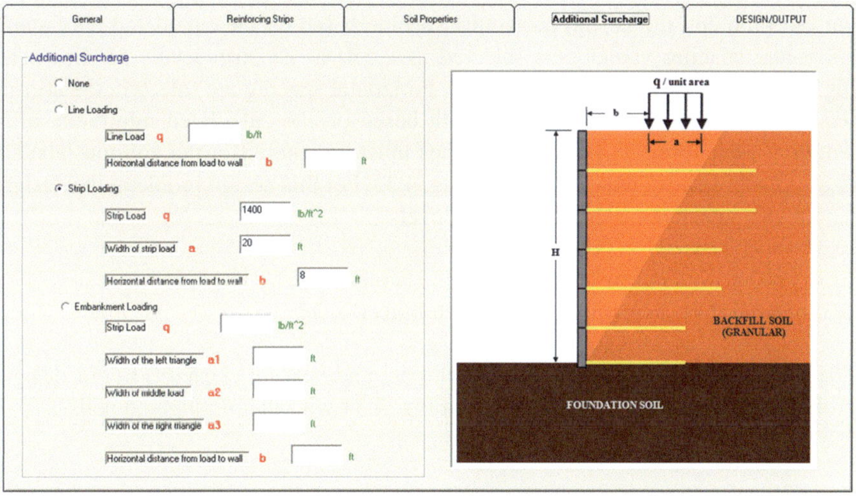

Fig. 4.31 Additional Surcharge Loading section

Additional Surcharge Section

For this design problem, a strip surcharge loading is applied on top of soil. Input data for this loading is entered as shown in Fig. 4.31.

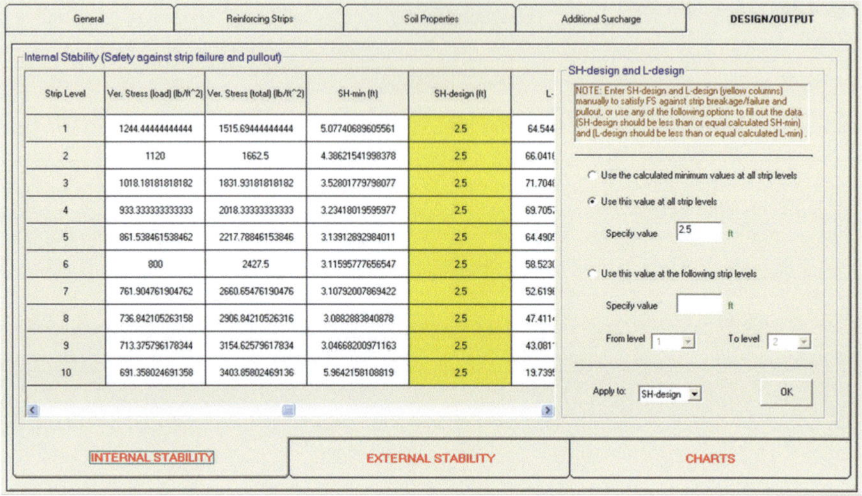

Fig. 4.32 DESIGN/OUTPUT section (design of horizontal spacing—internal stability)

DESIGN/OUTPUT Section (Internal Stability)

For our problem, the design horizontal spacing based on the calculated maximum horizontal spacing values was selected as 2.5 ft at all strip levels as shown in Fig. 4.32. After the design values for the horizontal spacing are selected, the user is required to enter the design strip length based on the calculated minimum strip length at each level. These design values must be entered in the column labeled (L_{design}) with yellow cells as shown in Fig. 4.33. For this design problem, the design length was selected as 72 ft at all strip levels.

DESIGN/OUTPUT Section (External Stability)

The three safety factors (overturning, sliding, and bearing capacity) in our design problem were all satisfied and shown in Fig. 4.34 with the highlighted cells.

DESIGN/OUTPUT Section (CHARTS)

In this third screen (CHARTS) of the DESIGN/OUTPUT section, one can view any required output results in table and chart formats. For example, the total horizontal stress (soil + load) with depth is plotted as depicted in Fig. 4.35. Also, distributions of safety factors against pullout are shown in Fig. 4.36.

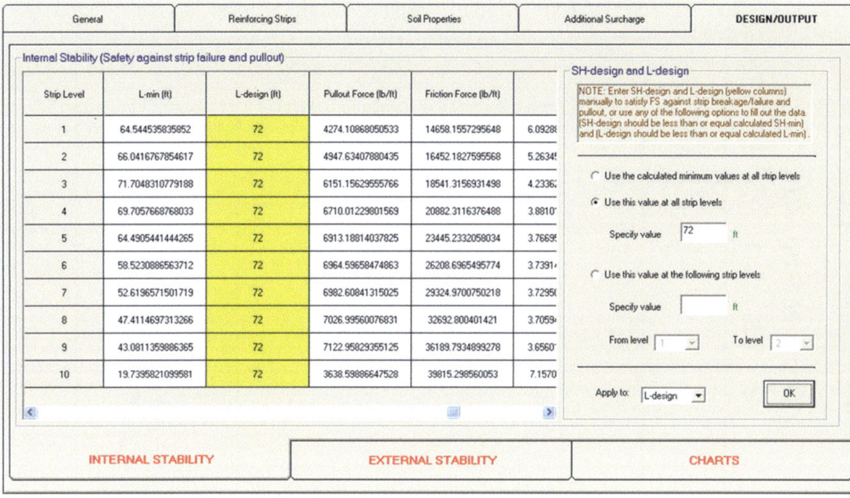

Fig. 4.33 DESIGN/OUTPUT section (design of strip length—internal stability)

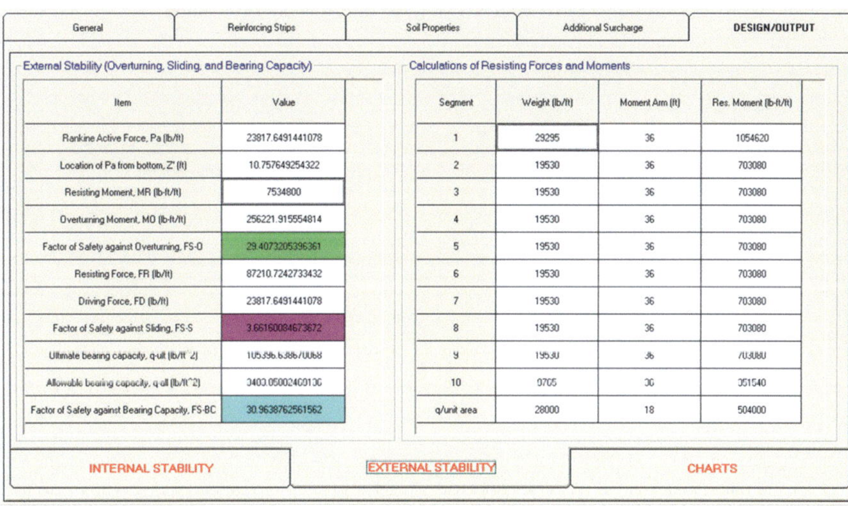

Fig. 4.34 DESIGN/OUTPUT section (external stability)

4.4.4 MSE Wall with Line Loading and Varying Strip Length

Design an MSE wall that is 25 ft high with galvanized steel reinforcement strips to retain a granular backfill behind it. Physical properties of the backfill soil, the foundation soil, and the strip reinforcement are listed in Table 4.29. An additional surcharge strip loading is applied on top of soil with details as provided in Table 4.30.

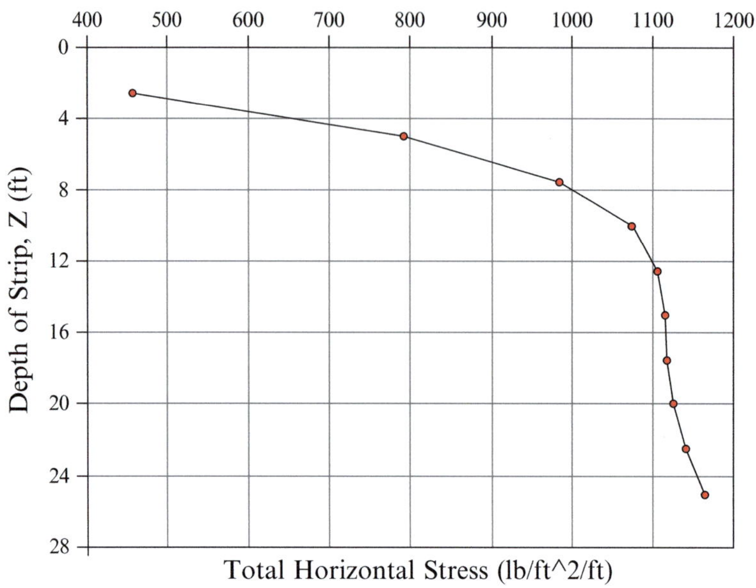

Fig. 4.35 Distribution of total horizontal stresses due to soil and strip loading

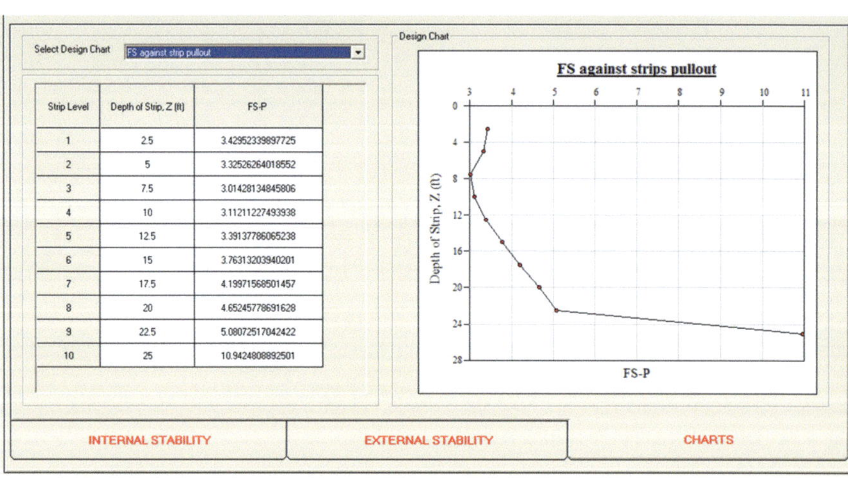

Fig. 4.36 Distribution of safety factor against pullout for the selected design parameters

Table 4.29 Physical properties and design parameters

Item	Property	Value	Unit
Backfill soil	ϕ_1	32	°
	γ_1	108.5	lb/ft^3
Foundation soil	ϕ_2	30	°
	γ_2	112	lb/ft^3
	c_2	500	lb/ft^2
Reinforcing strips	w	3	in.
	t	0.25	in.
	f_y	5,000,000	lb/ft^2
	ϕ_μ	18	°

Table 4.30 Details of strip loading

Property	Value	Unit
q (line)	650	lb/ft
b	4	ft

4.4.4.1 Hand Solution

To design the 25 ft high MSE wall, one must follow the steps below:

Step 1:
The following are given in the problem statement:

$H = 25$ ft; $\gamma_1 = 108.5\,\text{lb/ft}^3$; $\phi_1' = 32^\circ$; $\gamma_2 = 112\,\text{lb/ft}^3$; $\phi_2' = 30^\circ$; $c_2' = 500\,\text{lb/ft}^2$; $q = 650$ lb/ft; and $b = 4$ ft.

Step 2:
The following minimum values for the safety factors are considered to satisfy internal and external stability requirements in the design of the 25 ft high MSE wall:

For internal stability:

$$FS_{\text{B-min}} = 3.0 \text{ and } FS_{\text{P-min}} = 3.0.$$

For external stability:

$$FS_{\text{O-min}} = 3.5;\ FS_{\text{S-min}} = 3.5;\ \text{and } FS_{\text{BC-min}} = 4.0.$$

Step 3:
A suggestion for the dimensions of the reinforcing strips is given in the problem statement as follows:

$$w = 3 \text{ in. and } t = 0.25 \text{ in.}$$

Also,
$f_y = 5,000,000\,\text{lb/ft}^2$ and $\phi_\mu' = 18^\circ$.

Table 4.31 Suggested depths for strip levels

Strip level (i)	Depth, Z_i (ft)
1	2.5
2	5.0
3	7.5
4	10.0
5	12.5
6	15.0
7	17.5
8	20.0
9	22.5
10	25.0

Step 4:
Let's assume ten strip levels along the wall height (vertically) equally spaced with depths as in Table 4.31 below:

Step 5:
Using Eqs. (4.1) through (4.3), one can calculate $S_{v(i)}$ as follows:

At the first strip level ($i = 1$),

$$S_{v(1)} = \frac{2.5 + 5.0}{2} = 3.75 \text{ ft}$$

Also, at the third strip level ($i = 3$),

$$S_{v(3)} = \frac{Z_4 - Z_2}{2} = \frac{10 - 5}{2} = 2.5 \text{ ft}$$

At the last strip level ($i = 10$),

$$S_{v(10)} = \frac{2 \times 25 - Z_{10} - Z_9}{2} = \frac{2 \times 25 - 25.0 - 22.5}{2} = 1.25 \text{ ft}$$

The effective vertical distances, $S_{v(i)}$, are summarized in Table 4.32 below:

Step 6:
The total vertical stresses are calculated using Eq. (4.4) as follows:

$$\sigma'_{v(T)} = \sigma'_{v(soil)} + \sigma'_{v(load)}$$

The vertical stresses due to soil ($\sigma'_{v(soil)}$) can be found using Eq. (4.6):

Table 4.32 Effective vertical distances ($S_{v(i)}$)

Strip level (i)	Z_i (ft)	$S_{v(i)}$ (ft)
1	2.5	3.75
2	5.0	2.5
3	7.5	2.5
4	10.0	2.5
5	12.5	2.5
6	15.0	2.5
7	17.5	2.5
8	20.0	2.5
9	22.5	2.5
10	25.0	1.25

Table 4.33 Total vertical effective stresses ($\sigma'_{v(T)}$) at each strip level

Strip level (i)	Z_i (ft)	$\sigma'_{v(soil)}$ (lb/ft^2)	$\sigma'_{v(load)}$ (lb/ft^2)	$\sigma'_{v(T)}$ (lb/ft^2)
1	2.5	271.25	13.06	284.31
2	5.0	542.5	30.77	573.27
3	7.5	813.75	33.44	847.19
4	10.0	1,085	30.75	1,115.75
5	12.5	1,356.25	27.24	1,383.49
6	15.0	1,627.5	24.05	1,651.55
7	17.5	1,898.75	21.36	1,920.11
8	20.0	2,170	19.13	2,189.13
9	22.5	2,441.25	17.28	2,458.53
10	25.0	2,712.5	15.74	2,728.24

$$\sigma'_{v(soil)} = 108.5 \times Z_i$$

However, the vertical stresses due to the strip loading ($\sigma'_{v(load)}$) can be found using Eq. (4.9):

$$\sigma'_{v(load)} = \frac{2 \times 650 \times Z_i^3}{\pi\left(4^2 + Z_i^2\right)^2}$$

Calculations of the total vertical effective stresses at each strip level are summarized in Table 4.33.

Additionally, the total horizontal stresses are calculated using Eq. (4.5) as follows:

$$\sigma'_{h(T)} = \sigma'_{h(soil)} + \sigma'_{h(load)}$$

The horizontal stresses due to soil can be found using Eq. (4.7):

$$\sigma'_{h(soil)} = \tan^2\left(45 - \frac{32}{2}\right) \times 108.5 \times Z_i = 0.307 \times 108.5 \times Z_i = 33.3 \times Z_i$$

However, the total horizontal stresses due to the line loading on top of soil can be obtained as follows:

First, using Eq. (4.13), m_1 can be calculated as

$$m_1 = \frac{4}{25} = 0.16$$

Using Eq. (4.14), m_2 can be calculated as

$$m_2 = \frac{Z_i}{25}$$

Since $m_1 = 0.16 < 0.4$ Eq. (4.12) is used to determine $\sigma'_{h(load)}$:

$$\sigma'_{h(load)} = \frac{0.203 \times 650 \times \left(\frac{Z_i}{25}\right)}{25 \times \left(0.16 + \left(\frac{Z_i}{25}\right)^2\right)^2}$$

Now, $\sigma'_{h(load)}$ can be obtained using Eq. (4.17):

$$\sigma'_{h(T)} = 33.3 \times Z_i + \frac{0.203 \times 650 \times \left(\frac{Z_i}{25}\right)}{25 \times \left(0.16 + \left(\frac{Z_i}{25}\right)^2\right)^2}$$

Calculations of the total horizontal effective stresses at each strip level are summarized in Table 4.34.

Table 4.34 Total horizontal effective stresses ($\sigma'_{h(T)}$) at each strip level

Strip level (i)	Z_i (ft)	$\sigma'_{h(soil)}$ (lb/ft^2)	$\sigma'_{h(load)}$ (lb/ft^2)	$\sigma'_{h(T)}$ (lb/ft^2)
1	2.5	83.34	18.26	101.61
2	5.0	166.69	26.39	193.08
3	7.5	250.03	25.33	275.37
4	10.0	333.38	20.62	353.99
5	12.5	416.72	15.70	432.42
6	15.0	500.06	11.71	511.77
7	17.5	583.41	8.74	592.15
8	20.0	666.75	6.60	673.35
9	22.5	750.09	5.05	755.14
10	25.0	833.44	3.92	837.36

Table 4.35 Maximum horizontal spacing ($S_{H\text{-}max}$) between the strips at each strip level

Strip level (i)	Z_i (ft)	$S_{H\text{-}max}$ (ft)
1	2.5	22.8
2	5.0	18.0
3	7.5	12.6
4	10.0	9.8
5	12.5	8.0
6	15.0	6.8
7	17.5	5.9
8	20.0	5.2
9	22.5	4.6
10	25.0	8.3

Step 7:
The maximum horizontal spacing ($S_{H\text{-}max}$) between the strips at each strip level to satisfy $FS_{B\text{-min}} = 3.0$ can be calculated using Eq. (4.34) as follows:

At $Z_i = 2.5$ ft (first strip level, $i = 1$):

$$S_{\text{H-max}} = \frac{\frac{3}{12} \times \frac{0.25}{12} \times 5,000,000}{101.5 \times 3.75 \times 3} = 22.8 \text{ ft}$$

At $Z_i = 25$ ft (last strip level, $i = 10$):

$$S_{\text{H-max}} = \frac{\frac{3}{12} \times \frac{0.25}{12} \times 5,000,000}{837.36 \times 1.25 \times 3} = 8.3 \text{ ft}$$

Calculations of the maximum horizontal spacing between the reinforcing strips at each strip level are summarized in Table 4.35.

Let's use a design horizontal spacing between the strips at all strip levels a value of 3.5 ft. Therefore,

$S_{\text{H-design}} = 3.5$ ft at all strip levels.

Step 8:
The maximum breakage force the reinforcing strip can withstand is calculated using Eq. (4.36) as follows:

$$F_{\text{B-max}} = \frac{3}{12} \times \frac{0.25}{12} \times 5,000,000 = 26,041.7 \text{ lb}$$

Also, the actual applied breakage/pullout force and the actual factor of safety against breakage at each strip level can be determined using Eqs. (4.35) and (4.37), respectively:

At $Z_i = 7.5$ ft (third strip level, $i = 3$):

Table 4.36 Actual breakage/pullout forces and safety factors against breakage at each strip level

Strip level (i)	Z_i (ft)	F_{BP} (lb)	FS_B
1	2.5	1,333.59	19.5
2	5.0	1,689.43	15.4
3	7.5	2,409.45	10.8
4	10.0	3,097.44	8.4
5	12.5	3,783.66	6.9
6	15.0	4,478.03	5.8
7	17.5	5,181.33	5.0
8	20.0	5,891.80	4.4
9	22.5	6,607.51	3.9
10	25.0	3,663.46	7.1

Using Eq. (4.35),

$$F_{BP} = 275.37 \times 2.5 \times 3.5 = 2,409.5 \text{ lb}$$

Using Eq. (4.37),

$$FS_B = \frac{26,041}{2,409.5} = 10.8$$

Calculations of the actual breakage/pullout forces and safety factors against breakage at each strip level are summarized in Table 4.36.

Step 9:
The minimum strip length (L_{min}) at each level to satisfy $FS_{P\text{-}min} = 3.0$ can be calculated using Eq. (4.38) as follows:

For example, at $Z_i = 7.5$ ft (third strip level, $i = 3$):

$$L_{min} = \frac{(25 - 7.5)}{\tan\left(45 + \frac{32^o}{2}\right)} + \frac{3.0 \times 2409.45}{2 \times \frac{3}{12} \times 847.19 \times \tan 18^o} = 62.2 \text{ ft}$$

Calculations of the minimum strip length at each strip level are summarized in Table 4.37.

Let's use varying strip design lengths as suggested below:

L_{design} = 100 ft for strip level 1.
L_{design} = 66 ft for strip levels 2–6.
L_{design} = 55 ft for strip levels 7–10.

Step 10:
The maximum pullout force any reinforcing strip can withstand due to friction can be calculated using Eq. (4.39) as follows:

Table 4.37 Minimum strip length (L_{min}) at each strip level

Strip level (i)	Z_i (ft)	L_{min} (ft)
1	2.5	99.1
2	5.0	65.5
3	7.5	62.2
4	10.0	59.6
5	12.5	57.4
6	15.0	55.6
7	17.5	54.0
8	20.0	52.5
9	22.5	51.0
10	25.0	24.8

Table 4.38 Actual maximum pullout/friction forces and safety factors against pullout at each strip level

Strip level (i)	Z_i (ft)	$F_{P\text{-}max}$ (lb)	FS_P
1	2.5	4,042.83	3.0
2	5.0	5,114.31	3.0
3	7.5	7,748.78	3.2
4	10.0	10,456.33	3.4
5	12.5	13,276.92	3.5
6	15.0	16,221.18	3.6
7	17.5	15,859.87	3.1
8	20.0	18,574.82	3.2
9	22.5	21,414.20	3.2
10	25.0	24,377.59	6.7

For example, at $Z_i = 7.5$ m (third strip level, $i = 3$):

Using Eq. (4.39),

$$F_{\text{P-max}} = \left(2 \times \frac{3}{12} \times 847.19 \times \tan 18^{\circ}\right) \times \left(66 - \frac{25 - 7.5}{\tan\left(45 + \frac{32^{\circ}}{2}\right)}\right) = 7,748.8 \text{ lb}$$

Using Eq. (4.40),

$$FS_P = \frac{7,748.8}{2,409.45} = 3.2$$

Calculations of the actual maximum pullout/friction forces and safety factors against pullout at each strip level are summarized in Table 4.38.

Step 11:

Check external stability to avoid overturning, sliding, and bearing capacity failure using Eqs. (4.41), (4.48), and (4.55), respectively. Overturning and resisting moments can be calculated using Eqs. (4.42) and (4.43), respectively. Additionally, driving and resisting forces can be calculated using Eqs. (4.49) and (4.50), respectively.

– Check overturning (Eq. (4.41))

To determine the safety factor against overturning and compare it with the minimum safety factor, first, we need to calculate the overturning and resisting moments as follows:

Equation (4.42) can be used to determine the overturning moment (M_O):

$$M_O = \sum_{i=1}^{10} \left[\sigma'_{h(T)} \times S_{v(i)} \times (25 - Z_i) \right] = \underbrace{101.6 \times 3.75 \times (25 - 2.5)}_{i=1} + \cdots + \underbrace{354 \times 2.5 \times (25 - 10)}_{i=4} + \cdots$$

Thus,

$$M_O = 92{,}217.6 \text{ ft lb}$$

Equation (4.43) can be used to determine the resisting moment (M_R):

$$M_R = \underbrace{\left(\frac{108.5}{2} \sum_{i=1}^{10} \left[S_{v(i)} \left(L_{\text{design}} \right)^2 \right] \right)}_{M_{R(\text{soil})}} + \underbrace{(650 \times 4)}_{M_{R(\text{load})}}$$

$$M_R = \underbrace{\frac{108.5 \times 100^2}{2} \times 3.75}_{i=1} + \cdots + \underbrace{\frac{108.5 \times 66^2}{2} \times 2.5}_{i=4} + \cdots + 2{,}600$$

Therefore,

$$M_R = 6{,}426{,}817.2 \text{ ft lb}$$

So, using Eq. (4.41),

$$\text{FS}_O = \frac{6{,}426{,}817.2}{92{,}217.6} = 70 \geq \text{FS}_{\text{O-min}} = 3.5 \text{ O.K.}$$

– Check sliding (Eq. (4.48))

To determine the safety factor against sliding and compare it with the minimum safety factor for sliding, first, we need to calculate the driving and resisting forces as follows:

Equation 4.49 can be used to determine the driving force (F_D):

$$F_D = \sum_{i=1}^{10} \left[\sigma'_{h(T)} \times S_{v(i)}\right] = \underbrace{101.6 \times 3.75}_{i=1} + \cdots + \underbrace{354 \times 2.5}_{i=4} + \cdots$$

Thus,

$$F_D = 10,773.5 \text{ lb/ft}$$

Equation (4.50) can be used to determine the resisting moment (F_R):

$$F_R = \underbrace{\left(108.5 \times \tan\left(\frac{2 \times 32}{3}\right)\right)\left(\sum_{i=1}^{10}\left[S_{v(i)} \times L_{\text{design}}\right]\right)}_{F_{R(\text{soil})} - \text{Eq. (4.51)}} + \underbrace{\left(650 \times \tan\left(\frac{2 \times 32}{3}\right)\right)}_{F_{R(\text{load})} - \text{Eq. (4.52)}}$$

$$F_R = \left(42.375 \times \sum_{i=1}^{10}\left[S_{v(i)} \times L_{\text{design}}\right]\right) + (253.86)$$

Therefore,

$$F_R = 71,497 \text{ lb/ft}$$

So, using Eq. (4.48),

$$\text{FS}_S = \frac{71,497}{10,773.5} = 6.6 \geq \text{FS}_{\text{S-min}} = 3.5 \text{ O.K.}$$

– Check bearing capacity (Eq. (4.55))

Bearing capacity factors can be obtained as follows:

$N_c =$ 30.10 using Eq. (1.3) with $\phi' = 30°$.
$N_\gamma = 21.22$ using Eq. (1.4) with $\phi' = 30°$.

Also, the total vertical stress at depth ($Z = 25$ ft) can be obtained from Eq. (4.4) as follows (see also Step 6):

$$\sigma'_{v(T)} = \underbrace{(108.5 \times 25)}_{\sigma'_{v(\text{soil})}} + \underbrace{\frac{2 \times 650 \times 25^3}{\pi\left(4^2 + 25^2\right)^2}}_{\sigma'_{v(\text{load})}}$$

Thus,

$$\sigma'_{v(T)} = 2,728.2 \text{ lb/ft}^2$$

Therefore, Eq. (4.55) can be used to determine the safety factor against bearing capacity and compare it with the minimum safety factor for bearing capacity as follows:

$$\mathrm{FS_{BC}} = \frac{500 \times 30.10 + \frac{1}{2} \times 112 \times 21.22 \times 55}{2,728.2} = 30.8 \geq \mathrm{FS_{BC\text{-}min}} = 4.0 \text{ O.K.}$$

4.4.4.2 foundationPro Solution

Similar to the design problem in the previous section (Sect. 4.4.3), the first three sections in the MSE Wall-1 applications are the same (General, Reinforcing Strips, and Soil Properties). However, one should include a line loading on top of soil instead of a strip loading.

Additional Surcharge Section

For this design problem, a line loading is applied on top of soil. Input data for this loading is entered as shown in Fig. 4.37.

DESIGN/OUTPUT Section (Internal Stability)

For our problem, the design horizontal spacing based on the calculated maximum horizontal spacing values was considered constant and was selected as 3.5 ft at all

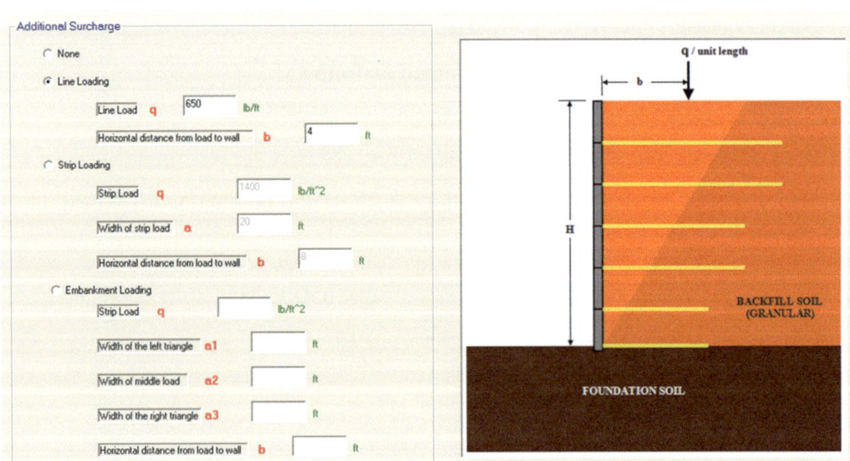

Fig. 4.37 Additional Surcharge Loading section

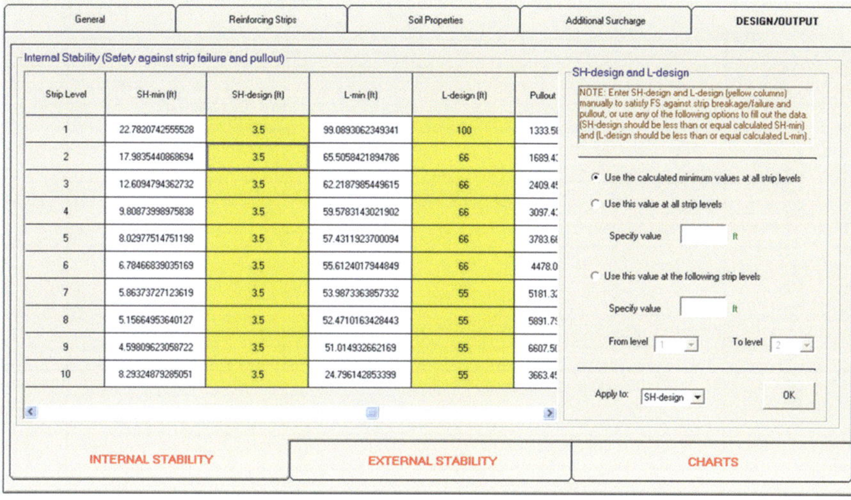

Strip Level	SH-min (ft)	SH-design (ft)	L-min (ft)	L-design (ft)	Pullout
1	22.7820742555528	3.5	99.0893062349341	100	1333.5
2	17.9835440868694	3.5	65.5058421894786	66	1689.4
3	12.6094794362732	3.5	62.2187985449615	66	2409.4
4	9.80873998975838	3.5	59.5783143021902	66	3097.4
5	8.02977514751198	3.5	57.4311923700094	66	3783.6
6	6.78466839035169	3.5	55.6124017944849	66	4478.0
7	5.86373727123619	3.5	53.9873363857332	55	5181.3
8	5.15664953640127	3.5	52.4710163428443	55	5891.7
9	4.59809623058722	3.5	51.014932662169	55	6607.5
10	8.29324879285051	3.5	24.796142853399	55	3663.4

Fig. 4.38 Design of horizontal spacing and strip length—internal stability

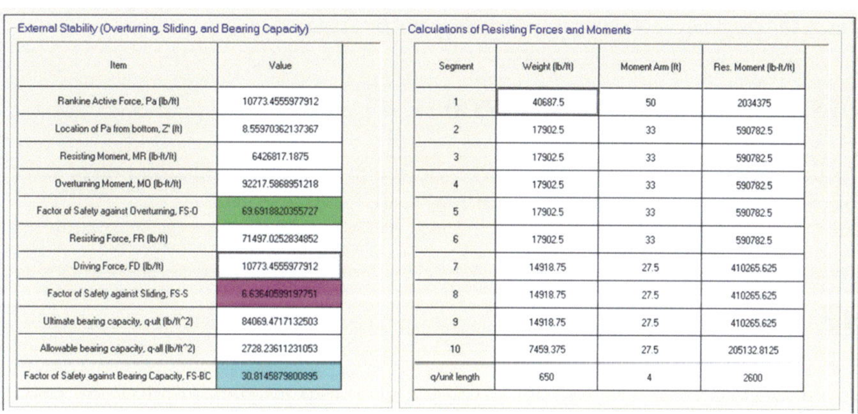

Item	Value
Rankine Active Force, Pa (lb/ft)	10773.4555977912
Location of Pa from bottom, Z' (ft)	8.55970362137367
Resisting Moment, MR (lb-ft/ft)	6426817.1875
Overturning Moment, MO (lb-ft/ft)	92217.5868951218
Factor of Safety against Overturning, FS-O	69.6918820355727
Resisting Force, FR (lb/ft)	71497.0252834852
Driving Force, FD (lb/ft)	10773.4555977912
Factor of Safety against Sliding, FS-S	6.63640599197751
Ultimate bearing capacity, q-ult (lb/ft^2)	84069.4717132503
Allowable bearing capacity, q-all (lb/ft^2)	2728.23611231053
Factor of Safety against Bearing Capacity, FS-BC	30.8145879800895

Segment	Weight (lb/ft)	Moment Arm (ft)	Res. Moment (lb-ft/ft)
1	40687.5	50	2034375
2	17902.5	33	590782.5
3	17902.5	33	590782.5
4	17902.5	33	590782.5
5	17902.5	33	590782.5
6	17902.5	33	590782.5
7	14918.75	27.5	410265.625
8	14918.75	27.5	410265.625
9	14918.75	27.5	410265.625
10	7459.375	27.5	205132.8125
q/unit length	650	4	2600

Fig. 4.39 DESIGN/OUTPUT section (external stability)

strip levels as shown in Fig. 4.38. After the design values for the horizontal spacing are selected, the design length based on the calculated minimum strip length at each level was considered varying and selected as depicted in Fig. 4.38. These design values were entered in the column labeled (L_{design}) with yellow cells as shown in the figure. The three design lengths that were considered are 100 ft (for level 1), 66 ft (for levels 2–6), and 55 ft (for levels 7–10).

DESIGN/OUTPUT Section (External Stability)

The three safety factors (overturning, sliding, and bearing capacity) in our design problem were all satisfied and shown in Fig. 4.39 as in the highlighted cells.

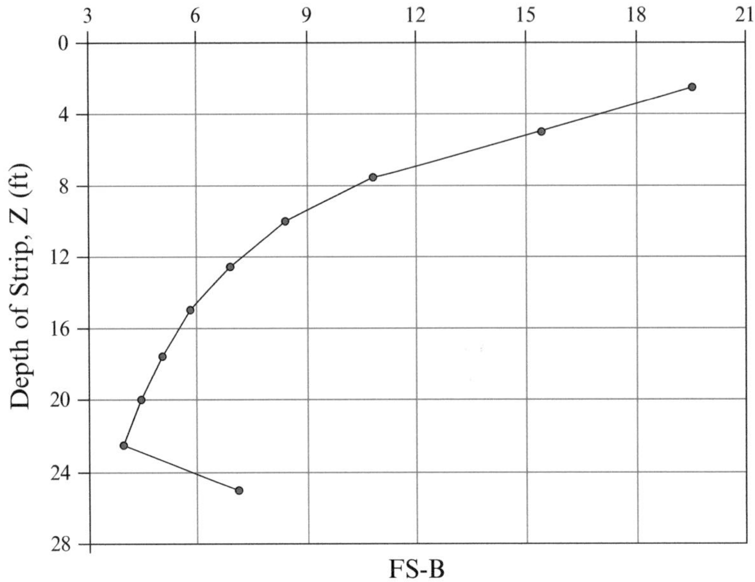

Fig. 4.40 Distribution of safety factors against strip breakage

DESIGN/OUTPUT Section (CHARTS)

As in examples of the charts that can be viewed in this section, distribution of safety factors against breakage is plotted in Fig. 4.40.

4.5 Suggested Projects

In this section, you will find some suggested design projects to allow the reader practice the concepts and the ideas discussed in the previous sections. These suggested projects will cover a variety of MSE wall designs by varying strip length and spacing (vertical and horizontal) under various loading conditions (strip load, embankment load, etc.).

4.5.1 Suggested Projects: MSE Wall with Applied Strip Loading

Design an MSE wall that is 35 ft high with galvanized steel reinforcement strips to retain a granular backfill behind it. Physical properties of the backfill soil, the foundation soil, and the strip reinforcement are listed in Table 4.39. An additional

Table 4.39 Physical properties and design parameters

Item	Property	Value	Unit
Backfill soil	ϕ_1	35	°
	γ_1	110	lb/ft^3
Foundation soil	ϕ_2	28	°
	γ_2	115	lb/ft^3
	c_2	734	lb/ft^2
Reinforcing strips	w	2.5	in.
	t	0.3	in.
	f_y	5,000,000	lb/ft^2
	ϕ_μ	19	°

Table 4.40 Details of strip loading

Property	Value	Unit
q (strip)	1,000	lb/ft^2
a	15	ft
b	6	ft

Table 4.41 Physical properties and design parameters

Item	Property	Value	Unit
Backfill soil	ϕ_1	33	°
	γ_1	17.8	kN/m^3
Foundation soil	ϕ_2	30	°
	γ_2	18.9	kN/m^3
	c_2	35	kN/m^2
Reinforcing strips	w	60	mm
	t	4	mm
	f_y	240,000	kN/m^2
	ϕ_μ	16	°

surcharge strip loading is applied on top of soil with details as provided in Table 4.40. Use a total of 14 reinforcing strips (vertically) along the wall height with depths (in ft) as follows: 2, 4, 6, 8, 10, 12.5, 15, 17.5, 20, 22.5, 25, 28, 31, and 34. Use a single value for the design horizontal spacing between the reinforcing strips at all levels. Also, use a safety factor of 3 for internal stability and 3.5 for external stability.

4.5.2 *Suggested Projects: MSE Wall with Embankment Loading*

Design an MSE wall that is 8 m high with galvanized steel reinforcement strips to retain a granular backfill behind it. Physical properties of the backfill soil, the foundation soil, and the strips reinforcement are listed in Table 4.41. An additional

Table 4.42 Details of strip loading

Property	Value	Unit
q (embankment)	75	kN/m^2
b	3	m
a_1	4	m
a_2	8	m
a_3	3	m

embankment loading is applied on top of soil with details as provided in Table 4.42. Use a total of 12 reinforcing strips (vertically) along the wall height with depths (in m) as follows: 0.75, 1.5, 2.25, 3, 3.75, 4.25, 5, 6, 7, 8, 9, and 10. Feel free to consider varying horizontal spacing and length of strips. Use a safety factor of 3 for internal stability and 4.5 for external stability.

References

Das B. Principles of foundation engineering. Boston, MA: Cengage Learning; 2010.
Das B. Advanced soil mechanics. London: Taylor and Francis; 1997.
Laba J, Kennedy J. Reinforced earth retaining wall analysis and design. Can Geotech J. 1986;23 (3):317–26.

Further Reading

Budhu M. Soil mechanics and foundations. New York: Wiley; 2000.